100
CITIES
5000
IDEAS

100 CITIES 5000 IDEAS

WHERE TO GO · WHEN TO GO · WHAT TO SEE · WHAT TO DO

JOE YOGERST

NATIONAL GEOGRAPHIC

WASHINGTON, D.C.

Contents

Introduction   7

Map of the World   8

United States, Canada & the Caribbean   10

Latin America   80

Australia & Oceania   126

Africa   148

Middle East & Central Asia   180

Eastern & Southern Asia   210

Europe   280

Acknowledgments   386

Illustrations Credits   387

Index   390

Pages 2-3: Take a walk at sunset along the marina promenade in Porto Arabia (p. 208), where you can take in views of the Persian Gulf and Doha West Bay.

Opposite: Enjoy the ultimate Italian shopping experience at Mercato Centrale in Florence (p. 316), where you can find fresh cheeses, meats, and other delicious fare.

INTRODUCTION

"Observing that nobody in the history of man had ever seen and described the entire urban world, I resolved to do it myself," wrote the late, great historian and author Jan Morris when asked why she was so obsessed with writing about cities. While it's impossible to replicate her literary quest, *100 Cities, 5,000 Ideas* offers an array of urban areas around the globe that should be on every traveler's bucket list.

Like Morris, I've managed to achieve my own long and lively rapport with urban areas. There's nothing I like more than wandering aimlessly through a new city, map and phone in hand but trusting more in dead reckoning to get around. Discovering where a side street leads or what's on the other side of a park or piazza. Ambling along a famous boulevard in the dead of night when most residents (and other travelers) are snug in their beds. Finding an obscure but captivating museum, a hole-in-the-wall restaurant with divine cuisine, or a neighborhood festival that only the locals know about.

Many of my fondest memories are urban adventures both planned and off-the-cuff. Sleeping atop the Great Pyramid of Giza with the dazzling lights of Cairo at my feet—and a stray black-and-white cat as my unexpected companion. Catching Keith Urban and Steven Tyler jamming together in a tiny Nashville honky-tonk. A Mexican family inviting me to share midnight tamales and tequila in an Oaxaca cemetery during Día de los Muertos (Day of the Dead). My first-time bungee jumping off the Harbour Bridge in Auckland and the relief I felt on that first return bounce.

But it's not just travel. I'm also drawn to cities as permanent abodes. The smallest place I have ever called home was Eugene, Oregon, which had just over 100,000 people when I attended journalism school at the University of Oregon. Otherwise, my home has been one large metropolis after another: Los Angeles, San Francisco, Johannesburg, London, Hong Kong, Singapore, and even Las Vegas count among the places I have lived, worked, and written since leaving home at the age of 18.

Like honeybees or schools of fish, is gathering with many others of our species part of our DNA? I'm beginning to think so after raising children who seem as obsessed with big cities as their father. I remember my oldest daughter, who must have been six or seven at the time, standing in the middle of Times Square and proclaiming, "This is my kind of town!" That's pretty much how I feel about cities, too.

And I'm not the only one. The percentage of the human race living in urban areas has steadily increased. In 1800, less than 10 percent of the global population lived in cities. By 1960, that figure had risen to 34 percent. By 2019, it was 55 percent. Even more astonishing is a United Nations estimate that Earth's urban population will reach 68 percent by the middle of this century.

Although we couldn't feature every city that begs attention (there are literally thousands of worthy candidates) in this book, we made an effort to include a wide variety of metropolitan personalities. Big and small, old and new, expensive and frugal (via Mercer's Cost of Living City Rankings), far off the beaten track and familiar even to those who have never visited (thanks to numerous books and movies).

Some cities simply can't be ignored—all-time greats like London, Tokyo, Sydney, and the Big Apple. Others are urban upstarts, places like Dubai, Auckland, Ulaanbaatar, and Nairobi. These destinations have exploded in both population and travel potential over the past half century. Still others are among my personal favorites: spunky San Juan in the Caribbean, canal-laced Bruges in Belgium, exotic Luang Prabang on the Mekong River, and multicultural Montréal (it's not just French!).

Looking to the future, I'm excited to see how cities will continue to evolve as places to stay and visit. A version of this book in the 22nd century might feature an entirely different lineup of "must-see" urban areas. But for the time being, we give you the cities that every traveler should visit now.

Tour Liberty Island and its famous statue on a weekend in the United States' most visited city, New York (p. 64).

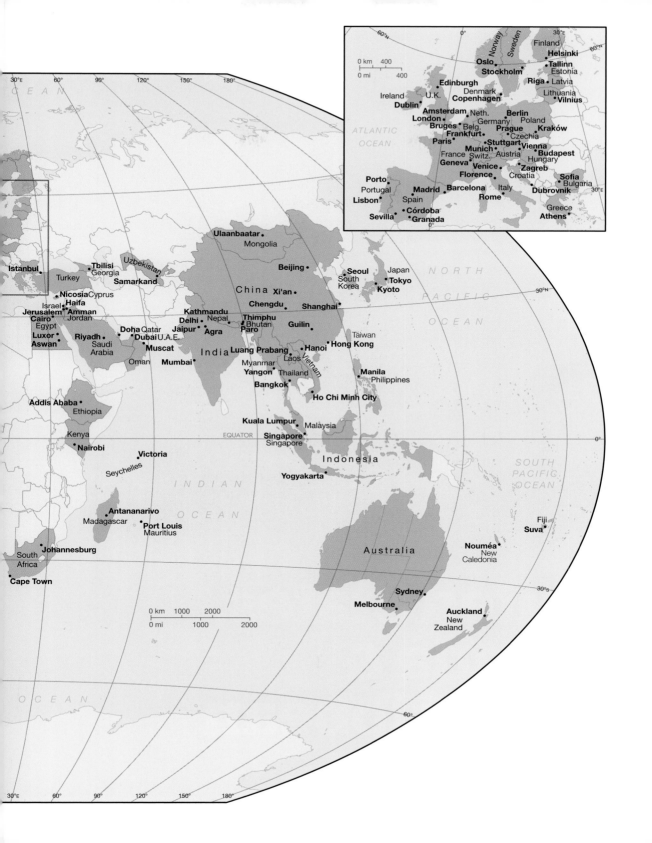

United States, Canada & the Caribbean

Manhattan's (p. 64) skyline lights up at night behind the Triborough Bridge.

Washington, D.C.
District of Columbia, U.S.A.

From its humble start as one of the first preplanned capitals of the Western world, Washington, D.C., has grown into a powerhouse of politics, culture, entertainment, and even nature in its own unique way.

THE BIG PICTURE

Founded: 1790

Population: 6.2 million

Size: 5,564 square miles (14,411 sq km)

Language: English

Currency: U.S. dollar

Time Zone: GMT -4 and -5 (EDT and EST)

Global Cost of Living Rank: 51

Also Known As: District of Columbia, the District, Inside the Beltway

It took years for people to think of Washington, D.C., as a proper city. Thomas Jefferson allegedly called it "that Indian swamp in the wilderness," but moved there anyway when he was elected president. After an 1842 visit, the great British novelist Charles Dickens complained about "spacious avenues that begin in nothing and lead nowhere." Another British wag, passing through the city during Lincoln's presidency, declared that an apparently disheveled capital was "still under the empire of King Mud."

In retrospect, none of those observations was off the mark. It did take Washington more than a century to get its urban act together. But nowadays there is no disputing that D.C. is one of the world's great cities.

Despite that tremendous growth since World War II, America's national capital still revolves around the carefully planned city that military engineer Pierre Charles L'Enfant laid out in the 1790s. His spacious avenues from nowhere now lead to a lot of significant somewheres, such as the **White House**, the **U.S. Capitol Building**, and the **Supreme Court**—the three pillars of American democracy (executive, legislative, judicial) and its separation of powers.

Now designated a national historic site, once muddy **Pennsylvania Avenue** stretches between the Capitol and the White House with other landmarks along the way, like the **National Archives** (where the original Declaration of Independence, Louisiana Purchase Treaty, and Emancipation Proclamation are on display), the rather intimidating **FBI Headquarters**, and **Freedom Plaza** with its huge, marble map of L'Enfant's original urban plan. The plaza derives its name from the fact that Martin Luther King, Jr., penned his famous "I Have a Dream" speech in the historic **Willard Hotel** overlooking the square.

Reverend King delivered that speech in 1963, on the steps of the **Lincoln Memorial**, which anchors

The color guard marches at the National World War II Memorial on the Mall.

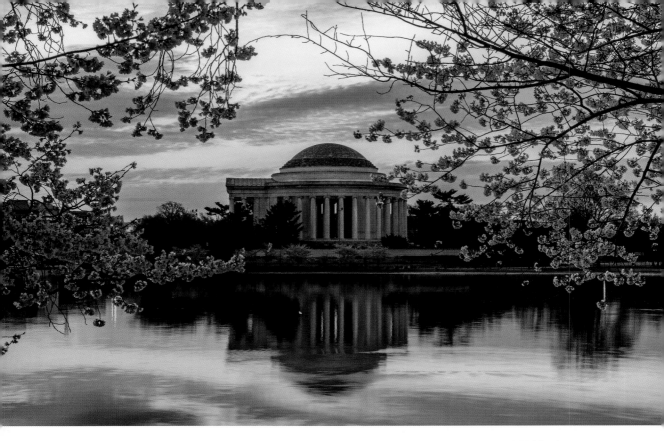

Visit D.C. in late March to early April for the Cherry Blossom Festival, where pink buds bloom around the Tidal Basin.

the western end of the **National Mall**. Spanning just under two miles (3 km) from stem to stern, the Mall has evolved from its bucolic days as a horse and cattle pasture into the most visited unit of the National Park Service and the world's single largest museum cluster.

In addition to the **National Gallery of Art**, the park is flanked by numerous Smithsonian museums ranging from vintage collections such as the **National Museum of Natural History**, the **Arts and Industries Building**, and the **Freer Gallery of Art**, to modern creations like the **National Air and Space Museum**, **National Museum of the American Indian**, and the latest arrival, the **National Museum of African American History and**

Culture, which opened in September 2016.

The **Washington Monument**, which honors the city's namesake president, towers over other tributes like the **Korean War Veterans Memorial**, **Vietnam Veterans Memorial**, **National World War II Memorial**, and **U.S. Holocaust Memorial Museum**. Around the western shore of the nearby Tidal Basin are memorials to **Martin Luther King, Jr.**, **Franklin Delano Roosevelt**, and **Thomas Jefferson**.

Although it's perhaps most famed for presidential inaugurations and political demonstrations, the Mall also flaunts a lighter side that includes a number of free public art and entertainment events. Each

summer, programs offer screenings of outdoor movies on the Mall, and the National Gallery of Art stages the **Jazz in the Garden** concert series.

Beyond the Mall, the district is home to more than 50 other museums, historic sites, and heritage homes. Although traditional paintings are still its stock-in-trade, the **National Portrait Gallery** also displays pop art and interpretive installations. **Ford's Theatre** and the **Petersen Boarding House** spin tales of John Wilkes Booth and Abraham Lincoln.

Hillwood Estate, Museum and Garden and the **Anderson House** afford a glimpse of American socialite life during the Gilded Age and early 20th century. Despite their rather ominous names, the

National Law Enforcement Museum and **International Spy Museum** offer family fun and education. Embellished with 112 gargoyles and 215 stained-glass windows, the **National Cathedral** is the last resting place of Woodrow Wilson, Helen Keller, and other luminaries.

Exhibits on science, exploration, and photography are the forte of the **National Geographic Museum** in the upscale **Dupont Circle** neighborhood that also includes a **Strivers' Section Historic District** steeped in African American lore. From the famed roundabout, the city's **Embassy Row** stretches northwest along **Massachusetts Avenue.** Each May, a number of the 177 foreign missions organize free open houses with entertainment as part of **Passport DC**, a month-long festival that showcases the city's international community.

Wedged between 27th and 37th Streets, **Georgetown Historic District** is the city's oldest neighborhood and home to the scenic **Georgetown University** campus. But old doesn't mean stodgy, because Georgetown also flaunts one of the city's hippest dining and drinking scenes. It's also the start of the 185-mile (298 km) **Chesapeake and Ohio (C&O) Canal** alongside the **Potomac River** between Washington, D.C., and Cumberland, Maryland. An outstanding example of the nation's canal-building heyday, the **C&O Canal Towpath** offers plenty of scope for walking, running, or biking.

The *Spirit of St. Louis,* lunar module LM-2, and SpaceShipOne are just three of the aircraft on display at the Smithsonian National Air and Space Museum.

ALTER EGO

Baltimore

Given its compact size, it's easy to forget that for much of the 1800s, **Baltimore** was the second most populous city in the United States behind New York—and leagues ahead of Washington, D.C.

Although it no longer carries megacity status, Baltimore's place in American history was sealed in 1814, when the British attack on **Fort McHenry** inspired a Francis Scott Key poem titled "The Star-Spangled Banner," which later became the national anthem.

Visitors invade the fort today for ranger talks and living history demonstrations. Arrayed around the nearby **Inner Harbor** are the **National Aquarium, Maryland Science Center**, and the **American Visionary Art Museum** with its masterpieces by self-taught artists.

The **Baltimore Museum of Industry** recalls the city's glory days as an industrial powerhouse, while historic ships like the U.S.S. *Constellation* docked along the waterfront reflect its ongoing status as a great seaport. Meanwhile, the **Fell's Point Historic District** has morphed from a shore-leave haven for drunken sailors into an area of hip restaurants, bars, and stores.

Away from the harbor, the **Baltimore Museum of Art** boasts the world's largest collection of Matisses, while works at the **Walters Art Museum** span the third millennium B.C. to the early 20th century.

Across the Potomac, **Arlington National Cemetery** is marked by row after row of stark white headstones, the graves of thousands who gave their lives in America's various wars. **President Kennedy's grave**, the **Tomb of the Unknown Soldier**, and the **U.S. Marine Corps War Memorial** are the cemetery's most visited sites. The Virginia side of the D.C. metro area also offers the **Alexandria Historic District**, the Civil War battlefields at **Manassas** and **Fredericksburg**, and outdoor summer concerts at **Wolf Trap National Park for the Performing Arts**, as well as the historic planes and spacecraft seen at the **Steven F. Udvar-Hazy Center** near Dulles Airport.

Connecticut Avenue, another one of the spokes off Dupont Circle, makes a beeline for the **Smithsonian National Zoological Park** (aka the Washington Zoo) and spacious **Rock Creek Park**. The woods and dales where President Teddy Roosevelt once took his friends and family on "scrambles" are now part of a 2,000-acre (809 ha) park administered by the National Park Service. The district's largest park renders various trails and sports facilities, plus a nature center, planetarium, ranger programs, and summer concerts and plays in the **Carter Barron Amphitheatre**.

Among the capital's many green spaces are the **U.S. Botanic Garden** beside the Capitol Building and the **U.S. National Arboretum** on the city's east side. The latter features numerous themed gardens, the **National Bonsai and Penjing Museum** of miniature trees, and the **National Grove of State Trees**. To interact with the locals, head to **Meridian Hill Park** on any summer Sunday and join into the drum circle that's taken place there for more than half a century. ■

Nashville & Memphis

Tennessee, U.S.A.

How many musical roots can you cram into 200 miles (322 km)? In the case of Nashville and Memphis, the answer is quite a few. These are the musical melting pots where the likes of country, blues, rock-and-roll, and half a dozen other genres were largely born and raised.

Located no more than a three-hour drive from one another along Interstate 40, the twin Tennessee towns of Nashville and Memphis have probably done more to nurture American music than all other cities combined.

NASHVILLE

Country music wasn't born in Nashville, but it certainly grew up there. Its roots are in the Ozarks and Appalachians, the cotton fields of Dixie, and the honky-tonks of Texas and Bakersfield, California. But it didn't become a nationwide and global phenomenon until "Music City" came along.

Some other city may have eventually claimed that title if Nashville radio station WSM hadn't launched a nationwide broadcast of the **Grand Ole Opry** in 1925. America's airwaves carried dozens of country music shows at that time. But the Opry managed to outlast them all

THE BIG PICTURE

Founded: Nashville, 1779; Memphis, 1819

Population: Nashville, 1.15 million; Memphis, 1.13 million

Size: Nashville, 563 square miles (1,458 sq km); Memphis, 497 square miles (1,287 sq km)

Language: English

Currency: U.S. dollar

Time Zone: GMT -5 and -6 (CDT and CST)

Global Cost of Living Rank: Unlisted

Also Known As: Nashville—Music City, Athens of the South; Memphis—Bluff City, Home of the Blues

and is now the nation's longest-running radio show.

The second event that solidified Nashville's music props came in 1942, when the Opry moved its weekly broadcast to a former gospel tabernacle called the **Ryman Auditorium**. Soon it was *the* place for country music stars and emerging talents to perform. Nowadays, tours of the redbrick "Mother Church of Country Music" bring the story to life.

Ryman's rise was complemented by an eruption of honky-tonks along nearby **Lower Broadway**, with legendary live music joints such as **Tootsie's Orchid Lounge**, **Legends Corner**, and **The Stage**, where many big names got their starts. Across the street, **Ernest Tubb Record Shop** still hosts live acts in a vinyl store founded in 1947 by its namesake.

A block farther south, the **Music City Walk of Fame Park** features pavement tributes to dozens of music stars as a prelude to the **Country Music Hall of Fame and Museum**, which covers the entire

The legendary Ryman Auditorium still hosts live performances and radio broadcasts.

Discover the anatomy of the guitar—as well as exhibits on country music legends—at the Country Music Hall of Fame and Museum.

history of American country music—from frontier fiddle tunes and folk songs and the gospel hymns of African American enslaved people through Hank Williams, Johnny Cash, and Taylor Swift. The museum offers public activities, from music workshops and family programs to films and live performances in the **CMA Theater**.

The museum is also the place to buy tickets and catch the shuttle bus to the historic **RCA Studio B**. Located along **Music Row** in the city's Edgehill Village neighborhood, the studio produced more than 45,000 songs between 1957 and 1977, including hits by Elvis, Roy Orbison, Dolly Parton, Charlie Pride, and Waylon Jennings.

There are plenty of other places around town to catch live tunes. And not all of it is country. Home to the Nashville Symphony, the state-of-the-art **Schermerhorn Symphony Center** is one of the world's most acoustically sophisticated classical musical venues. Big-name acts tend to play the new **Ascend Amphitheater** in **Riverfront Park** or downtown's **Bridgestone Arena** (also home to the Nashville Predators of NHL ice hockey).

The Gulch, an old industrial neighborhood on the edge of downtown, has been revitalized into an eats and entertainment scene that includes music clubs like the **Station Inn** and **Cannery Row**. Wedged between the **Vanderbilt University** campus and the city's famous full-scale replica of the

BACKGROUND CHECK

• Although he lived much of his life in Nashville, Andrew Jackson was one of three men who founded the city of Memphis in 1819.

• Memphis is named after an ancient Egyptian city on the River Nile, home to the pharaohs who built the great pyramids.

• Local high society initially objected to Nashville's country music fame because they considered it lowbrow for a city they considered to be the cultural "Athens of the South."

• The thoroughbred bloodline established at Belle Meade Plantation during the 19th century continues to produce champions (think Seabiscuit, Secretariat, and Barbaro) and dominate American horse racing.

• The Seeing Eye—the first guide-dog training school in the United States—was founded in Nashville in 1929.

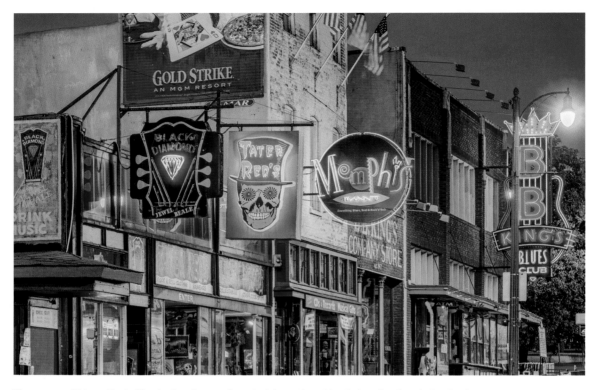

The mecca of blues, Beale Street offers live music, entertainment, and food along its vibrant storefronts.

Parthenon, an outdoor venue called **Musicians Corner** presents two dozen free concerts between May and September among the leafy confines of **Centennial Park**.

Nowadays the Grand Ole Opry splits its time between the Ryman and a modern **Grand Ole Opry House** unveiled in 1974, about six miles (9.7 km) up the **Cumberland River** from downtown Nashville. The new theater is the keystone of an extensive **Opryland** complex that includes the **Opry Mills** outlet mall, **Madame Tussauds Nashville** wax museum, **SoundWaves** water park, **Nashville Palace** live music venue, and a Dukes of Hazzard museum and swag shop called **Cooter's Nashville**.

For a peek at what Nashville was like before tunes took over, take a guided tour of **Belle Meade Plantation** (which also boasts a winery), antebellum **Belmont Mansion**, and Andrew Jackson's home at the **Hermitage**. Or take a drive along the **Natchez Trace Parkway** through the gorgeous Tennessee countryside southwest of Nashville.

MEMPHIS

Located along the Mississippi River at a strategic location between the Deep South and Midwest, Memphis has long served as a meeting place and mixing ground for diverse cultural forces, including music.

Beale Street in downtown Memphis was a gathering spot for African American musicians as early as the 1860s, a trend that continued into the early 20th century, when the city served as a hub for traditional blues and gospel, as well as a newfangled thing called jazz. The advent of radio shows and records stoked a Memphis music industry that created legends like B.B. King and Muddy Waters.

All these years later, Beale Street is still the focus of the local music scene, a seven-block stretch running up from the river that harbors **Rum Boogie**, **Silky O'Sullivan's** pub, the **Blues City Cafe**, and **B.B. King's Blues Club**, among others. Down the street, the historic **Orpheum Theatre** first opened in 1928. Together with a modern annex called the **Halloran Centre**, the grand old dame of Memphis song presents a wide variety of musical acts throughout the year, from country headliners and blues legends to Elvis tributes and the Memphis Songwriters Series. Many of those who have played the Orpheum are enshrined in the **Sidewalk of Stars** outside the theater.

Near the eastern end of Beale Street, the **W. C. Handy House Museum** preserves the modest shotgun shack where the "Father of the Blues" lived in the early 20th century. Just off the avenue, the **Memphis Music Hall of Fame** honors an ever growing number of local musical greats and showcases an oddball collection of mementos of the musical past that includes half of a flashy Cadillac Eldorado once owned by Jerry Lee Lewis and an espionage-worthy briefcase phone that belonged to Elvis. Over on the other side of Beale, the Smithsonian-affiliated **Memphis Rock 'n' Soul Museum** spins tales of how the city provided a venue for Black and white musicians to mingle and merge various strains into two of America's iconic sounds.

The city's other musical cornerstone is **Sun Studio**, where producer Sam Phillips birthed rock-and-roll in the 1950s by recording future stars like Elvis Presley, Roy Orbison, Jerry Lee Lewis, and Carl Perkins. After soaring to the top of the music charts, Elvis purchased **Graceland**, the southside estate where the King of Rock-and-Roll died in 1977. Tours of the mansion and the **Meditation Garden** where Elvis is buried are the main attraction. However, the complex also includes separate museums that house the King's automobiles, motorcycles, private jets, and musical memorabilia.

Memphis was also a mover and shaker in the civil rights movement. The **National Civil Rights Museum** chronicles that five-century struggle via artifacts, films, oral histories, and interactive exhibits. The museum campus includes the **Lorraine Motel**, where Dr. Martin Luther King, Jr., was assassinated in 1968, as well as the boardinghouse where assassin James Earl Ray fired the fatal shot.

The **Slave Haven Underground Railroad Museum** zeroes in on the abolitionist movement inside an antebellum home used to help runaway enslaved people flee the South. ∎

Martin Luther King, Jr.'s room at the Lorraine Motel, where he was assassinated, is preserved at the National Civil Rights Museum.

Los Angeles
California, U.S.A.

Los Angeles and its endless suburbs may have pioneered California's legendary car culture, but there's a lot more to the City of the Angels than cloverleaf interchanges, high-occupancy vehicle (HOV) lanes, and Sig alerts.

THE BIG PICTURE

Founded: 1781

Population: 15.4 million

Size: 2,452 square miles (6,351 sq km)

Language: English

Currency: U.S. dollar

Time Zone: GMT -7 and -8 (PDT and PST)

Global Cost of Living Rank: 20

Also Known As: L.A., City of Angels, Tinseltown, La La Land

Outsiders tend to assume that Los Angeles is all about freeways. Although it's true that the nation's first freeway opened between downtown L.A. and Pasadena in 1940, and that millions of Angelenos travel on freeways each day, surface streets offer deeper exploration that is key to understanding the expansive city.

Sunset Boulevard is the quintessential surface street. Meandering 22 miles (35 km) from the Pacific Ocean to downtown, it passes through an astounding variety of L.A. neighborhoods and along the way personifies many a local cliché.

It starts with surfers and sunbathers on the **Malibu** coast before twisting through the hills of the posh **Pacific Palisades** and **Will Rogers State Historic Park**. Hovering high above the boulevard in Brentwood is the **Getty Center**, both an architectural wonder and one of the world's great art museums. The museum boasts artwork that dates from the eighth century through modern day, dramatic architecture, beautiful gardens, and some of the most famous views of downtown Los Angeles.

Leaping over the San Diego Freeway, Sunset provides the northern boundary of the **University of California, Los Angeles** (UCLA) campus noted for its striking neo-Romanesque buildings, numerous museums, and remarkable athletic success. Climbing up the hillside above the campus is **Bel Air**, where many movie and music stars have lived, from Elizabeth Taylor and Clint Eastwood to Joni Mitchell and Beyoncé.

Descending into the flatlands, the boulevard makes a beeline across **Beverly Hills**, flanked by fabulous mansions and the glamorous **Beverly Hills Hotel** ("Hotel California" of the famous Eagles song). The only manse open to the public is **Greystone Mansion and Gardens**, a 1928 Tudor Revival castle built for the oil-rich Doheny family and now a popular moviemaking spot.

Beyond Beverly is the neon-spangled Sunset Strip and its legendary music clubs: **Whiskey A Go Go**, the **Rainbow Room**, the **Viper Room,** and the **Troubadour**. Tucked between the gargantuan movie and music billboards is **Sunset Tower**, a classic art deco building completed in 1931, and now a boutique hotel. Farther along is the

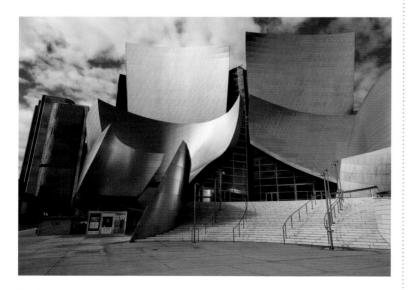

The Walt Disney Concert Hall was designed by architect Frank Gehry.

Griffith Observatory, a popular film location, offers views of downtown L.A. along with its world-renowned exhibits.

Chateau Marmont, a favorite entertainment industry "let's do lunch" location and refuge for renowned writers doing their Hollywood thing (F. Scott Fitzgerald, Dorothy Parker, Hunter S. Thompson, just to name a few).

By the time the boulevard reaches the globe-topped **Crossroads of the World** monument, it's deep into **Hollywood** (the neighborhood, not the frame of mind). **Hollywood Boulevard** and all of its landmarks—the **Chinese Theatre**, **Walk of Fame**, **Hollywood Bowl**, and **Dolby Theatre** where the Oscars are presented each year—are two blocks north of Sunset.

After a fleeting glimpse of the **Hollywood Freeway** (the world's second freeway), the boulevard runs through **Little Armenia** and **Thai Town** with their tasty restaurants, through trendy **Silver Lake** (where Walt Disney built his first animation studio in the 1930s) and **Echo Park**

GATHERINGS

• **Coachella:** One of the world's largest music festivals features chart-topping and emerging artists (past performers include Beyoncé, Childish Gambino, Radiohead, Lorde, and others) across two consecutive April weekends in Indio with more than 250,000 in attendance each year; coachella.com.

• **L.A. Pride Parade:** The world's first gay pride parade (launched in 1970) draws around 200,000 people to West Hollywood each June to champion equity, diversity, and inclusion in a lively display of colorful floats, costumes, and numerous events en route; lapride.org.

• **Mariachi USA:** Bands from around the U.S. and Mexico gather at the Hollywood Bowl in June for a festival that also features folkloric dancing and fireworks; mariachiusa.com.

• **Pageant of the Masters:** Notable works of art are brought to life with actors, music, and technology on summer nights at an amphitheater in Laguna Beach; foapom.com.

Find your favorite celebrity among more than 2,300 stars on the Hollywood Walk of Fame.

(where the West Coast motion picture industry was born before World War I). Sunset finally peters out in **Chinatown**, close to **Dodger Stadium** and the **Olvera Street** neighborhood where Spanish settlers founded Los Angeles in 1781.

By the 1850s, Pueblo de los Ángeles was rapidly expanding to the south, an area that grew into today's **downtown L.A.** The advent of freeways nearly killed downtown, allowing citizens to live, work, and play anywhere they wanted in the Los Angeles Basin and the inland valleys beyond.

However, like many older city centers, the L.A. version is repopulating with younger people who want to "live in a real city" rather than faceless suburbia, and visitors are lured by **The Broad** and **Geffen Contemporary at MOCA** art museums, **Walt Disney Concert Hall**, the revitalized **Arts District** and **Grand Central Market**, **Crypto .com Arena** (home of the NBA's Lakers and NHL's Kings), the **Grammy Museum** at L.A. Live, the **California African American Museum** in **Exposition Park**, the amazing **Skyslide** at the top of the U.S. Bank Tower, and the daily flea market that unfolds along **Santee Alley**.

Another epic surface street starts its journey to the ocean from the heart of downtown. **Wilshire Boulevard** started life as Calle de los Indios ("Street of the Indians") in the Spanish days because it stretched along a route that Native Americans had used for centuries to reach the tar pits and the coast. It wasn't until the auto appeared in the 1890s that Wilshire took on its present name.

The boulevard starts with a giant exclamation point: the 73-story **Wilshire Grand Center**, the tallest building west of the Mississippi River at 1,100 feet (335 m). Perched at the summit of the blue-glass tower are four restaurants and **Spire 73**, the highest open-air bar in the Western Hemisphere.

Leaving downtown behind, Wilshire dives into the **Westlake** district, **MacArthur Park** of the famous 1960s song, and then

ALTER EGO

SAN DIEGO

Around 120 miles (193 km) down the coast from downtown L.A., "America's Finest City," **San Diego**, revolves around the 70 miles (112.7 km) of sand (including clothing-optional Black's Beach), every imaginable water sport, and 340 parks to hike, bike, and chill.

As the first European settlement in California (1769), San Diego's history is on display at **Old Town State Historic Park**, the **Mission San Diego de Alcalá**, **Presidio Park**, and the **Gaslamp Quarter** (which doubles as the city's modern nightlife hub).

Balboa Park boasts 17 museums, many of them housed in buildings from the city's 1915 and 1935 international expositions, as well as the **San Diego Zoo**. Other family-friendly attractions include **Legoland** in Carlsbad and the zoo's sprawling **Safari Park** near Escondido.

Arrayed around big San Diego Bay are the **U.S.S. Midway Aircraft Carrier Museum**, **Cabrillo National Monument**, and the **Maritime Museum of San Diego**'s armada that includes the *Star of India* clipper ship, a reproduction of Cabrillo's Spanish galleon, and two submarines.

With more than 150 breweries, San Diego is America's undisputed capital of craft beer. And being so close to the border, Mexican food is pervasive in places like **Barrio Logan**, **Old Town**, and **Eden Gardens**.

Koreatown, one of the city's most vibrant ethnic neighborhoods. The **Robert F. Kennedy Community Schools** campus is located on the site of the demolished **Ambassador Hotel**, where the presidential candidate was assassinated in 1968.

A 15-block stretch of Wilshire between La Brea and Fairfax Avenues is the **Miracle Mile**, originally named for its many shops and department stores behind classic art deco Streamline Moderne facades. Nowadays the marvel is world-class museums: the esteemed **Los Angeles County Museum of Art** (LACMA), the **Petersen**

Automotive Museum wrapped in massive steel ribbons, the **Page Museum** of prehistoric natural history and adjacent **La Brea Tar Pits**, and the new **Academy Museum of Motion Pictures** inside the old May Company department store.

Wilshire's journey through the Beverly Hills shopping district features **Neiman Marcus**, **Saks Fifth Avenue**, **Tiffany**, and a side street called **Rodeo Drive**. Those skyscrapers off to the south are **Century City** and the historic **20th Century Fox** movie studios. **Westwood** throws up another high-rise forest, the edgy **Hammer Museum** of

contemporary art, and **Westwood Village Memorial Park**, a tiny cemetery where Marilyn Monroe, Dean Martin, Jack Lemmon, Roy Orbison, Truman Capote, Natalie Wood, and many other celebrities are buried.

Crossing the pedestrian-friendly **Third Street Promenade** with its abundant bars, restaurants, and shops, Wilshire Boulevard finally reaches the coast at the palm-fringed palisades of **Santa Monica**. South along the shore are **Santa Monica Pier**, **Muscle Beach**, and the eccentric **Venice Beach Boardwalk**, an oceanfront path that extends

Take an illuminated walk through "Urban Light" by artist Chris Burden on your visit to the Los Angeles County Museum of Art.

You can hike to the landmark Hollywood Sign, originally created in 1923 as an advertisement piece.

four miles (6.4 km) down to **Marina del Rey.**

L.A.'s various beach communities are connected by another fabled thoroughfare—the **Pacific Coast Highway** (PCH), a southern extension of California Highway 1 that runs up and down Los Angeles County.

The northern end of the PCH is anchored by **Leo Carrillo State Beach** and **Neptune's Net**, a popular seafood eatery established in 1956 by a retired NASA rocket scientist. The highway's ramble through Malibu features plenty of other sandy strands, from **Zuma** and **Point Dume**, where plenty of movies have been filmed (like the shocking final scene of the original *Planet of the Apes*) to **Surfrider**

Beach beside the pier in downtown Malibu.

Getty Villa and its priceless collection of ancient Greek, Roman, and Etruscan treasures mark the transition from the beach strip into the urban portion of the PCH. The highway takes on several pseudonyms as it heads south: Lincoln Boulevard in Santa Monica and Sepulveda Boulevard through **Manhattan Beach** (where the Beach Boys were born and raised), **Hermosa Beach** (the best place in L.A. to catch professional beach volleyball), and **Redondo Beach** with its sailboat charters, romantic gondolas, and restaurant-laden pleasure pier.

The PCH curves around to the **Wilmington Harbor** area and

attractions such as the **Battleship U.S.S. *Iowa* Museum** and **San Pedro Fish Market**, before crossing the Los Angeles River into **Long Beach**.

This second biggest city in L.A. County is home to the excellent **Aquarium of the Pacific**, the historic ocean liner ***Queen Mary***, and the **Museum of Latin American Art** (MOLAA), the nation's only collection devoted to Latino and Latin American modern and contemporary art. Long Beach also offers plenty of boating: sportfishing charters, whale-watching cruises, and ferries to Avalon village on **Catalina Island.**

The last of the Los Angeles area's four epic surface streets is Colorado Boulevard in the San Gabriel

HIDDEN TREASURES

- **Hollywood Forever Cemetery:** The final resting place of many showbiz icons (Mickey Rooney, Rudolph Valentino, Charlie Chaplin). During summer months, Angelenos picnic and watch outdoor films on the lawn.

- **California Institute of Abnormal Arts:** Underground art, music, burlesque, and independent films at a Burbank club cum museum that displays freak show memorabilia.

- **Underground tunnels:** Around 11 miles (18 km) of abandoned tunnels

beneath downtown L.A. were originally used to move goods, prisoners, horses, and bootleg booze. Cartwheel Art offers 2.5-hour guided tours of "Underground L.A.," as well as speakeasy, mural, graffiti, and paranormal tours.

- **Phantasma Gloria at RandylandLA:** Echo Park resident Randlett Lawrence continues his decade-old project to build a massive glass Skycatcher sculpture around his home and bathe it in colorful reflections.

Valley east of Hollywood and downtown. Although it's the road "The Little Old Lady (from Pasadena)" cruised in the early 1960's pop song, its fame derives from its status as **Historic Route 66** and the path of the annual **Rose Parade** on New Year's Day.

Colorado starts its westward passage in **Arcadia** near the **L.A. County Arboretum and Botanic Garden** (where many of the early Tarzan movies were filmed) and the sprawling **626 Night Market**, which unfolds more than 10 weekends between May and September at **Santa Anita Park**.

Making a beeline across **Pasadena**, the boulevard runs close to **Caltech** (self-guided walking tours) and the extraordinary **Huntington Library, Art Museum, and Botanical Gardens**, where visitors can gaze on everything from an original Chaucer manuscript and "The Blue Boy" by Thomas Gainsborough to a rare (and exceptionally stinky) titan arum "corpse flower."

Making quick work of **Old Pasadena** and its many bars and restaurants, the boulevard passes

the **Norton Simon Museum** of European and Asian art. The historic **Colorado Street Bridge** (opened in 1912) takes the route high above **Arroyo Seco** (home of the **Rose Bowl**) before a roadside marker at 1500 W. Colorado—the spot where the cheeseburger was invented in 1924.

The boulevard passes through

easygoing **Eagle Rock** (where President Barack Obama first attended college) before gliding into **Glendale**, an Armenian American enclave where the **Museum of Neon Art** and the vast **Glendale Galleria** mall offer roadside amusement.

Colorado Boulevard would probably run all the way to the ocean if not for **Griffith Park** blocking the way. In addition to 53 miles (85 km) of hiking, biking, and equestrian trails, the city's biggest and best park features landmarks such as the **Griffith Observatory**, the **Autry Museum of the American West**, the **L.A. Zoo**, and the beloved **Hollywood Sign**. Formerly the Hollywoodland Sign, this American landmark, which stretches across 350 feet (107 m) with letters that stand 45 feet (14 m) tall, was first erected in 1923. You can reach a scenic overlook of the sign and Los Angeles from a trailhead at the **Bronson Canyon** entrance of Griffith Park. ■

Catch flips and tricks at the skateboarding park along the boardwalk at Venice Beach.

Philadelphia
Pennsylvania, U.S.A.

Nearly 350 years since it was founded, the "City of Brotherly Love" endures as one of North America's oldest and most intriguing cities, a medley of American independence, ethnic neighborhoods, and pop culture highlights perched along the western bank of the Delaware River.

THE BIG PICTURE

Founded: 1682

Population: 5.8 million

Size: 2,060 square miles (5,335 sq km)

Language: English

Currency: U.S. dollar

Time Zone: GMT -4 or -5 (EDT or EST)

Global Cost of Living Rank: Unlisted

Also Known As: Philly, City of Brotherly Love, Workshop of the World, America's Birthplace, Cradle of Liberty

Founded in 1682 by William Penn in an area already occupied by Swedish and Dutch settlers and the Indigenous Lenape people, Philadelphia was the largest city in the 13 colonies (and second largest in the British Empire) by the time the American Revolution rolled around.

So it comes as no surprise that so much history unfolded in and around Philadelphia: the creation of the Declaration of Independence and the U.S. Constitution, Washington crossing the Delaware, the ghastly winter at Valley Forge. For 10 years it served as the U.S. capital. Powered by textiles and railroads, Philly was one of the main sparks of the American industrial revolution. For many years it was at the forefront of labor, women's rights, and African American rights.

Arrayed along four blocks between downtown and the river, **Independence National Historical Park** safeguards many of the buildings where the United States was born including **Independence Hall, Congress Hall**, the **Supreme Court Chamber**, the **First Bank of the United States**, and many more. The nation's most famous bell is now housed inside the glass-wrapped **Liberty Bell Center**, while the adjacent Old City district is flush with historical nuggets like the **National Constitution Center**, the **Betsy Ross House**, and the **Museum of the American Revolution**.

But it's not all history. Downtown Philly nurtures a forest of high-rise towers, including **One Liberty Place** with its 57th-floor observation deck. The new **Comcast Technology Center**—one of the tallest buildings in the Western Hemisphere at 1,121 feet (342 meters)—features a Steven Spielberg virtual reality experience inside a giant silver ball called the **Universal Sphere** as well as a penthouse bar and restaurant.

The city is also endowed with fine art. The **Barnes Foundation** boasts an impressive array of Impressionist art, a collection complemented by the **Rodin Museum**. With more than 200 galleries, the **Philadelphia Museum of Art** is renowned for both the quantity and quality of its artistic treasures—although perhaps best known

The Liberty Bell—on display near Independence Hall—was cracked during a Presidents' Day celebration in 1846.

From the 72 "Rocky Steps" at the entrance to the Philadelphia Museum of Art, you can take in a view of downtown.

nowadays for its monumental **"Rocky Steps"** and statue of Sylvester Stallone's movie boxer. Among the city's other compelling collections are the cutting-edge **Franklin Institute** of science, the **Penn Museum** of archaeology and anthropology, and the kid-friendly **Please Touch Museum**.

Philadelphia also has its macabre side. The eerie **Eastern State Penitentiary**—where the likes of mobster Al Capone and bank robber "Slick Willie" Sutton were once imprisoned—is now filled with tourists rather than convicts. The remains of Albert Einstein's brain and the corpse of the "Soap Lady" are just two of the medical oddities on display at the **Mütter Museum**. The **Edgar Allen Poe National Historic Site** (where the bard lived in

the early 1840s) and the abandoned **Mount Moriah Cemetery** round out the city's creepy attractions.

Several historic ships are moored along the Philadelphia riverfront including the **S.S. *United States*** (the fastest passenger liner of all time), the

submarine **U.S.S. *Becuna***, and the 1890's battleship **U.S.S. *Olympia***, a veteran of the Spanish-American War. The place to party along the Delaware is **Penn's Landing** with its summer beer garden and concert series and wintertime ice-skating rink. ■

HIDDEN TREASURES

• **Reading Terminal Market:** Philadelphians love to "meet at the pig," the market's metallic mascot, before exploring the fresh produce and flower stalls, gourmet eateries, and kitchenware outlets.

• **Mummers Museum:** Find displays of outrageous costumes and other paraphernalia from the New Year's Day parade staged in Philadelphia since 1901.

• **Wanamaker organ:** The world's largest pipe instrument (28,750 pipes) is played for concerts inside Macy's in downtown Philly.

• **Elfreth's Alley:** Constructed in 1702, the cobblestone alleyway is considered the nation's oldest residential street and is flanked by 32 historic federal and Georgian homes.

Seattle
Washington, U.S.A.

On the cusp of the future for more than half a century, Seattle continues to change the world as a whole and reshape its own charismatic cityscape.

From home computing and online shopping to jet planes and frothy flavored coffee, Seattle helped shape how much of the world lives today. We should have seen it coming in 1962, when the city on Puget Sound hosted an international expo. Most people called it the Seattle World's Fair, but its official name was the Century 21 Exposition.

The centerpiece of that fair was the **Space Needle**, a 604-foot (184 m) observation tower that's become as much a symbol of Seattle as the Eiffel Tower is of Paris or Golden Gate Bridge is of San Francisco. Refurbished in 2018, the Needle's viewing deck features new floor-to-ceiling glass windows and translucent benches that literally hang off the edge.

Surrounding the tower, the old expo grounds are now an arts, education, and entertainment district called **Seattle Center**. Among its eclectic attractions are the **Pacific Science Center**, a whimsical blend of art and plants called **Chihuly Garden and Glass**, and the **McCaw Hall** home of the **Seattle Opera** and **Pacific Northwest Ballet**.

But the center's most talked-about tenant is the **Museum of Pop Culture** (MoPOP) inside an eccentric metallic structure that architect

THE BIG PICTURE

Founded: 1851

Population: 3.4 million

Size: 1,239 square miles (3,209 sq km)

Languages: English

Currency: U.S. dollar

Time Zone: GMT -7 or -8 (PDT or PST)

Global Cost of Living Rank: 67

Also Known As: Emerald City, Jet City, Queen City of the Pacific Northwest

Frank Gehry claims was inspired by the Stratocaster guitar of Seattle-born Jimi Hendrix. In addition to the world's largest collection of artifacts on Hendrix and other Seattle musicians, the building houses the **Science Fiction and Fantasy Hall of Fame**.

Another fair relic, the **Seattle Center Monorail** whisks riders to downtown Seattle along an elevated route that stretches for nearly a mile (1.6 km). The city center skyscrapers aren't nearly as interesting as low-level buildings like the artistically rendered **Central Library**—which begs a visit even if you're not borrowing a book—and the **Seattle Art Museum** (SAM) with its limestone facade and iconic **Hammering Man** sculpture.

Directly opposite the art museum, the **Harbor Steps** lead down to the Seattle waterfront and harbor views on the slowly spinning **Seattle Great Wheel**; puffins, moon jellies, and spawning salmon are just a few creatures at the **Seattle Aquarium**; and fresh-off-the-boat seafood and craft stalls can be enjoyed at the historic **Pike Place Market** (opened in 1907).

Huddled around **Pioneer Square**, Seattle's old town preserves remnants

The Amazon Spheres—part of the company's world headquarters—include a "greenhouse" with a larger-than-life bird's nest.

Mount Rainier stands tall in the distance behind Seattle's metal-and-glass skyline, as seen from Kerry Park.

of the city's Wild West days as a rowdy timber town and gateway to the Alaskan frontier. **Klondike Gold Rush National Historical Park** offers a visitors center with historical exhibits and ranger-guided walking tours in summer. **Seattle Underground** and its creepy, sunken gold rush–era structures (fodder for several cinematic horror stories) can be explored on commercial guided tours. And be sure to stroll down **Yesler Way**—the street that coined the term "skid row" back in the day when they skidded freshly cut logs down its steep slope to the waterfront.

Yesler Way leads uphill to the **Chinatown Historic District**. Founded in the 1850s by the city's first Chinese immigrants, the neighborhood has grown to embrace

many Asian ethnic groups. **Hing Hay Park** provides a venue for Lunar New Year celebrations, a summer dragon festival, and other cultural events, while the **Wing Luke Museum of the Asian Pacific American Experience** spotlights

Seattle's multicultural heritage.

Chinatown segues into the **Central District**, the city's oldest Black neighborhood. Musical legends Quincy Jones and Jimi Hendrix are among those who grew up here. Nowadays the neighborhood is

ALTER EGO

Portland

Although Seattle marches boldly into the future, **Portland** revels in what it already has: amazing gardens, a scenic riverside location, and quick escapes to the Cascade Range, **Columbia Gorge**, and Pacific coast.

And did we mention beer? The city's trendy **Pearl District** is a cradle of the craft beer movement that has swept the entire planet. The metro area boasts more than 120

breweries that produce more than 18,000 different types of beer.

Oregon's largest burg is also flush with green spaces: rose and rhododendron gardens, Chinese and Japanese gardens, **Hoyt Arboretum**, sprawling **Forest Park**, and nearby **Mount Hood National Forest**. Hundreds of miles of hiking and biking trails are also on offer—with a well-deserved brew or two at the end of your trek.

Children look at salmon swimming upstream at Ballard Locks.

home to the **Northwest African American Museum** and the **Langston Hughes Performing Arts Institute**. Located in a 1915 building that started life as a synagogue, the latter presents a wide variety of dance, drama, music, and movies.

Seattle's older neighborhoods offer a stark contrast to modern downtown's stunning architecture. The highest building in the Pacific Northwest, the 76-floor **Columbia Center** (933 feet/284 m) is crowned by the **Sky View Observatory & Bar** with incredible views of the city, the Sound, and the snowcapped Cascades. Among downtown's other architectural icons are the futuristic **Seattle Central Library** with its

unique glass-and-steel exoskeleton and the **Arctic Building**, a 1916 beaux arts social club founded by miners who struck it big in the Klondike.

Nearly surrounded by water, Seattle sprawls along a large isthmus between **Lake Washington** in the east and Puget Sound to the west. The big water bodies are connected by the **Lake Washington Ship Canal**, originally built (between 1883 and 1934) as a means to convey logs and timber to the Sound but is nowadays an amazing scenic and cultural asset.

Bookended by **Discovery Park** and the gorgeous **University of Washington** campus, the canal

provides waterfront venues for the **National Nordic Museum** of Scandinavian American culture, the excellent **Museum of History & Industry** (MOHAI) on **Lake Union**, and the **Center for Wooden Boats**, which preserves historic watercraft and also rents boats for cruising the canal. If paddling is more your thing, nearby **Moss Bay** offers rental kayak and paddleboards, as well as guided kayak tours of the waterway.

It's possible to hike sizable sections of the Lake Washington Ship Canal via shoreline trails like the **Cheshiahud-Lake Union Loop**, the **South Ship Canal Trail**, and the **Burke-Gilman Trail**. The waterway is also flanked by plenty of green spaces: **Washington Park Arboretum** near the university, **Gas Works Park** with its awesome skyline views, and the historic **Ballard (Hiram M. Chittenden) Locks**, which offers a visitors center, botanical garden, waterfront promenade, and fish ladder. Home port for much of Seattle's fishing fleet, **Wild Salmon Seafood Market** on the canal's south side tenders an amazing array of fresh-off-the-boat delicacies.

Rather than erecting skyscrapers emblazoned with their corporate logos, Seattle's business giants are leaving a much different legacy. The Amazon **Spheres** shelter 40,000 cloud forest plants from around the globe. The **Bill & Melinda Gates Foundation Discovery Center** is dedicated to innovation and bold ideas. More than 150 planes and spacecraft are housed inside the **Museum of Flight** at Boeing Field. And you can still sip a coffee at the original **Starbucks,** opened in 1971 on the edge of Pike Place Market.

Departing from docks along the downtown waterfront, the ubiquitous **Washington State Ferries** link the city center with spots along Puget Sound that make great day trips.

Bainbridge Island (40 minutes by ferry) offers wineries and restaurants along **Winslow Avenue** as well as regional arts and crafts at the modern **Bainbridge Island Museum of Art** and the poignant **Japanese American Exclusion Memorial** that honors Washington State residents of Japanese descent relocated to internment camps during World War II. **Bremerton** (one hour by ferry) flaunts its maritime heritage with sights like the **Puget Sound Navy Museum,**

ARTBEAT

- **Best Music:** *Electric Ladyland* by Jimi Hendrix, *Dreamboat Annie* by Heart, *Nevermind* by Nirvana, *Superunknown* by Sound Garden, *There Is Nothing Left to Lose* by Foo Fighters, *The Woods* by Sleater-Kinney.

- **Best Movies:** *It Happened at the World's Fair* (1963), *The Night Strangler* (1973), *The Parallax View* (1974), *An Officer and a Gentleman* (1982), *Say Anything* (1989), *Singles* (1992), *Sleepless in Seattle* (1993), *10 Things I Hate About You* (1999), *The Ring* (2002).

- **Best TV Shows:** *Twin Peaks* (1990–91), *Frasier* (1993–2004), *Grey's Anatomy* (2005–present), *The Killing* (2011–14).

- **Best Books:** *Disclosure* by Michael Crichton, *Snow Falling on Cedars* by David Guterson, *Waxwings* by Jonathan Raban, *The Art of Racing in the Rain* by Garth Stein, *Truth Like the Sun* by Jim Lynch, *Where'd You Go, Bernadette* by Maria Semple.

U.S.S. *Turner Joy* museum ship, and the **Puget Sound Naval Shipyard Memorial**.

Seattle also enables breaks from the city on day trips to nearby wilderness areas. **Mount Rainier National Park** is a two- to three-hour drive southeast of the city center. **Deception Pass State Park** on Puget Sound lies 90 minutes to the north, while the Staircase area of **Olympic National Park** is just over two hours to the west via Tacoma and Olympia. ■

The original Starbucks location opened at Pike Place Market in 1971.

Chicago
Illinois, U.S.A.

Chicago is still suffused with the braggadocio that provoked its Windy City nickname. But residents have an awful lot to boast about: world-class museums and championship sports teams, record-breaking skyscrapers and legendary pizza, and a pair of extraordinary streets.

THE BIG PICTURE

Founded: 1780s

Population: 9.3 million

Size: 2,647 square miles (6,856 sq km)

Languages: English

Currency: U.S. dollar

Time Zone: GMT -5 and -6 (CDT and CST)

Global Cost of Living Rank: 45

Also Known As: Windy City, Second City, Chicagoland, City of Big Shoulders, Paris of the Prairie, the Third Coast

Very few large urban areas can be defined by just two streets. But **Michigan Avenue** and **Lake Shore Drive** certainly give it a try when it comes to Chicago. That's because the celebrated downtown street and waterfront road are indelibly connected to many of the events and locations that molded Chicago into the "stormy, husky, brawling, City of the Big Shoulders" that poet Carl Sandburg lauded.

The Windy City actually got its start where modern-day Michigan Avenue crosses the **Chicago River**.

Although French explorer Jean Baptiste Point du Sable was the first person of European descent to settle the shoreline that would grow into America's third largest city, the establishment of **Fort Dearborn** in 1803 fairly well guaranteed a permanent presence. The only reminders of the sturdy wooden stockade are three tributes on the facade of the **London Guarantee Building** that now occupies the spot.

Nowadays the pastoral shores of olden days are barely imaginable, the river engulfed in a canyon of skyscrapers like doubled-towered **Marina City** and dyed green on St. Patrick's Day to honor Chicago's Irish American citizens. Departing from a floating dock beside Michigan Avenue, **Chicago's First Lady Cruises** offers guided sightseeing, photography, architecture, and even yoga-session river tours.

For those who would rather stroll, the **Chicago Riverwalk** runs about a mile and a half (2.4 km) along the south bank between Lake Michigan and the Lake Avenue Bridge with two beer gardens, several cafés, and **Sweet Home Gelato** along the way. Yet another way to explore the river is joining a guided paddle or hiring your own boat at **Urban Kayaks** near the lake end of the Riverwalk.

Michigan Avenue leaps the river on the **DuSable Bridge** (built 1920) to a stretch of the boulevard called the **Miracle Mile** because of its fabulous shops, hotels, and landmark skyscrapers. Foremost among the latter are the neo-Renaissance **Wrigley Building** that chewing gum funded (completed in 1924) and the neo-Gothic **Tribune Tower** newspaper building (completed in 1925), its facade decorated with architectural fragments from

See dinosaurs among other natural history treasures at the Field Museum.

One of the most visited sites in Chicago, the Cloud Gate in Millennium Park was designed by artist Sir Anish Kapoor.

other global landmarks. Opened in 1929 as the Shriners' Medinah Athletic Club, the **InterContinental Chicago** hotel surpasses its neighbors in sheer flamboyance—especially the 14th-floor indoor swimming pool where Johnny Weissmuller trained for his Olympic gold medals and *Tarzan* stardom.

Another six blocks up Michigan Avenue, the stone **Water Tower** is one of the few structures that survived the Chicago Fire of 1871 and remains a symbol of Chicago's resurgence after the blaze. Nowadays, it houses a showroom for local art and photography called the **City Gallery**. Across the street, the nearby **Chicago Sports Museum** offers virtual reality challenges and a cool Curses & Superstitions gallery. No. 875 North Michigan Avenue is better known as the 100-story **John Hancock Center**, the planet's second tallest building when it was finished in 1968 and still among the 20 highest. High-speed elevators rocket visitors to a **Chicago 360 Observation Deck** that floats

1,000 feet (305 m) above the city.

South of the Chicago River, Michigan Avenue quickly reaches **Grant Park**, the city's most celebrated green space and a showpiece of innovative modern outdoor architecture. Founded in 1839 with a mandate to forever remain vacant of buildings, Chicago's "Back Yard" flaunts iconic structures like the big

silver **Cloud Gate** sculpture, aka "The Bean," and the **Jay Pritzker Pavilion**, which hosts alfresco symphony, rock concerts, and other events. The ban on park buildings was broken in the early 1890s but for good reason: creation of the **Art Institute of Chicago** and its world-class Impressionist and American collections.

ARTBEAT

- **Best Movies:** *Scarface* (1932), *His Girl Friday* (1940), *The Blues Brothers* (1980), *Ferris Bueller's Day Off* (1986), *The Untouchables* (1987), *Eight Men Out* (1988), *Hoop Dreams* (1994), *High Fidelity* (2000), *Barbershop* (2002).

- **Best Songs:** "That Toddlin' Town" by Frank Sinatra, "In the Ghetto" by Elvis Presley, "Bad, Bad Leroy Brown" by Jim Croce, "Chicago/We Can Change the World" by Graham Nash, "Chi-City" by Common, "Homecoming" by Kanye West.

- **Best Books:** *Sister Carrie* by Theodore Dreiser, *The Jungle* by Upton Sinclair, *The Man with the Golden Arm* by Nelson Algren, *The Adventures of Augie March* by Saul Bellow, *The Devil in the White City* by Erik Larson, *The House on Mango Street* by Sandra Cisneros, *Native Son* by Richard Wright, *The Time Traveler's Wife* by Audrey Niffenegger.

- **Best Plays:** *Chicago* by Maurine Dallas Watkins, *The Front Page* by Ben Hecht and Charles MacArthur, *A Raisin in the Sun* by Lorraine Hansberry, *Hellcab* by Will Kern, *Glengarry Glen Ross* and *American Buffalo* by David Mamet.

Opposite the art museum, the intersection of Michigan Avenue and Adams Street marks the eastern end of **Historic Route 66**. The nearby Adams/Wabash Station is the closest place to catch the celebrated **Chicago "L"** elevated train that loops around downtown. South of Grant Park, Michigan Avenue cruises through a recently revitalized **Motor Row District**, where car dealerships that sprang up in the 1920s have found new life as trendy restaurants, hip music joints, and even a drag club called **Lips**.

Lake Shore Drive is a whole different urban animal, a stylish waterfront route that stretches 15 miles (24 km) down the Lake Michigan shore between **Kathy Osterman Beach** and Jackson Park. Constructed in the 1880s as a carriage road for the wealthy residents of Chicago's fabled Gold Coast, it was gradually reworked into an expressway after the age of autos arrived.

The drive's northern reaches are dominated by parks and recreation, places like **Montrose Beach**, **Marovitz Golf Course**, and **Belmont Harbor** with its Lake Michigan fishing charters and sailboat rentals. Just a 10-minute walk west of Lake Shore Drive, **Wrigley Field** is home to the Chicago Cubs of major league baseball.

The "800-pound gorilla" of the north shore—in both a literal and figurative sense—is **Lincoln Park**. Created right after the Civil War and named for the recently deceased president, the green space is most renowned for the **Lincoln Park Zoo**, one of the world's more progressive menageries with modern habitats

The Chicago River sits in a valley of skyscrapers and offers opportunities for sightseeing cruises.

---

like the **Regenstein Center for African Apes**, where western lowland gorillas are captive bred.

Among the park's other tenants are the botanical **Lincoln Park Conservatory**, **Chicago History Museum**, and **Peggy Notebaert Nature Museum**, with its living **Butterfly Haven** and an indoor **Wilderness Walk** that re-creates the prairie, savanna, and dune environments once found along Chicago's lakeshore. Near the park's southwest corner, **The Second City** comedy club fostered Bill Murray, Tina Fey, John Belushi, Dan Aykroyd, Stephen Colbert, Nia Vardalos, and many other famous funny people.

The bike- and hike-friendly **Lakefront Trail** (7.6 miles/12.2 km) runs the entire length of northern Lake Shore Drive from Osterman Beach to the magnificent **Navy Pier**. During World War II, the 3,300-foot (1,006 m) wharf was used as the U.S. Navy's Great Lakes training facility. After years of neglect, it was transformed into a theme park with carnival rides and restaurants, as well as the **Chicago Children's Museum**, **Chicago Shakespeare Theater,** and the indoor **Crystal Gardens**.

Lake Shore Drive continues along the eastern edge of Grant Park to the **Museum Campus**, a waterfront venue for four very different attractions: the **Field Museum** of natural history (dinosaurs!), **Adler Planetarium**, **Shedd Aquarium**, and gigantic **Soldier Field** stadium, which hosts both the Chicago Bears of professional football and epic outdoor concerts.

Beyond the Museum Campus, Lake Shore Drive and the Lakeshore Trail continue for another six miles to **Jackson Park**, the least visited of Chicago's trio of big city parks but far from the least interesting. Its lagoons, gardens, and islands are remnants of the Columbian Exposition of 1893, a world's fair that transformed both Chicago and the world as a whole.

The only remaining expo building is the Palace of Fine Arts, which now houses the **Museum of Science and Industry**. The park's other seminal features were added after the fair: the 18-hole **Jackson Park Golf Course** (opened in 1900) and the **63rd Street Beach House** (finished in 1919). If all goes as planned, they will be joined by the **Barack Obama Presidential Center**, the 44th president's official library and museum. ■

New Orleans
Louisiana, U.S.A.

Louisiana's largest city is renowned for its cultural melting pot and around-the-clock party scene. What's not so well known is that New Orleans also boasts a unique blend of neighborhoods, lesser known areas that nearly match the French Quarter when it comes to merriment, memories, and moodiness.

THE BIG PICTURE

Founded: 1718

Population: 970,000

Size: 251 square miles (650 sq km)

Language: English

Currency: U.S. dollar

Time Zone: GMT -5 and -6 (CDT, CST)

Global Cost of Living Rank: Unlisted

Also Known As: The Big Easy, Crescent City, NOLA

Mardi Gras, voodoo, Bourbon Street, and gumbo—New Orleans is saddled with more clichés than any other American city. Although all these are undoubtedly local institutions, that doesn't mean they have to form the core of a visit to the Big Easy.

Take the **French Quarter**, for example. Although you can certainly while away days quaffing hurricane cocktails in the dive bars or trying to channel the spirit of Marie Laveau, the Vieux Carré also has its more refined side.

Catch an early morning Mass at **St. Louis Cathedral**, followed by an alfresco breakfast in **Jackson Square** with coffee and beignets from the original **Café Du Monde** (opened in 1862) across the street. Take a guided tour of the restored **Hermann-Grima** and **Gallier Historic Houses** for a glimpse of New Orleans during its 19th-century heyday.

The Quarter also boasts captivating collections like the African American artifacts on display at the **Backstreet Cultural Museum** (located in an old funeral home),

the Mardi Gras and Hurricane Katrina exhibits at the **Presbytère** state museum, and the **New Orleans Pharmacy Museum** with its ancient elixirs, surgical instruments, and *gris-gris* potions in an apothecary shop established in 1823.

Many locals will swear the real New Orleans lies in the neighborhoods that ring the French Quarter, like the **Marigny** district, where **Frenchmen Street** is flanked by music clubs and the **New Orleans Jazz Museum** in the **Old U.S. Mint** building. Located in an antebellum church, the **Marigny Opera House** presents a year-round slate of dance, music, and offbeat theater.

A longtime bastion of Creole and African American culture, the gritty **Tremé** district offers historic **Congo Square** in **Louis Armstrong Park**, the celebrated **St. Louis Cemetery No. 1** (where the "voodoo queen" Marie Laveau is buried), the **Free People of Color Museum** (Le Musée de f.p.c.), and the **Edgar Degas House**, where the famous French Impressionist lived and painted in 1872–73.

The city's central business district (CBD) is full of really big things like the **Superdome** indoor stadium, the

Dine on classic beignets and chicory coffee at historic Café Du Monde.

A sight to be seen by day or night, Jackson Square boasts the St. Louis Cathedral, first opened in 1794.

National WWII Museum, and the **Audubon Aquarium of the Americas**. Inside, flit through the **Audubon Butterfly Garden & Insectarium**. Pedestrians cannot walk across the towering **Crescent City Connection** bridges over the **Mississippi River**, but they can hop a ferry to **Algiers** with its handicraft workshops, waterfront walk, and **Jazz Walk of Fame**.

From the CBD, you can also ride the vintage **St. Charles Streetcar Line** (operating since 1835) to the **Garden District** and its lovingly preserved 19th-century mansions. Or reach the district by browsing your way along funky **Magazine Street** and its six miles (9.7 km) of eateries, art galleries, and antique shops.

New Orleans' compact size makes it easy to explore that city's hinterland.

Ride a replica paddle wheeler down the Mississippi to **Chalmette Battlefield**, where Andrew Jackson vanquished the British, or explore the bayou trails of **Barataria Preserve**, part of **Jean Lafitte National**

Historical Park. More wilderness awaits on the north side of **Lake Pontchartrain**, where **Big Branch Marsh National Wildlife Refuge** offers hiking through piney woods and wetlands. ∎

ARTBEAT

- **Best Movies:** *A Streetcar Named Desire* (1951), *King Creole* (1958), *The Cincinnati Kid* (1965), *The Big Easy* (1986), *Angel Heart* (1987), *Bad Lieutenant: Port of Call New Orleans* (2009).

- **Best Books:** *Interview with a Vampire* by Anne Rice; *Other Voices, Other Rooms* by Truman Capote; *The Awakening* by Kate Chopin; *A Confederacy of Dunces* by John Kennedy Toole; *The Great Deluge* by Douglas Brinkley; *Nine Lives:*

Mystery, Magic, Death, and Life in New Orleans by Dan Baum.

- **Best TV Shows:** *Treme* (2010–13), *When the Levees Broke: A Requiem in Four Acts* miniseries (2006), *NCIS: New Orleans* (2014–21).

- **Best Music:** *The Definitive Collection* by Louis Armstrong, *Southern Nights* by Allen Toussaint, *Warm Your Heart* by Aaron Neville, *Black Codes (From the Underground)* by Wynton Marsalis, *The Essential Preservation Hall Jazz Band*.

Vancouver
British Columbia, Canada

Vancouver never gets the credit it deserves as one of the world's most beautiful cities. Surrounded by water on three sides and with a backdrop of mountains that rise straight up from the sea, Canada's far western metropolis renders the perfect blend of nature and nurture.

It wasn't the spectacular scenery that lured the first European settlers to the Fraser River Valley area in the 1850s—it was the shiny yellow metal they could pan from local streams. Yet that gold rush provided the seed that would grow into modern Vancouver, a vibrant multicultural city at the crossroads of Canada and the Pacific Basin.

Unlike many other urbanites who seem oblivious to their natural surroundings, those who have come to call Vancouver home have gone out of their way to complement Mother Nature's gifts by cultivating awesome parks, building spectacular bridges, and fostering a skyline that doubles as future Earth in the many science-fiction television shows and movies filmed in Vancouver.

Bold modern architecture is definitely one of its fortes. Like the **Bloedel Conservatory** in **Queen Elizabeth Park**. A huge dome comprising 1,400 acrylic panels, the conservatory is home to more than 500 plant species as well as tropical birds and fish. In the plaza outside is the sculpture **"Knife**

THE BIG PICTURE

Founded: 1862

Population: 2.34 million

Size: 338 square miles (875 sq km)

Language: English

Currency: Canadian dollar

Time Zone: GMT -8 and -7 (PST or PDT)

Global Cost of Living Rank: 93

Also Known As: Hollywood North, Raincouver, Vancity

Edge Two Piece"** by Henry Moore.

Along the city's northern waterfront, **Canada Place** and its five giant fabric sails started life as the Canada Pavilion for the Expo 86 world's fair and is now the city's convention center. Inside is **FlyOver Canada**, a wind-, mist-, and aroma-enhanced virtual reality experience that simulates an aerial journey from coast to coast. Among the other modern architectural treasures are the Colosseumesque **Vancouver Public Library** on Liberty Square, the **Bill Reid Gallery** and its unsurpassed collection of Northwest coast Indigenous art, and another relic of Expo 86 called **Science World**, an interactive museum tucked inside a giant geodesic dome.

Vancouver channels its Wild West roots in **Gastown**, a historic neighborhood wedged between downtown and the harbor. Named after "Gassy Jack" Deighton, a British steamboat captain who allegedly opened the town's first saloon, the historic neighborhood flaunts plenty of spots to eat, drink, and make merry. On the south side of Gastown, the flamboyant **Millennium Gateway** and super-skinny **Sam Kee Building** (opened 1913) announced the start of **Vancouver Chinatown**.

On the tip of Vancouver Island, Butchart Gardens has been celebrating blooms for 100 years.

Vancouver Lookout and the sails of Canada Place mark the harbor front of the robust city.

Vancouver's other blast from the past is **Granville Island,** which floats in False Creek on the peninsula's south side, where bygone factories have been converted into trendy bars, restaurants, art galleries, and performance spaces like the **Arts Club Theatre** and the **Improv Centre**. Farmers from around British Columbia hawk their flowers and fresh produce at **Granville Island Public Market,** which also boasts gourmet food stalls. The **"Giants" of Granville** are six old grain silos transformed into cartoonish murals by graffiti artists.

World-renowned **Stanley Park** crowns the end of the peninsula. Named after the same Canadian governor-general as the Stanley Cup of ice hockey championship fame, the giant green space offers 27 miles (43 km) of hiking and biking trails, including a **Seawall** path that connects beaches, viewpoints, and landmarks along the waterfront. Among the park's human-made attractions are **Vancouver Aquarium,** the cluster of **Brockton Point Totem Poles** on the spot where a Squamish Nation village once stood, and the spectacular **Lions Gate Bridge** (opened in 1938), which leaps Burrard Inlet to the North Shore. ■

ALTER EGO

Victoria

As Vancouver marches boldly into the future, **Victoria** continues to cherish its past as a remote outpost of the British Empire.

In truth, the British Columbia capital—located at the southern tip of Vancouver Island—is a thoroughly Canadian city. But many of its landmarks date to bygone days.

Arrayed around the busy **Inner Harbour** are the domed **B.C.** **Legislative Buildings** (finished 1897), the majestic **Empress Hotel** (opened 1904), the wooden **Church of Our Lord** (opened 1870), and **Thunderbird Park** with its vintage totem poles and historic pioneer-era structures.

Butchart Gardens, the city's top attraction, traces its birth to 1909, when the wife of a local cement tycoon began developing a "sunken garden" in a former quarry.

Montréal
Québec, Canada

Montréal may be the largest French Canadian city—and a wellspring of francophone movies, music, books, and other culture—but La Belle Ville is far from being a homogeneous metropolis.

THE BIG PICTURE

Founded: 1642

Population: 3.6 million

Size: 500 square miles (1,295 sq km)

Languages: French, English

Currency: Canadian dollar

Time Zone: GMT -4 or -5 (EDT, EST)

Global Cost of Living Rank: 129

Also Known As: Ville-Marie (French colonial), La Belle Ville, La Métropole, City of Saints, City of a Hundred Steeples

Like just about every large city in the Western world these days, Montréal supports a wide range of people from all around the globe, more than 200 total ethnic groups. Even among the two-thirds of the population that count French as their mother tongue, many of those are descended from kinfolk who immigrated from Haiti, Lebanon, Morocco, and other former parts of the French Empire.

And like in many other places, these groups carved out their own enclaves, neighborhoods that flavor the metropolis with foods, faiths, languages, and other customs from places far away and very different from the French Canadian heartland.

The city's oldest neighborhood is most definitely French. **Vieux-Montréal** (Old Montréal) hugs a stretch of the St. Lawrence River where Samuel de Champlain established a trading post in 1605. Derelict for decades, the historic buildings of the old town have found new life in the 21st century as boutique hotels, sidewalk cafés, and new cultural attractions. The neighborhood is anchored by **Place Jacques-Cartier** and its entourage of buskers, and the cobblestone **Rue**

Saint-Paul—first paved in 1672 and now the city's oldest street.

Tales of olden days are spun at the cutting-edge **Pointe-à-Callière** museum of history and archaeology, a few blocks from the neo-Gothic bulk of **Notre-Dame Basilica** and its extravagant interior. The silver-domed **Bonsecours Market** (opened in 1847) has morphed from selling fresh vegetables and fish to an elegant emporium of Québécois crafts and fashion.

Along the river, the **Old Port** that once welcomed freighters and passenger ships now swarms with families headed for the **Montréal Science Centre** with its IMAX cinema, the towering **La Grande Roue** Ferris wheel, an indoor adventure maze called **SOS Labyrinthe**, and a dock for the **Saute-Moutons** jet boats that run the wild and crazy Lachine Rapids on the St. Lawrence River.

Starting from the waterfront, **Boulevard Saint-Laurent** cuts across the heart of the city. Nicknamed "the Main" because of its significance to Montréal's evolution, the avenue was declared a national historic site in 2002. Many of the city's most distinctive neighborhoods

Pedestrians make their way to shops and restaurants along cobblestone streets in Old Montréal.

A tour on La Grande Roue Ferris wheel offers a stunning view of the city, including the Bonsecours Market.

are located along its seven-mile (11 km) length, including a new **Quartier des Spectacles** arts and entertainment area, as well as **Chinatown**, **Little Portugal**, **Little Italy**, and the **Historic Jewish Quarter**.

Hometown hero Mordecai Richler both satirized and romanticized life in the Jewish Quarter through books like *The Apprenticeship of Duddy Kravitz*. Though most of the Yiddish-speaking residents have relocated elsewhere in the metro area, the old neighborhood maintains vestiges like **Schwartz's Deli** (opened in 1928), **Segal's Market** (opened in 1927), **Moishes Steakhouse** (opened in 1938), and the **Museum of Jewish Montréal**.

The boulevard continues through an artsy area called **Mile End** with vintage stores, ethnic restaurants, and cocktail bars frequented by artists, musicians, filmmakers, and other creative types. Reflecting Mile End's mixed bag, the **Rialto Theatre** offers a smorgasbord of music,

LAY YOUR HEAD

• **Hotel William Gray:** An 18th-century mansion in the old town provides an atmospheric venue for this chic modern abode beside Place Jacques-Cartier; restaurants, bar, spa, fitness center, library/lounge; from $168; hotelwilliamgray.com.

• **Hotel Kutuma:** Modern African-themed decor doesn't seem a bit out of place at this boutique in the funky Plateau-Mont-Royal neighborhood; restaurant, bar, terrace, parking; from $125; kutuma.com.

• **Le Mount Stephen:** Suites with panoramic ceilings are part of the buzz at this downtown hotel tucked inside an opulent 1920s social club mansion; restaurant, bar, spa, fitness center, robotic Japanese commodes; from $203; lemountstephen.com.

• **Hotel 10:** Located along Boulevard St. Laurent between the old Greek and Jewish neighborhoods, this hip boutique blends a 1915 art nouveau building and modern tower; restaurant, bar, nightclub, fitness center, roof terrace; from $104; hotel10montreal.com.

Stunning gilded frescoes are worth the visit to Notre-Dame Basilica.

drama, dance, comedy, cabaret, and burlesque shows inside a marvelous old movie palace erected in the 1920s. Mile End's cinematic cred is bolstered by the fact that William Shatner (aka Captain Kirk) was born and raised there.

Little Italy revolves around **Jean-Talon Market,** where as many as 300 farmers from the Québec countryside hawk their tasty wares. The market also features specialty stores like **Charcuterie Balkani** (artisanal meats), **Fromagerie Qui Lait Cru!?!** (300 cheese varieties), **Épices de Cru** (global spices), and **La Librairie Gourmande** (cookbooks). Rue Jean-Talon leads to the nearby **La Petite-Patrie** neighborhood known for chic shopping along **Rue Saint-Hubert** and an eclectic eating scene that features Vietnamese, Mediterranean, Latin America, Portuguese, Japanese, and Indian restaurants around a single intersection (Jean-Talon at St. Denis).

The heavily forested peak that rises above these neighborhoods is **Mont-Royal,** the city's namesake and largest and most beloved green space. The park is rife with hikers, bikers, and joggers in summer, cross-country skiers and tobogganers

LOCAL FLAVOR

• **Verses Bistro:** Nouvelle Québécois dishes like salmon gravlax with maple syrup and deer haunch in a black beer sauce are the forte of this chic eatery in the Hotel Nelligan; *100 Saint-Paul Street West, Old Montreal;* versesrestaurant.com.

• **Les Enfants Terribles:** Food with a view on the 44th floor of Place Ville Marie skyscraper; indoor and outdoor tables, popular Sunday brunch, and killer cocktails; *Boulevard Robert-Bourassa, downtown Montreal;* jesuisunenfantterrible.com.

• **St-Viateur Bagel Shop:** Smaller and sweeter than its New York cousin, the bagel developed by Montréal's Jewish community is baked in wood-fired ovens at this beloved neighborhood bakery; *263 St-Viateur Ouest, Mile End;* stviateurbagel.com.

• **Olive et Gourmando:** Foodies flock to this gourmet café turned deli at the southern end of the old town for breakfast, lunch, and snacks; *351 Saint-Paul St. West, Old Montreal;* oliveetgourmando.com.

• **Dinette Triple Crown:** Southern soul food in Little Italy; order a picnic basket with fried chicken, pulled pork, hush puppies, and creamy coleslaw, and eat in the park across the street; *6704 Clark Street, Little Italy;* dinettetriplecrown.com.

in winter. Those who don't wish to work up a sweat can also drive to **Belvédère Kondiaronk** and **Voie Camillien-Houde** with their spectacular city views. From "Musical Sunday" concerts at the **Mont Royal Chalet** (May to October) to **City Lights Snowshoe Excursions** during the winter months, the park hosts many special events.

But it's not the only thing on the mountaintop. Massive **Mont-Royal Cemetery** offers guided history, botany, and birding walking tours, as well as summer **Shakespeare in the Park** in a setting the Bard would most appreciate. **Saint Joseph's Oratory**, Canada's largest church, dominates another part of the summit. The **Oratory Museum** exhibits sacred art from Québec and beyond, while the church hosts a year-round slate of concert series featuring everything from Bach to Christmas carols.

Several of Montréal's outlying districts are also intriguing. Founded by enslaved African Americans fleeing to Canada via the Underground Railroad and immigrants from the Caribbean, **Little Burgundy** is Canada's most acclaimed Black neighborhood. Its famous jazz clubs are long gone, but places like the renovated **Atwater Market**, the new **Arsenal Art Contemporain**, and the ensemble of antique shops and cafés along **Rue Notre-Dame** have brought new life to the old quarter. **Lachine Canal**, which demarcates one side of Little Burgundy, has been transformed into an elongated eight-mile (13 km) national historic park with hiking, biking, kayaking, fishing, and winter recreation.

North of the city center, the **Hochelaga-Maisonneuve** area renders another huge green space, with a much different character than Mont-Royal. Among the highlights of **Parc Maisonneuve** are the **Montreal Botanical Garden**, the creepy-crawly **Montreal Insectarium** (more than a quarter million bugs), **Rio Tinto Alcan Planetarium**, and the **Olympic Stadium**, where the 1976 Summer Games played out. A funicular whisks visitors to the top of the stadium's **Montréal Tower**—the world's tallest leaning tower—for a lofty look at the city.

Montréal's other landmark event—the Expo 67 world's fair—largely transpired on two St. Lawrence River islands that make up **Parc Jean-Drapeau**. Following the expo, the geodesic dome of the U.S. Pavilion was transformed into the **Biosphere** environmental museum, and the French and Québec pavilions made into the **Montreal Casino**. The islands also harbor **La Ronde** theme park and its multiple roller coasters, the 19th-century **Fort de l'Île Sainte-Hélène**, and an amphitheater for shows and concerts in Parc Jean-Drapeau. ∎

Find Montréal-style bagels (sweeter than New York's) at St-Viateur Bagel Shop.

Toronto
Ontario, Canada

Sprawling along the north shore of Lake Ontario, Canada's largest city is at once the most iconic maple-leaf metropolis and a reflection of the nation's ethnic and cultural diversity.

THE BIG PICTURE

Founded: 1750

Population: 6.2 million

Size: 888 square miles (2,300 sq km)

Language: English

Currency: Canadian dollar

Time Zone: GMT -4 or -5 (EDT or EST)

Global Cost of Living Rank: 98

Also Known As: Fort Rouillé, York, the 6ix

By any definition, Toronto got off to a rough start. It was ridiculed as "Muddy York" during English colonial days owing to its sorry state whenever it rained. In the 1850s it lost out in the race to become Canada's capital to much smaller Ottawa. By the turn of the 20th century, downtown Toronto had twice burned to the ground.

It wasn't until the 1970s that Toronto finally surpassed Montréal as the nation's most populous metro area. And the Ontario city has never looked back, growing into a business, cultural, and political force, and a sports powerhouse (the first Canadian city to win a National Basketball Association title).

The tallest freestanding structure in the Western Hemisphere (and tallest in the world for many years), the **CN Tower** looms 1,814.3 feet (553 m) above downtown Toronto. The summit features a viewing deck with glass flooring and panoramic windows, a revolving restaurant, and the **EdgeWalk** adrenaline adventure in which participants walk hands-free around the *outside* of the tower's main pod.

Ripley's Aquarium of Canada and the **Toronto Railway Museum** populate the park at the tower's base. Just a few blocks to the south is **Toronto Harbour** with its waterfront trails, canoe and kayak rentals, and ferries to the parklands, beaches, and amusement centers on the **Toronto Islands** in Lake Ontario.

Heading inland from the CN Tower, the **PATH Skywalk** over the railroad tracks leads to **downtown Toronto** and its forest of skyscrapers. Beneath the surface, a 19-mile (31 km) maze of pedestrian walkways facilitates visits to the **Hockey Hall of Fame**, **Roy Thomson Hall** (symphony, pop concerts, film festivals), and **Nathan Phillips Square** with its diametrically opposed **City Halls**: The old one (1899) is Romanesque Revival; the new one (1965) is boldly and unashamedly postmodern.

Scattered around the fringe of downtown are the **Art Gallery of Ontario** (AGO) and 200-year-old **St. Lawrence Market** with its 120 food and craft stalls, as well as ethnic neighborhoods like **Chinatown**, **Little Italy**, and **Koreatown**, all

Find outdoor cafés, galleries, and shops in the Distillery Historic District.

From the panoramic windows in the viewing deck of the CN Tower, take in the Toronto skyline and Lake Ontario in the distance.

rightly renowned for their ethnic cuisines.

East of downtown along the lakefront, the **Distillery Historic District** safeguards the largest ensemble of Victorian-era industrial buildings left in North America. The district takes its name from the fact it was once home to Gooderham & Worts Distillery, the nation's leading maker of alcoholic beverages. The vintage buildings now host trendy restaurants, art galleries, clothing boutiques, **Mill Street Brewery**, and the **Soulpepper Theatre Company**.

The nearby **Docks** area is slowly transitioning from maritime to entertainment with the addition of an indoor golf driving range, outdoor go-kart track, recreational axe-throwing facility, and a performance space. Farther out along the east coast is **Woodbine Beach**, the

city's best bet for beach volleyball or a dip in Lake Ontario.

The upscale **Yorkville** neighborhood presents a whole different side of Toronto, from fine dining and posh shopping along the **"Mink Mile"** (Bloor Street) to concerts at the **Royal Conservatory of Music** and 30 galleries of artifacts from around the globe at the massive **Royal Ontario Museum** (ROM)—Canada's largest museum of any kind.

Toronto's location makes it a convenient launchpad for day trips to scenic wonders in the region. The Canadian side of **Niagara Falls** is about a 90-minute drive around the western end of Lake Ontario, while **Algonquin Provincial Park**—one of eastern Canada's finest nature areas—is a three-hour drive to the north across the Ontario province countryside. ■

Québec
Québec, Canada

The only remaining walled city north of the Rio Grande blends aspects of old France and modern Canada into a delectable blend of sidewalk cafés, one-off boutiques, outdoor activities, some of the nation's best museums, and four seasons that generate an ever changing cityscape.

THE BIG PICTURE

Founded: 1608

Population: 720,000

Size: 165 square miles (427 sq km)

Language: French

Currency: Canadian dollar

Time Zone: GMT -4 and -5 (EDT and EST)

Global Cost of Living Rank: Unlisted

Also Known As: La Vieille Capitale (the Old Capital), Ville Ancien Cent Clochers (City of a Hundred Steeples)

First-time visitors to Québec are often overcome by an urge to pinch themselves as a reminder that they're exploring a North American city rather than an ancient French ville. A feeling conjured by a blend of cobblestone streets and Old World architecture, an aroma of fresh baguettes each morning and French wine over lunch or dinner, but primarily the fact that more than 90 percent of residents are francophone.

On bluffs along the north bank of the **St. Lawrence River**, Québec's historic old town is divided into two levels—**Haute-Ville** (Upper Town) and **Basse-Ville** (Lower Town), where Nouvelle-France (New France) first took root in the early 16th century after Jacques Cartier's epic voyage across the Atlantic. **Saint-Jean-Baptiste**, **Montcalm**, and **Saint-Roch** were the first neighborhoods outside the city walls as the city spread along the **Plains of Abraham** and the riverside lowlands.

How you wind up passing time in Québec City largely depends on the season. Summer is a time for hiking, cycling, river trips, and other outdoor activities, as well as celebrations like the **Festival d'été de Québec** (an 11-day music festival in July),
the **Fêtes de la Nouvelle-France** (an August event inspired by 17th- and 18th-century New France), and **Grands Feux Loto-Québec** (an incredible August fireworks display).

The popular sidewalk cafés remain open in autumn, but temperatures are cooler and nature lovers rejoice as the city's green spaces blaze with orange, red, and golden foliage. Winter lures cross-country skiers, snowshoers, and snowman builders to the parks and frozen-over river. The **Hôtel de Glace** (Ice Hotel) opens in suburban **Saint-Gabriel-de-Valcartier** and a second festival season arrives, highlighted by the **Marché de Noël Allemand** (German Christmas market in November–December) and the big, boisterous **Carnaval de Québec** in February.

In addition to the **Québec Exquis** food festival in May, the spring thaw brings the city's beautiful gardens back to life. The mix of snowmelt and rain makes this the best time to visit already spectacular **Montmorency Falls**, eight miles (13 km) north of the city center.

Still, any season is right for visiting the **Haute-Ville**, the labyrinth of narrow lanes and cobblestone squares

Chill out, literally, at the spectacular Hôtel de Glace, made entirely of ice.

The famous Fairmont Le Château Frontenac stands tall in the center of downtown Québec City.

tucked inside 17th-century walls that are both a UNESCO World Heritage site *and* Canadian national park. Enclosed by roughly three miles (4.8 km) of ramparts, it's the only remaining fortified city north of Mexico. The entrances are accessible via the **Québec National Historic Site** and the star-shaped **Citadelle de Québec**, an active military base and the secondary official residence of the governor-general of Canada, as well as a military museum.

A broad promenade called the **Terrasse Dufferin** renders an underground archaeological dig and incredible views across the St. Lawrence Valley. Rising behind is **Fairmont Le Château Frontenac**, a stupendous castle-like hotel opened in 1893, and it's worth a breeze through the majestic lobby even if you're not staying there.

The **Funiculaire du Vieux-Québec** and aptly named **Escalier Casse-Cou** (Breakneck Steps) descend to the riverside Basse-Ville,

LAY YOUR HEAD

• **Auberge Saint-Antoine:** Custom-designed rooms in an upscale, family-owned "museum hotel" that flows through several historic buildings near the St. Lawrence River; restaurant, bar, fitness center, cinema, archaeological tours, electric car charging; from $150; saint-antoine.com.

• **Le Germain:** Located in the Old Port area, this stylish boutique hotel offers 60 rooms in a century-old bank building; restaurant, bar, gym; from $133; germainhotels.com.

• **Hotel Manoir d'Auteuil:** Views across the Parc de l'Esplanade to the city walls highlight a hotel located inside a historic 1835 building in old town; breakfast, bar; from $97; manoirdauteuil.com.

• **Auberge Aux Deux Lions:** Three-star hotel with 15 unique rooms and suites in the Montcalm district; restaurant, bar, garden, communal kitchen; from $77; aubergeauxdeux lions.com.

where Samuel de Champlain founded the city in 1608. The early French town revolves around the **Place Royale** and the lovely **Notre-Dame-des-Victoires** church, built between 1687 and 1723 atop the remains of Champlain's trading post.

Around a corner from the square, the **Fresque des Québécois** (Québec City Mural) on Rue Notre-Dame tells the story of Québec City through 25 historical figures and artists painted on the side of a five-story building. With its sundry boutiques and art galleries, pedestrian-friendly **Quartier Petit Champlain** is considered North America's oldest shopping district. Near the southern end of the street is the **Fresque du Petit-Champlain**, another giant mural that depicts working-class life during French colonial days.

The nearby **Old Port**—one of the globe's five busiest harbors in the early 19th century—now shelters yachts and Royal Canadian Navy vessels. Many of the old warehouses along **Rue Saint-Paul** have been converted into boutiques and restaurants like the **Café du Monde** with its panoramic river views. Canada's Indigenous First Nations people, early French settlers, and modern Québécois take center stage at the **Musée de la Civilization**, while scenic cruises depart from the **Quai Chouinard**.

Once upon a time, the **Saint-Roch** area along the St. Charles River was the city's shipyard. After falling on hard times in the late 20th century, the traditional working-class neighborhood has rebounded on the back of hip sidewalk cafés, craft breweries, and one-off stores like **Benjo Toys**. A turnaround so

Take a stroll underneath the colorful art installation of Umbrella Alley in Rue du Cul de Sac.

complete that locals now called it Nouvo St-Roch. The area's history is alive and well thanks to historic **Église Saint-Roch** and the **L'îlot des Palais** archaeology museum housed in an old redbrick brewery.

From the old town's **Porte Saint-Louis** gateway, the busy **Grand Allee** runs along the crest of Parliament Hill beneath imposing structures like the **Parlement du Québec** provincial assembly building (completed in 1886) and the 31-story **Édifice Marie-Guyart**, topped by the **Observatoire de la Capitale** with its 360-degree views of Québec City and beyond. Down the block is yet another notable mural—**Fresque BMO de la Capitale Nationale du Québec**—a colorful take on Québec's political history.

A one-block stretch of the **Grand Allee** between **Place George V** and Rue de la Chevrotière is packed with upscale restaurants in Victorian-era mansions. Farther along is the **Musée National des Beaux-Arts du Québec**, which packs an array of art from the 16th century through present day into four buildings including the new glass-wrapped **Pierre Lassonde Pavilion**.

The nearby **Saint-Jean-Baptiste** neighborhood remains more boho than gentrified, especially the cluster of restaurants, bakeries, and nightlife along **Rue Saint-Jean** and around **Passage Olympia**. Among the area's many intriguing stores are **Épicerie J.A. Moisan** (a grocery store founded in 1871) and **Érico Chocolaterie**, which doubles as a sweet shop and small chocolate museum.

Saint-Jean-Baptiste also supports two performing arts venues: **Grand Théâtre de Québec** and **Théâtre Périscope**, the latter in a former synagogue. The neighborhood is also ground zero for the annual **Fete Arc-en-Ciel** (Rainbow Festival) celebration in September.

Rambling across the riverside bluffs south of the old town are the fabled **Plains of Abraham**, where the British defeated the French in 1759 during the Seven Years War as a prelude to seizing complete and permanent control of New France. The **Plains of Abraham Museum** recounts that pivotal battle in Canadian (and world) history. The only battles nowadays are games waged on the ball fields and lawns that sprawl across the city's largest park. ■

LOCAL FLAVOR

• **Chez Muffy:** French Canadian farm-to-fork menu and 12,000-bottle wine cellar inside a renovated 1822 maritime warehouse; *10 Rue Saint-Antoine, Old Port;* saint-antoine.com/chez-muffy.

• **Restaurant Champlain:** The chef's surprise five-course "Experience Modat" tasting menu and premium "wine discovery" pairing will leave you as breathless as the views from this elegant Château Frontenac restaurant; *1 rue des Carrière, Haute-Ville;* restaurantchamplain.com.

• **La Boîte à Pain:** More than 40 varieties of bread, plus soups, quiches, salads, and croissants at a popular local bakery with four locations around Québec City; boiteapain.com.

• **Aux Anciens Canadiens:** A 17th-century house provides an apropos setting for an eatery that serves savory deep-dish *tourtière* (meat pies) stuffed with pork, beef, veal, rabbit, salmon, or wild game; *34 rue Saint-Louis, Haute-Ville;* auxancienscanadiens.qc.ca.

Las Vegas
Nevada, U.S.A.

Beyond the rhinestones and roulette tables, Nevada's neon-splashed metropolis is actually an eclectic urban area with a wide variety of activities for those who like their fun both indoors and out.

THE BIG PICTURE

Founded: 1855

Population: 2.4 million

Size: 417 square miles (1,080 sq km)

Languages: English

Currency: U.S. dollar

Time Zone: GMT -7 and -8 (PDT and PST)

Global Cost of Living Rank: Unlisted

Also Known As: Sin City, Lost Wages, Entertainment Capital of the World

"I don't get no respect!" was comedian Rodney Dangerfield's catchphrase all those years he headlined casino showrooms on the Strip. Las Vegas could easily make the claim: It's never received the props it deserves as an all-around travel destination rather than just "Sin City."

One of the best kept secrets is the fact that Las Vegas ranks among America's best museum cities. The downtown **Mob Museum** offers an incredible perspective on American organized crime and law enforcement efforts to extinguish it. The collection is located in the historic former **U.S. Courthouse** where Senator Estes Kefauver staged a 1950 hearing to investigate organized crime, and the courtroom where the inquest played out is preserved as part of the museum.

Just a few blocks away, the **Neon Museum** highlights another blast from the Vegas past—more than 200 casino, nightclub, and restaurant signs rescued from the literal scrap heap of history and now safeguarded in a vast "boneyard" that visitors maneuver on guided tours. The museum visitors center is located inside the lobby of the old **La Concha Motel**, a masterpiece of the wacky Googie architecture that spangled American highways in the 1950s.

That's just getting started. The **National Atomic Testing Museum** focuses on the testing of nuclear bombs in the Nevada desert near Las Vegas, while the **Pinball Hall of Fame** features more than 2,000 working machines that patrons can play. Located in the new Art District on the edge of downtown, the **Burlesque Hall of Fame** explores the world of American erotic dance from the 1860s through modern times.

Two entirely different collections are dedicated to Las Vegas legend Liberace: the **Liberace Garage**, which showcases his rhinestone-covered Duesenberg, a Bradley GT painted with genuine gold flakes, and other vehicles; and the **Liberace Museum Collection** of his pianos, outrageous stage costumes, and other mementos now housed at **Thriller Villa**, Michael Jackson's former home. Wayne Newton, another Vegas icon, has his sprawling **Casa de Shenandoah** estate in the city, complete with a priceless art collection and beloved Arabian horses.

Elvis Presley doesn't have his own museum in Vegas. But "the King" is remembered in several other ways, including a statue in the lobby of **Westgate Hotel** (called the

The Insanity Ride hangs 1,149 feet (350 m) above the Vegas Strip.

The Vegas Strip glows bright at night, with the famous Bellagio fountains as a centerpiece.

International in the 1970s when Elvis lived and performed there for seven years), **Pink Cadillac Tours** in a vintage ragtop Caddy driven by an Elvis impersonator, and Elvis-themed nuptials at **Graceland Wedding Chapel**.

Several casino-hotels boast their own art and history collections, most notably the **Bellagio Gallery of Fine Art** (where rotating exhibits mainly feature famous Impressionists and 20th-century masters), **Titanic: The Artifact Exhibition** at the Luxor Hotel (more than 250 objects recovered from the infamous shipwreck), and the anatomically focused **Real Bodies** at Bally's. The **Aria Fine Art Collection** includes large-scale works by the likes of Henry Moore, Nancy Rubins, and Frank Stella dispersed around the casino-hotel, as well as a satellite sculpture installation called **Seven Magic Mountains** found in the

desert roughly 10 miles (16 km) south of the Strip.

For those who relish full-on glitz and glamour, nothing beats **the Strip**. Officially dubbed Las Vegas Boulevard, the street was born in the 1930s, as the Nevada end of the cross-desert highway to Los Angeles. The long-gone El Rancho (opened

in 1941) was the boulevard's first casino, but the Strip really began its rise to fame in 1946, when gangster Bugsy Siegel opened **The Flamingo**, a tropical art deco–style casino-hotel.

Even if you don't gamble, the Strip is a gas that features a cluster of gourmet eateries run by celebrity chefs; shows starring world-renowned

GATHERINGS

• **Electronic Daisy Festival:** The world's premier electronic dance music (EDM) event takes over Las Vegas Motor Speedway for three days in May with carnival rides, giant art installations, and wacky costumes as part of the fun; lasvegas.electricdaisycarnival.com.

• **Life Is Beautiful:** Art, music, food, and learning are the four cornerstones of a September fest that engulfs 18 blocks of downtown Las Vegas; lifeisbeautiful.com.

• **Rise Lantern Festival:** World peace and wonderful music are the vibes at this October event in the Mojave Desert that culminates in the world's largest lantern release; risefestival.com.

• **National Finals Rodeo:** America's super bowl of cow roping, bull riding, and bronco busting takes place each December at the Thomas & Mack Center along with a holiday gift fair called Cowboy Christmas; nfrexperience.com.

singers, magicians, and comedians—and as many as six **Cirque du Soleil** extravaganzas at any one time; shopping that runs the gamut from top-shelf luxury goods to kitschy souvenirs; and probably the best people-watching on planet Earth.

If walking the Strip gets to be too much (especially during the triple-digit temperatures of the Nevada summer), hop onto the **Las Vegas Monorail**. The 3.9-mile (6.3 km) line features seven stations between the **Sahara** hotel at the north end of the Strip and the colossal **MGM Grand**. For those who want to continue southward, a series of air-conditioned pedestrian bridges connects the MGM with five other major resort hotels.

In addition to the aforementioned hotel museums, the big digs along the Strip also feature action attractions like the carnival thrill rides atop the 1,149-foot (350 m) **Stratosphere Tower**, the **Big Apple Coaster** that undulates across the top of New York-New York Hotel and Casino, **Shark Reef Aquarium** and the artificial wave pool at the **Mandalay Bay**, and **High Roller** at the **LINQ Promenade**—the world's tallest Ferris wheel at 550 feet (168 m) above ground level.

There's more action of the gambling and adrenaline-sports kind along **Fremont Street** in downtown Las Vegas. Under the watchful eye of **Vegas Vic**, the much photographed neon cowboy, "Glitter Gulch" boasts historic casinos like the **Golden Gate** (opened 1906), **Pioneer Club** (opened 1942), and **Golden Nugget** (opened 1946). Flying high above the street is the **SlotZilla Zip Line** and higher still is a barrel-shaped canopy that hosts **Viva Vision,** where sound-and-light shows unfold on the world's second largest video screen.

Las Vegas is also a great base for day trips into the **Mojave Desert** that engulfs the city. Just beyond the western suburbs, **Red Rock Canyon National Conservation Area** offers hiking, mountain biking, rock climbing, and horseback riding along routes leading to lofty

Find old-time gambling machines and other treasures at the Mob Museum.

Take a ride outside the city through the majestic views of Red Rock Canyon National Conservation Area.

viewpoints and oasis canyons in a pristine Bureau of Land Management (BLM) preserve. The often snowcapped peak rising beyond the ruddy cliffs is **Mount Charleston**, an 11,916-foot (3,632 m) mountain with hiking through forested areas during the warmer months and the **Lee Canyon** snow sports resort in winter.

Over on the other side of town, **Valley of Fire State Park** is another geological wonderland, with strange rock formations like the Elephant, the Fire Wave, and the Silica Dome, as well as Native American petroglyphs.

The area's largest park, **Lake Mead National Recreation Area**, offers 1.5 million acres (607,000 ha) of aquatic adventure and desert wilderness along the Colorado River corridor. **Willow Beach Harbor** rents motorboats,

pontoon boats, canoes, and kayaks in blocks ranging from two to eight hours, while **Desert Adventures** is one of several outfitters with guided one-day paddle trips down the deep **Black Canyon of the Colorado**.

Lake Mead is also home to the **Hoover Dam**. Completed in 1936, the giant concrete barrier is both a marvel of 20th-century engineering and a tour de force of art deco architecture. Guided tours of the dam and power plant descend deep into the belly of the beast. The best view of the dam is from the acrophobia-inducing pedestrian walkway on the **O'Callaghan–Tillman Bridge**.

It's also possible to visit the **Grand Canyon** on flightseeing excursions from Las Vegas, including helicopter tours that touch down in the gorge. ■

Santa Fe
New Mexico, U.S.A.

Santa Fe is the urban equivalent of an archaeological site, with layers of cultural and historical strata that reveal an enchanting blend of Native American, Spanish colonial, Wild West, bohemian artists, and most recently, the epitome of Southwest chic.

THE BIG PICTURE

Founded: 1610

Population: 85,000

Size: 37 square miles (96 sq km)

Language: English

Currency: U.S. dollar

Time Zone: GMT -6 and -7 (MDT and MST)

Global Cost of Living Rank: Unlisted

Also Known As: La Villa Real de la Santa Fe de San Francisco de Asís (Spanish colonial), Oghá P'o'oge (Tewa)

Santa Fe has always been a lodestone, a place that attracts people searching for that proverbial greener grass. "Actually touch the country, and you will never be the same again," poet D. H. Lawrence wrote of New Mexico in the 1930s.

In Spanish days the grass was literal: establishing a foothold of Iberian civilization amid the unrelenting desert. After the U.S. annexation of the 1840s, it was American settlers hoping for a new life in the West. By the 20th century, the attraction had turned metaphorical—artists and writers seeking inspiration, young people searching for nirvana, city slickers enthralled with the wide-open spaces.

Three entirely different museums delve into the region's Indigenous culture. Crowning the summit of the city's **Museum Hill**, the **Museum of Indian Arts & Culture** focuses on art, anthropology, culture, and history, while the **Wheelwright Museum of the American Indian** zeroes in on modern and historic Native American art. Meanwhile, the **Museum of Contemporary Native Arts** on Cathedral Place showcases contemporary intertribal art via exhibitions and special events.

Beyond the city limits, the **Rio Grand Valley** offers Native American sites like the modern **Poeh Cultural Center** at Pojoaque Pueblo, **La Cienequilla Petroglyphs**, and the 11,000-year-old ruins of **Bandelier National Monument**.

Santa Fe Plaza flaunts Spanish roots that stretch all the way back to the early 17th century, when soldiers, priests, and civilians arriving from Mexico founded the city. The adobe **Palace of the Governors**—where Lew Wallace finished writing *Ben-Hur* while he was the territorial governor in the 1870s—now houses the **New Mexico History Museum**. Arrayed around the plaza are Mexican and Southwest eateries, art galleries, and stores selling cowboy boots, sheepskin coats, and other Western wear.

Two blocks east of the plaza, **St. Francis Cathedral** is the latest iteration in a series of Catholic churches that have stood on the site since the 1620s. Of special note are the French rose window, the nation's oldest Madonna statue, and bronze front doors that portray the cathedral's history. Santa Fe's oldest church,

Find Native American crafts and clothing, like concho belts, in shops throughout the city.

The Poeh Cultural Center highlights the arts and culture of the Puebloan peoples, especially the Tewas.

Mission San Miguel was erected shortly after the city was founded and has been restored to look much as it did in Spanish times. The mission and the nearby **Oldest House** (built in 1610) are the nexus of a **Barrio de Analco Historic District** that harbors six colonial-era structures.

Santa Fe's Wild West era is best expressed by **Rancho de las Golondrinas** on the southern edge of town, a living history museum dedicated to New Mexico's 18th- and 19th-century ranching heritage. Back in town, the **Railyard Arts District** has transformed from a train depot and service area into an eclectic array of art galleries, farmers market, drinking establishments, and the **Violet Crown** indie movie house.

The **Georgia O'Keeffe Museum** preserves artwork and mementos of the most famous of the creative minds who flocked to Santa Fe in the early 20th century. Depending on your fondness for luxury shopping and spas, **Canyon Road** represents either the best or worst of modern Santa Fe. But there's always the **Santa Fe Opera** (are the structure or the shows more dazzling?) and an oddball interactive art installation called **Meow Wolf**. ■

URBAN RELIC

Taos Pueblo

Continuously inhabited for more than 1,000 years, **Taos Pueblo** in northern New Mexico is one of the oldest examples of urban living in the Americas.

The pueblo comprises two main structures—Hlaukwima (South House) and Hlauuma (North House)—where around 150 permanent residents live.

Constructed entirely of adobe with wooden beams for roof support, the pueblo was declared a UNESCO World Heritage site in 1992.

The community hosts many events throughout the year including a Deer or Buffalo Dance in January, Taos Pueblo Powwow in July, San Geronimo Feast Day in September, and Procession of the Virgin Mary on Christmas Eve.

Boston
Massachusetts, U.S.A.

Many of the ideas that shaped the United States of America—and by extension the entire world—were hatched in the minds of men and women from Greater Boston, an urban area that gradually drifted away from its staunch Puritan roots to become a beacon of progress.

From Titletown to Beantown, Boston has earned many monikers over the years. Perhaps it's time for another: Heroesville USA or something of the sort. Because there isn't another city that has produced so many iconic Americans.

From bygone greats like John Adams, Ben Franklin, Louisa May Alcott, and Henry David Thoreau, to modern-day heroes like John F. Kennedy, Julia Child, Tom Brady, and Mark Wahlberg, the list of prominent Bostonians goes on and on.

As they left their mark on the city as a whole, they also touched certain Boston neighborhoods like the **North End**, home to many of those responsible for the ideals and events that sparked the American Revolution. Marked by red bricks, the 2.5-mile (4 km) **Freedom Trail** links numerous North End landmarks associated with the heroes of that era, including **Paul Revere House** on North Square, **Faneuil Hall** where Sam Adams and others lobbied for liberty from their British oppressors,

THE BIG PICTURE

Founded: 1630

Population: 4.6 million

Size: 4,500 square miles (11,655 sq km)

Languages: English

Currency: U.S. dollar

Time Zone: GMT -4 and -5 (EDT and EST)

Global Cost of Living Rank: 50

Also Known As: Cradle of Liberty, Beantown, Titletown, Athens of America, the Hub, Trimountaine (early colonial)

the **Old North Church** of "One if by land, two if by sea" fame, the **Old South Meeting House** where Ben Franklin was baptized in 1706, and the spot where the **Boston Massacre** played out in 1770.

By the end of the 19th century, Italians and other immigrants had replaced the North End's longtime Anglo-American residents, transforming the neighborhood into Boston's Little Italy. **Hanover Street** and **Salem Street** are still crowded with Italian cafés, trattorias, and bakeries like **Mike's Pastry** (try the cannoli).

That's not to say that there's nothing modern. Like the rest of Boston, the North End has undergone some radical changes. The neighborhood was once split by the JFK Expressway, but a massive 16-year construction project dubbed the "Big Dig" sent the highway underground, replaced by the leafy **Rose Fitzgerald Kennedy Greenway**.

Meanwhile, the steamers that once docked at local finger piers have been replaced by swank condos, whale-watching tours, harbor cruise boats, and the excellent **New England Aquarium**. The North End waterfront is also the jumping-off spot for

Paddle a historic Swan Boat on a pond inside Boston's Public Garden.

The Boston Tea Party Ships and Museum offers an educational and enlightening adventure back in time to December 1773.

ferries to Salem, Provincetown, and **Boston Harbor Islands National Recreation Area**, a cluster of more than two dozen islands and peninsulas with beaches, picnic areas, hiking trails, historic forts and lighthouses, and wildlife areas.

The Freedom Trail continues through the high-rises of downtown Boston to **Beacon Hill**, another hub of American liberty. In 1795, Paul Revere laid the ceremonial cornerstone of the golden-domed **Massachusetts State House** on a plot that once belonged to John Hancock. Revere, Hancock, Sam Adams, and Boston Massacre victims are buried around the corner in the **Granary Burying Ground**. The redbrick trail ends on the **Boston Common**, famed as the nation's first public park and as a place where British

troops bivouacked during the Revolution. The greenery spills over into the **Boston Public Garden** with its Swan Boats and darling **"Make Way for the Ducklings"** sculpture.

Back in the day, Beacon Hill was also an epicenter for Boston's African American population and by extension a hotbed of abolitionism. That legacy is the focus of a 1.6-mile (2.6 km) **Black Heritage Trail** that includes the **Massachusetts 54th**

BACKGROUND CHECK

• Christmas was banned in Boston from 1659 to 1681 because the colony's Puritan leaders considered it a pagan holiday rather than Christ's birthday.

• Until Back Bay was landfilled during the late 19th century, the North End and Beacon Hill were confined to a virtual island attached to the mainland by a narrow isthmus.

• Boston was the site of the nation's first chocolate factory (1765), first

public beach (1896), and first subway rail line (1897).

• Ben Franklin and Louis Farrakhan are among the better-known teenage dropouts from Boston Latin School in the North End.

• Athletic teams at Tufts University in suburban Medford are called the Jumbos in honor of the famous elephant that belonged to circus showman (and Tufts benefactor) P. T. Barnum.

Find the Old State House and Custom House Tower nestled amid downtown Boston's modern skyscrapers.

Regiment Memorial, sites associated with the **Underground Railroad**, and the **Museum of African American History** in a former church and meetinghouse where William Lloyd Garrison founded the New England Anti-Slavery Society in 1831.

On the other side of Arlington Street are **Back Bay** and **Fenway**, an eclectic area along the south side of the **Charles River** that transitions from elegant brownstones into the city's tallest skyscrapers and favorite stadium.

Among the older landmarks are graceful **Trinity Church** (stunning stained-glass windows) and the Renaissance-style **Boston Public Library** (John Singer Sargent murals). You can catch the **Boston Pops** at the neighborhood's **Symphony Hall** and ogle world-renowned works at the **Museum of Fine Arts** (MFA). Modern **Copley Place** mall is surrounded by a high-rise forest that includes the 52-story **Prudential Tower**. Outside the mall, amphibious duck boats depart on fun, fact-filled tours that include a cruise on the Charles River.

Women were instrumental in creating some of the area's iconic institutions. The massive **First Church of Christ, Scientist** (which harbors one of the world's largest pipe organs) is the foremost architectural expression of a metaphysical religious movement founded by Mary Baker Eddy in 1879. Inside the adjoining library is the **Mapparium**, an incredible stained-glass globe viewed inside out from a glass bridge. Around the same time, an eccentric philanthropist and ardent Boston Red Sox fan was collecting enough priceless art to fill her namesake **Isabella Stewart Gardner Museum**.

Gardner's beloved Red Sox play nearby at **Fenway Park**. Opened in 1912, it's the oldest stadium in major league baseball and produced whole different kinds of local legends—Ted Williams, Carl Yastrzemski, "Big Papi" David Ortiz, and their teammates. A guided tour of Fenway—including a chance to stand atop the famous **"Green Monster"** left-field wall—is a must for any sports fan. Just down Brookline Avenue, a former Sears distribution center has found new life as **Time Out Market Boston**, a trendy gaggle of bars and gourmet eateries, a fitness club, an outdoor adventure store, and a multiplex cinema.

Across the Charles River, Cambridge is home to **Harvard University** and the **Massachusetts Institute of Technology** (MIT) and their own museum clusters. Down at the confluence of the river and Boston Harbor, the old **Charlestown** neighborhood anchors the north end of the Freedom

Trail with landmarks like **Bunker Hill** and the **U.S.S. *Constitution*,** a legendary three-masted frigate that served in the U.S. Navy for nearly a century.

South Boston is another neighborhood in transition, especially the **Seaport District**, where the waterfront is now flush with restaurants, craft breweries, and cultural institutions like the **Institute of Contemporary Art**, the **Boston Children's Museum**, and the **Boston Tea Party Museum & Ships**. Once dominated by working-class Irish Americans, "Southie" is undergoing a dramatic gentrification that includes a slew of hip bars, restaurants, and shops along West Broadway.

Perched at the bottom end of South Boston is a tribute to the most famous Irish American of them all: the **John F. Kennedy Presidential Library & Museum**. Farther down the shore is Quincy, where **Adams National Historical Park** preserves the heritage of two other American patriots who became presidents.

Boston's western suburbs include leafy **Lexington** and **Concord**, where the American Revolution started on April 19, 1775, with the "shot heard round the world." **Minute Man National Historic Park** offers battle relics like the **North Bridge** and **Paul Revere Capture**

Visit the Orchard House, Louisa May Alcott's childhood home, now a museum.

Site, as well as the 4.5-mile (7.2 km) **Battle Road Trail**. Concord also boasts American literary sites like the **Old Manse** home of Ralph Waldo Emerson and Nathaniel Hawthorne, Louisa May Alcott's **Orchard House**, and **Walden Pond**, where local lad Henry David Thoreau kick-started the global environmental movement. All four of them are buried in the town's **Sleepy Hollow Cemetery**.

More history awaits along Boston's **North Shore**. From the **House of the Seven Gables**, the **Salem Witch Trials Memorial**, and the whimsical **Bewitched Statue of Elizabeth Montgomery**, there are plenty of ways to explore Salem's dark side. But the seaside town is also home to the **Salem Maritime National Historic Site** and its historic wooden ships. The setting of the *Wicked Tuna* television series, **Gloucester** is an active fishing port with whale-watching tours, the **Rocky Neck Art Colony** of galleries and shops, and a long waterfront promenade that leads to the sandy beaches of **Stage Fort Park**. ∎

LAY YOUR HEAD

• **Four Seasons One Dalton:** Boston's latest, greatest five-star property offers posh digs in Back Bay with views to die for and incredible Japanese cuisine at Zuma; restaurants, bar, spa, fitness center, indoor pool; from $445; fourseasons.com/onedalton.

• **Inn at Hasting Park:** The Boston area's only Relais & Châteaux property, this chic boutique inn near Lexington Green flaunts designer-savvy guest rooms and epic cuisine; restaurant, bar; from $172; innathastingspark.com.

• **Beauport Hotel:** This seaside getaway on Gloucester Harbor offers a sandy beach, rooftop bar, and swimming pool; restaurants, fitness center, bike rental; from $188; beauporthotel.com.

• **The Liberty Hotel:** The old Charles Street Jail in Beacon Hill provides an evocative setting for this modern, waterfront luxury hotel; restaurants, bar, fitness center; from $158; marriott.com.

Havana
Cuba

It's cliché to call Havana a time capsule. But that's exactly what the Cuban capital is: a city that seems forever stuck in the 1950s or early 1960s in architecture, automobiles, and so much more. Which is why it's on the bucket list of so many travelers: one of those rare places where it actually feels like you're traveling back in time.

THE BIG PICTURE

Founded: 1515

Population: 2.1 million

Size: 120 square miles (311 sq km)

Language: Spanish

Currency: Cuban peso

Time Zone: GMT -4 or -5 (EDT or EST)

Global Cost of Living Rank: 134

Also Known As: Ciudad de las Columnas (City of Columns)

The first thing that anyone should do in Havana is undertake a casual walkabout of the old town and waterfront, taking in the sights, sounds, and smells of another era.

Duck in and out of the funky bars and art galleries along pedestrian-only **Calle Mercaderes**. Catch a concert in the cobblestone **Plaza de la Catedral** or street musicians in the ancient **Plaza de Armas**. Feel the spray on your face as waves break over the **Malecón** seawall along the Florida Straits. And ride a vintage Buick or Chevy taxi to the **Hotel Nacional** for martinis (shaken or stirred) in one of the historic bars.

Nearly all of **Habana Vieja** (Old Havana) is a UNESCO World Heritage site that includes many of the city's Spanish colonial relics, including the baroque 18th-century **Catedral de Habana**, the exquisite **Plaza Vieja** with its lively bars and outdoor cafés, the moat-enclosed **Castillo de la Real Fuerza**, and the hulking **El Morro** and its trademark lighthouse on the other side of the harbor entrance, a massive Renaissance-style fortress constructed between 1589 and 1630.

Fast-forward several hundred years to independent Cuba, as exemplified by the **Capitolio Nacional**, the huge domed structure unveiled in 1929 and modeled after the Panthéon church in Paris. Guided tours of the former legislative building include a glimpse of "La Republica," the world's third largest indoor statue. Havana's art deco masterpiece is **Edificio Bacardí**, opened in 1930 as headquarters of the famous rum company. Visitors can admire the lobby, but the best view of the building is from the roof terrace at the historic **Plaza Hotel** (opened in 1909) across the street.

The city entered its third phase in 1959, when Fidel Castro and his guerrillas swapped American imperialism for Soviet-style communism. Relics like the *Granma* yacht are displayed at the expansive **Museo de la Revolución** in the former Presidential Palace. Even though it seems Stalinist in form and function, the massive **Plaza de la Revolución** was designed by a

A monument to revolutionary Che Guevara looks down onto Revolution Square.

Explore Paseo de Martí in Habana Vieja (Old Havana) to see the colorful and historic architecture, as well as vintage cars.

Frenchman. Three heroes of the revolution—Camilo Cienfuegos, José Martí, and Che Guevara—are honored by supersize monuments around the square.

Havana boasts a very active arts scene as personified by the **Museo Nacional de Bellas Artes** (one of the world's largest and most significant collections of Latin American art) and the recently refurbished **Gran Teatro,** home of the national ballet company and annual ballet festival as well as various other dance events. The city's live music scene ranges from the venerable **Casa de la Música** to the cheesy (but enchanting) floor show at the Tropicana, while literature is the focus at **Finca Vigía,** where Ernest Hemingway

wrote *For Whom the Bell Tolls* and *The Old Man and the Sea*.

Havana also has its lighter side: the transformation of sugarcane into rum at the **Museo del Ron** in the Fundación Havana Club, Cuban-style baseball at **Estadio Latino-americano,** and the whimsical **Fusterlandia**, a fantasyland of out-door folk-art mosaics. ∎

ALTER EGO

Santiago de Cuba

The island's second largest city, **Santiago de Cuba** is celebrated as the spot where Castro launched the Cuban Revolution, the place where Teddy Roosevelt charged up San Juan Hill, and as one of the world's most musical cities.

Cuba's Afro-Caribbean culture is much more prominent in Santiago than the capital, a factor that plays into the city's art, dance, and music.

Joints like the legendary **Casa de la Trova, Iris Jazz Club,** and **Casa de las Tradiciones** present live musical acts from a wide variety of world beat genres.

The **Moncada Museum** details how the revolution started in 1953, with Castro's attack on the barracks. Roosevelt and his Rough Riders are immortalized by monuments on Loma de San Juan on the city's east side.

St. Louis & Kansas City

Missouri, U.S.A.

Missouri is very much a tale of two cities—St. Louis and Kansas City—that slowly evolved from frontier outposts and economic powerhouses into the eclectic Renaissance cities they are today.

Separated by a long stretch of the Missouri River and 250 miles (402 km) of Interstate 70, St. Louis and Kansas City have more in common than serving as Missouri's largest cities.

Both were gateways for the conquest of the American West, as well as trade and transportation hubs, followed by economic decline that left them languishing at the edge of the rust belt before urban renewal and new industries restored their civic vigor.

ST. LOUIS

High above downtown St. Louis looms the glimmering **Gateway Arch**, a symbol of both the city and America's manifest destiny. Upgraded to national park status in 2018, the world's tallest arch (630 feet/192 m) is the centerpiece of a complex that also includes the **Old Courthouse** (1864) and **Mississippi River** paddleboat tours.

A string of parks and plazas called **Gateway Mall** stretches 17 blocks through the heart of downtown between the arch and **Union Station**, an 1890s train terminus reworked in modern times into a luxury hotel (the station's **Grand Hall** is an art nouveau masterpiece).

Basically an open-air museum of American architecture, the mall is flanked by other iconic structures like the art deco-style **Stifel Theatre** (1934), the neo-Renaissance **City Hall** (1904), and the terra-cotta-colored **Wainwright Building**, unveiled in 1892 as one of the globe's first skyscrapers.

Reflecting the nation's westward march, the city's urban core continues in a westerly direction to **Forest Park**, venue for both the 1904 Louisiana Purchase Exposition and the Summer Olympics. Fairground attractions were repurposed into the **St. Louis Art Museum** and **St. Louis Zoo**, renowned for its research and conservation efforts. With its Victorian district and tropical Climatron greenhouse, the nearby **Missouri Botanical Garden** is also deemed one of the nation's finest.

THE BIG PICTURE

Founded: St. Louis, 1764; Kansas City, 1821

Population: St. Louis, 2.2 million; Kansas City, 1.7 million

Size: St. Louis, 978 square miles (2,533 sq km); Kansas City, 719 square miles (1,862 sq km)

Currency: U.S. dollar

Time Zone: GMT -5 or -6 (CDT or CST)

Language: English

Global Cost of Living Rank: St. Louis, 103; Kansas City, unlisted

Also known as: St. Louis—River City; Kansas City—KC, Cowtown

Named a national park in 2018, the Gateway Arch is a visual symbol of the gateway to the West and national expansion.

Walk through interactive displays at the National World War I Museum and Memorial in Kansas City.

KANSAS CITY

"KC" boasts its own revitalized **Union Station** (opened in 1914). With the decline of U.S. train travel in the late 20th century, the massive beaux arts structure was modified into a shopping, restaurant, and museum complex with a high-tech attraction called **Science City**.

Tucked inside **Union Hill** opposite the station is the **National World War I Museum and Memorial**, which details America's role in the global conflict through modern interactive exhibits in a large underground space. Crowning the hilltop is the lofty **Liberty Memorial**, an Egyptian Revival monument erected in 1926 to honor those who died in the war.

On the east side of Union Hill, **Crown Center** is the superstar of Kansas City urban renewal. Starting in the 1970s, the derelict area was reworked into a modern high-rise neighborhood with offices, apartments, and some of the city's best restaurants, as well as the **Legoland Discovery Center, Sea Life Kansas City** aquarium, and **Hallmark Cards Visitors Center**.

Home to the bustling **City Market** and the *Arabia* **Steamboat Museum**, the historic **River Market** area has also bounced back. South of downtown, the fashionable **Southmoreland** area boasts the largest concentration of creativity in the Midwest beyond Chicago. Among its venerable institutions are the **Kemper Museum of Contemporary Art** and the **Nelson-Atkins Museum of Art** with its old masters and whimsical sculpture garden. ■

ALTER EGO

Oklahoma City

Oklahoma's capital, **Oklahoma City** (OKC), is a much different sort of midwestern city, born of a crazy land rush and raised on bubbling crude. The **National Cowboy & Western Heritage Museum** and the bronze "Sooners" of the **Centennial Land Run Monument** reflect those pioneer days, while the vintage oil derricks around the **State Capitol** pay tribute to the petrol boom.

Downtown is framed by the solemn **Oklahoma City National Memorial & Museum** and buoyant **Bricktown** district, where bars, restaurants, and the **American Banjo Museum** surround a warren of picturesque canals.

Among OKC's offbeat offerings are the hipster **Midtown District** (cool bars), kayaking and rafting at **Riversport Adventures**, and the fascinating, bone-filled **Museum of Osteology**.

New York

New York, U.S.A.

Slip into an Empire State of mind in the Big Apple and its mind-blowing array of museums, parks, and other attractions scattered across a vast, audacious metropolis that started life as a modest 16th-century trading post and farming community.

THE BIG PICTURE

Founded: 1624

Population: 21 million

Size: 4,669 square miles (12,093 sq km)

Language: English

Currency: U.S. dollar

Time Zone: GMT -4 and -5 (EDT and EST)

Global Cost of Living Rank: 14

Also Known As: The Big Apple, Gotham, Empire City, Capital of the World

Start spreading the news: New York boasts more than 80 museums—three dozen in Manhattan alone. At any given time, 40 playhouses are active in the **Broadway Theater District**, complemented by almost 100 **Off-Broadway** stages. The five boroughs nurture around 1,700 green spaces ranging from neighborhood playgrounds to state parks, national parks, and a world-famous city park in the middle of Manhattan.

Even though it's not considered a surf, sand, and sun destination, the Big Apple offers 14 miles (23 km) of beach, many of those sandy strands in **Gateway National Recreation Area** along the city's Atlantic shore. More than 300 skyscrapers provide the city one of the world's most impressive skylines. And the metro area supports 13 professional sports teams, from legendary franchises like baseball's **New York Yankees** to upstarts like **NJ/NY Gotham FC** of women's professional soccer.

A typical visit to the nation's largest city might include all these treasures. Or you could examine New York City through an entirely different lens: A metaphorical archaeological tour starting at the oldest layers and working up to the present.

Two reflecting pools at the 9/11 Memorial and Museum cover the exact footprints of the original twin towers.

The waterfront walkways along Hudson River Park look out toward One World Trade Center, built near ground zero.

As contrary as it seems given all the people and traffic and buildings, the city still harbors a couple of its original 17th-century Dutch farms. **Wyckoff House** in **Brooklyn** is the city's oldest building of any kind, founded in 1652, and still growing corn, tomatoes, and other food crops. Just two years younger, the **Lent-Riker-Smith Homestead** near **LaGuardia Airport** is the oldest house in **Queens,** as well as the oldest in the entire city that's still a private residential home.

The solid stone **Conference House** at the far end of **Staten Island** was erected in the 1680s during the early years of British rule. Its 15 minutes of fame came in September 1776, when John Adams, Ben Franklin, and Edward Rutledge held peace talks with Lord Howe to end the colonial rebellion (spoiler

alert: The talks failed!). **Van Cortlandt House** in the **Bronx** spins its own revolutionary tales: General George Washington twice used the handsome three-story manse as his field headquarters.

Although **Manhattan** was the first of the boroughs to be settled by Europeans (1624), the tangible aspects of its Dutch and early British eras disappeared long ago as buildings were demolished and

BACKGROUND CHECK

• Sailing for the French crown, Giovanni da Verrazzano "discovered" New York Harbor in 1524, but France didn't settle the area.

• Following Henry Hudson's maritime expedition up the river that would bear his name, the Dutch founded New Amsterdam in 1624. Shortly after, they purchased Manhattan from the Native Canarsee people for 60 silver guilders (around $1,100 in modern dollars).

• British forces captured the Dutch colony in 1664 and quickly renamed it after the Duke of York.

• New York City served as the U.S. capital for two years (1789–90). As such, it hosted the first session of Congress and the Supreme Court, and President George Washington's first inauguration.

• New York City didn't extend beyond Manhattan and the Bronx until 1898, when the five boroughs united as a single municipal entity.

others built in their place. However, the heritage endures in thoroughfares like **Wall Street** (the Dutch colonial "Waalstraat") and **Broadway** ("Heeren Wegh"), which were developed along a Native American path called Wickquasgeck Trail that ran the length of the island.

Historians disagree about Manhattan's oldest structure. Is it **Fraunces Tavern** on Pearl Street, erected in 1719 but much modified over the years—and still a functioning pub where visitors can quaff oysters and ale? Or is it **St. Paul's Chapel** on lower Broadway, consecrated in 1766 and barely changed from the days when President Washington worshipped there during New York's brief stint as the U.S. capital city?

Much of the city's colonial and early American heritage was swept away by an industrial revolution that reached a fever pitch in the decades surrounding the Civil War. A surplus of jobs lured millions of immigrants via **Ellis Island,** transforming New York into the world's first global city. By the end of the century, the wealth generated by the rise of American capitalism had spawned a Gilded Age dominated by wealthy New York families like the Vanderbilts, Rockefellers, Carnegies, and Astors.

From **Central Park**, the **Statue of Liberty**, the **Brooklyn Bridge**, **Carnegie Hall**, **Grand Central Terminal**, and **Macy's** department store on Herald Square, many a Big Apple icon was created during the latter half of the 19th century. It was also a golden age for faith and fun: **St. Patrick's Cathedral** and the

Times Square is Manhattan's most iconic destination, with neon advertisements and swarms of taxis.

HIDDEN TREASURES

- **Third Rail Projects:** This Brooklyn-based troupe creates site-specific experiential theater with a high level of audience participation. Past performances at Lincoln Center include *The Grand Paradise, Behind the City*, and *Ghost Light*.

- **Ground Zero Museum Workshop:** As official photographer at ground zero for the FDNY fire unions, Gary Marlon Suson spent seven months documenting the World Trade Center site. His images—along with video footage, artifacts, and firsthand recordings from relief workers—are on display in a tiny museum that holds just 28 people at a time.

- **A Piece of the Berlin Wall:** A 12-by-20-foot (3.7-by-6.1-m) mural by German artists Thierry Noir and Kiddy Citny is rendered on five concrete slabs of the original Berlin Wall given to Manhattan's Battery Park City by the German capital.

- **Mmuseumm:** This curated display of artifacts housed in a Chinatown freight elevator (that fits three visitors at a time) specializes in "overlooked, dismissed, or ignored" items like the shoe thrown at President George W. Bush at the Minister's Palace in Baghdad.

Cathedral of St. John the Divine opened within a year of one another, while the **Metropolitan Museum of Art**, the **American Museum of Natural History**, and the **Bronx Zoo** all came into being.

By 1870, a still independent Brooklyn was the nation's third largest city behind neighboring New York (Manhattan in those days) and Philadelphia. And it was also a thriving era that created legendary

institutions like **Prospect Park**, the **Brooklyn Museum**, and **Coney Island**'s amusement zone.

Yet prosperity also had a steep downside: economic inequality. Neighborhoods like the **Bowery**, **Chinatown**, and **Little Italy**, where many immigrants lived, were squalid at the time. In addition to **Ellis Island** and the new **Statue of Liberty Museum**, the legacy of those who built the city with their own hands is preserved in places like the **Tenement Museum, Eldridge Street Synagogue**, and **Italian American Museum** on the **Lower East Side**, as well as the **Weeksville Heritage Center** in Brooklyn, which preserves one of the nation's first free Black communities.

Although it's now a popular food hall, **Chelsea Market** is located in the sort of workplace where so many of New York's underclass labored in those days, a massive National Biscuit Company factory opened in 1898 (and where the Oreo cookie was invented).

By the turn of the 20th century, the city was ready for skyscrapers. Chicago may have invented the high-rise, but New York literally took them to new heights in landmark structures like the **Flatiron Building** (finished in 1902) and **Met Life Tower** (completed in 1909) overlooking **Madison Square**, and the **Woolworth Building** (1913) in Lower Manhattan. The **Skyscraper Museum** in Battery Park City is a great place to learn more about New York's behemoths.

Yet it wasn't until after World War I—and the invention of a whole new type of architecture—that New York truly reached for the sky. **Art deco** was like nothing that came before, a design style that drew

Stroll through Central Park, one of the world's most famous green spaces.

on a range of predecessors to create a bold, streamlined look that typified the 1920s and '30s.

While the rest of the nation was distracted by Prohibition, the Roaring Twenties, and the Great Depression, New York went on a building spree that produced iconic art deco structures like the **Chrysler Building**, **Empire State Building**, and **Rockefeller Center**, as well as **Radio City Music Hall** and the posh **Waldorf-Astoria Hotel**. Reflecting the city's obsession with everything new, the **Museum of Modern Art (MoMA)** is another child of this era. But the **Metropolitan Museum of Art** (the Met) went medieval, creating **The Cloisters** in **Washington Heights**.

Harlem was also undergoing vast changes at this time, as the neighborhood on the north side of Central Park evolved from European migrants to African American residents relocating from the southern U.S. The "Harlem Renaissance" of the early 20th century spawned an artistic and intellectual awakening expressed through landmarks like the **Apollo Theater**, **Duke Ellington House**, and the **Schomburg Center for Research in Black Culture**.

By the end of World War II, New York was the undisputed champion of the urban world, a global attitude expressed by the rise of the **United Nations** along the East River and the New York World's Fair of 1964–65 in **Flushing Meadows** with its trademark **Unisphere** (the world's largest globe) and a New York City Pavilion that's now the **Queens Museum**.

With habitués like Bob Dylan, Joan Baez, Andy Warhol, and Jack Kerouac, **Greenwich Village** reached its boho peak in the mid-20th century. Having already broke ground with the city's first racially integrated nightclubs, the Village became a vanguard of gay rights after the Stonewall riots of 1969 outside the **Stonewall Inn**, now a historic landmark.

Performing arts flourished at the new **Lincoln Center for the Performing Arts**, where the **New York Philharmonic**, **Metropolitan Opera**, and **New York City Ballet** took up permanent residence. Frank Lloyd Wright's **Guggenheim Museum** looks just as radical now as the day it opened. But another landmark from that era is no longer with us: the 110-story World Trade Center.

The city rebounded from 9/11

with typical New York gumption. **One World Trade Center** is the now the tallest building in the Western Hemisphere at 1,776 feet (541 m), topped by the **One World Observatory** with the city's highest views. The plaza at the bottom preserves the original footprint of the twin towers and the **9/11 Memorial and Museum**, which honors the 2,977 victims killed by the 2001 terrorist attacks. Meanwhile, the cathedral-like **Oculus** atrium hovers above a new trade center transportation hub.

It may not be as high as One World, but the new **30 Hudson Yards** high-rise offers an open-air observation deck called **Edge** that extends 80 feet (24 m) into thin air from the 101st floor. Down at ground level is **The Shed**, a futuristic performance space.

A public park created along an old elevated railway line, the **High Line** connects the sprawling Hudson Yards campus with the latest iteration of the **Whitney Museum of American Art** in the **Meatpacking District**, another relic of the industrial revolution undergoing sweeping change.

Since the turn of the 21st century, the entire waterfront from 59th Street to the lower tip of Manhattan has gone green via **Hudson River Park**. The island's second biggest green space (after Central Park) features a 4.5-mile (7.2 km) promenade for hiking, biking, and jogging, as well as batting cages, soccer fields, and dog-friendly spaces. **Little Island**, its newest component, is an innovative "floating park" that hovers above the river on 132 concrete tulips. Anchoring the lower end of Manhattan are the pyramid-like **Museum of Jewish Heritage**, the

The Metropolitan Museum of Art (aka The Met) boasts many treasures, including Fabergé eggs, as the largest art museum in the U.S.

Staten Island Ferry terminal, the **National Museum of the American Indian** in the old Customs House, and the amazing new **Sea-Glass Carousel** in Battery Park.

Brooklyn's once decrepit waterfront is also undergoing a complete face-lift via the transfer of the **Brooklyn Navy Yard** from military to civilian use and conversion of old finger piers into a marvelous **Brooklyn Bridge Park** with picnic areas and alfresco eateries, a roller rink and beach, and summer outdoor movies set against a backdrop of the Manhattan skyline. ■

ALTER EGO

Queens

Founded as one of the 12 original counties of New York, Queens is now the largest of the five boroughs at 109 square miles (282 sq km) and boasts the second largest population, after Brooklyn. Just across the river from Manhattan, the Long Island destination is a hidden gem with loads to offer visitors.

Get to Queens from Manhattan or Brooklyn by subway or train and be ready to eat. The **Flushing** neighborhood boasts a larger **Chinatown** than the one in Manhattan, or head to **Jackson Heights** for authentic Indian fare. In **Astoria**, you'll find a smorgasbord of options, including Greek, Peruvian, Tibetan, Egyptian, Colombian, and Japanese cuisines.

There's a cultural side to Queens, too: Stop in at **MoMA PS1**, an extension of Manhattan's Museum of Modern Art, or visit the **Unisphere**, a stainless-steel globe that was the centerpiece of the 1964 World's Fair.

For recreation, watch a **Mets** game at **Citi Field**, spend the day at **Flushing Meadows Corona Park** (familiar to fans of *The Great Gatsby*), or, in late summer, watch one of tennis's biggest matches: The **U.S. Open** has been held at the **USTA Billie Jean King National Tennis Center** since 1978.

San Francisco
California, U.S.A.

San Francisco still offers plenty of old favorites. But with the rise of Silicon Valley and the tech economy, the City by the Bay also nurtures innovative new museums, cutting-edge parks, and neighborhoods that reflect the future of urban areas around the planet.

THE BIG PICTURE

Founded: 1776

Population: 6.4 million

Size: 1,109 square miles (2,872 sq km)

Language: English

Currency: U.S. dollar

Time Zone: GMT -7 and -8 (PDT and PST)

Global Cost of Living Rank: 25

Also Known As: The City, City by the Bay, Fog City

No city has impacted life in the 21st century more than San Francisco and the surrounding Bay Area. From personal computing, social media, and ride-sharing to eBay, Netflix, and Tesla, many of the things we take for granted in everyday life were born and raised in the City by the Bay.

It has also been a fulcrum of big ideas. Fueled by the California gold rush and the "Big Four" robber barons who followed, San Francisco sparked many aspects of modern American capitalism. But the 20th century spawned the age of counterculture: Save the Redwoods and the environmental movement, the sixties social revolution, and LGBTQ rights reflect the city's role as America's conscience.

Perhaps it was destiny rather than coincidence that San Francisco was founded in 1776, just five days before the 13 colonies adopted their Declaration of Independence. Like the nation it came to embody, the metropolis has never stopped evolving. Its power base shifted to suburban Silicon Valley in the 1980s and '90s, but "the City" has regained much of its muscle since the turn of the 21st century with an explosion of new corporate headquarters, cultural institutions, and neighborhoods.

The once dilapidated **South of Market** (SoMa) area kicked off the San Francisco Renaissance with a flurry of new attractions around the beautiful **Yerba Buena Gardens** (christened after the city's original Mexican name). The **San Francisco Museum of Modern Art** anchors a neighborhood that also includes the **Children's Creativity Museum** and its historic 1906 **LeRoy King Carousel**, the **Museum of the African Diaspora**, the **Contemporary Jewish Museum**, and the **Yerba Buena Center for the Arts** with its eclectic exhibits, movies, lectures, and live performances.

As software and internet companies flooded into the area, the revival spread to adjacent areas. The gritty waterfront **China Basin** was transformed by **Oracle Park**, the baseball stadium where the San Francisco Giants play their home games beside an inlet called **McCovey Cove** that's often full of kayakers waiting to fetch home-run balls.

Farther down the shore, **Mission Bay** is undergoing a similar makeover via the new **U.C. San Francisco Mission Bay Campus** and the

Explore Chinese delicacies and traditional teahouses in the city's beloved Chinatown.

From Golden Gate National Park you can take in the San Francisco skyline across the bay and its iconic Golden Gate Bridge.

state-of-the-art **Chase Center**, home of the Golden State Warriors NBA team, as well as a shoreline food-and-beverage hot spot called the **Mission Rock Resort**. The latest upstart is **Dogpatch**, a vintage waterfront industrial area that blends blue-collar and hipster attractions ranging from a popular boxing gym and small factories with craft breweries and the **Museum of Craft & Design** (MCD).

San Francisco's green scene has also undergone radical change in recent years with the transition of several large military bases to civilian use. Founded in September 1776, the **Presidio of San Francisco** remained a military base (Spanish, Mexican, American) until the 1990s, when it was transferred to

the National Park Service and folded into **Golden Gate National Recreation Area**. The Civil War–era **Fort Point National Historic Site** and the adjacent surf break, **Crissy**

Field's waterfront walkway, **Baker Beach** with its boulder-strewn views of the **Golden Gate Bridge**, and the **Presidio Pet Cemetery** were already open to the public.

BACKGROUND CHECK

• Though Manhattan's Central Park often gets the attention, Golden Gate Park is actually 20 percent larger than its East Coast cousin.

• The modern-day fortune cookie was invented in San Francisco by Japanese immigrant Makoto Hagiwara. The first version of the cookie was served at Golden Gate Park's Tea Garden. Today, you can see the cookie being made on a tour of the Golden Gate Fortune Cookie Factory in Chinatown.

• Many of San Francisco's historic homes and buildings are made from ship parts. After the city's port became too packed with abandoned vessels from the Gold Rush, the ships were stripped and repurposed as building materials.

• Before it was a prison, Alcatraz was actually a military fort. The U.S. Army used it for decades after it was completed in the 1850s. It only became a prison in 1934.

However, many of the Presidio's attractions are recent additions. The first ever visitors center for the iconic span, the **Golden Gate Bridge Welcome Center** includes a national park–style swag shop. The base shoreline provides a scenic venue for the **Planet Granite** rock-climbing facility, **House of Air** trampoline park, and the **Military Intelligence Service Historic Learning Center** museum in the very building where the U.S. Army's spy corps was founded in 1941.

The old base hospital has been replaced by the **Letterman Digital Arts Center**, home to **Lucasfilm** (of *Star Wars* and *Indiana Jones* fame), the legendary **Industrial Light & Magic** visual effects company, and the **Yoda Fountain**. Old military buildings around the **Main Parade Ground** are now occupied by the **Walt Disney Family Museum** and the **Presidio Officers' Club** history museum.

In the middle of the bay, the naval base on **Treasure Island** was also decommissioned in the 1990s and invaded by civilian uses, including the **Treasure Island Museum** in a wonderful art deco building and **Sottomarino Winery** inside a U.S. Navy submarine vessel called the U.S.S. *Buttercup*. Irish football, hurling, and camogie matches play out on the green fields of the **Gaelic Athletic Association** (GAA).

Treasure Island is accessed via the spectacular **San Francisco-Oakland Bay Bridge**, a double span that underwent its own makeover following the Loma Prieta earthquake of 1989. Opened in 2013, the new eastern span is the world's widest bridge. At around the same time,

Hop on a cable car down Hyde Street for city views and to spot Alcatraz Island.

the western span was illuminated by **"The Bay Lights,"** an after-dark art installation by Leo Villareal that features 25,000 LED bulbs.

On the other side of the bridge, **Naval Air Station Alameda** has had a much slower transition from uniforms to civvies, but the ex-base is finally coming into its own. Giant hangars that once housed A-4 Skyhawk fighter jets and husky Sea Stallion helicopters have been repurposed into **Faction Brewery**, **Hanger 1** vodka, **Rock Wall Wine**, and the **Alameda Vintage Fashion Faire**. The old base also affords killer views across the bay to the bridge and San Francisco skyline.

Most of San Francisco's oldies but goodies are still around. But even those have been reenergized by the perpetual motion that distinguishes today's Bay Area.

Chinatown has expanded way beyond its post–gold rush core, pushing into adjacent areas. But it's also less authentic than days past.

Many of the mom-and-pop stores and eateries that once infused the neighborhood have given way to art galleries and upscale restaurants with fancy tasting menus rather than à la carte Cantonese cuisine. Yet bygone Chinatown survives in **Tin How Temple** (opened in 1852), the Asian-inspired architecture of **Waverly Place**, and moody **Li Po Lounge**, serving mai tais and other cocktails since the 1930s.

Nowadays you would be hard-pressed to find Italian Americans living in **North Beach**. Wedged between **Coit Tower** and the **Transamerica Pyramid**, the neighborhood where "Joltin' Joe" DiMaggio grew up is now far more multicultural. **Saints Peter and Paul Church**—the historic "Italian Cathedral of the West"—now offers services in Italian, English, and Chinese. Yet the past lives on at eateries like **Fior d'Italia** (opened in 1886) and **Tommaso's Ristorante Italiano** (opened in 1935).

LOCAL FLAVOR

• **Mr. Pollo:** The reasonably priced four-course tasting menu at this mega-popular (and foodie-favorite) Colombian restaurant features salad, arepas, an entree, and dessert; *2823 Mission Street, Mission District, (860) 912-9168.*

• **Dragon Beaux:** Traditional dishes and modern, Asian fusion near Golden Gate Park, including dim sum, hot pots, dumplings, and Peking duck; *5700 Geary Boulevard, Richmond District;* dragonbeaux .com.

• **Palace Hotel:** Whether it's brunch in the jewel-box Garden Court or cocktails in the Pied Piper Bar, this meticulously restored landmark is a throwback to old San Francisco;

2 New Montgomery Street, Financial District; marriott.com.

• **Auntie April's:** No nonsense but delicious chicken, waffles, and soul food restaurant in southeast San Francisco; *4618 3rd Street, Bayview;* auntieaprilssf.com.

• **Dancing Yak:** Turquoise booths, purple walls, and copious art complement the traditional Nepalese dishes. Momos, samosas, tandoori and thali plates, goat curry, vegan choices, and cocktails with a Himalayan twist like bourbon infused with turmeric and peppercorn; *280 Valencia Street, Mission District;* dancingyaksf.com.

In addition to its time-honored blend of waterfront restaurants, kitschy souvenir shops, **Ghirardelli chocolate**, and street performers, **Fisherman's Wharf** has added two floating museums—the Liberty ship **S.S. *Jeremiah O'Brien*** and submarine **U.S.S. *Pampanito***—as well as the **Musée Mécanique** and its collection of more than 300 mechanical fortune-tellers, player pianos, toys, and carnival games. In addition to the standard **Cellhouse Audio Tour**, it's possible to explore **Alcatraz** prison on guided after-dark, behind-the-scenes, and garden tours.

From the turnaround near **Aquatic Park** and the **San Francisco Maritime National Historical Park**, visitors can hop one of the city's national landmark cable cars up and over **Russian Hill** and **Nob Hill**, before plunging into the skyscraper canyons of the **Financial District** and nearby **Embarcadero**.

Like everything else in San Francisco, the downtown area is also in the midst of change. The family-friendly **Exploratorium** science museum has relocated to Pier 15 and nowadays the historic **Ferry Terminal** does double duty as an aquatic transport hub and lively farmers market. But the biggest change is up above: the soaring, 61-story **Salesforce Tower** that became the Bay Area's tallest building when it opened in 2018. Poised at the foot of the giant, **Salesforce Park** is an elevated botanical garden with more than 16,000 plants and a year-round slate of open-air concerts, fitness classes, kid activities, and other events.

Almost always busy **Market Street** runs inland to the stately **City Hall** and other civic treasures like the **War Memorial Opera House**, home of the city's opera and ballet companies, the ultramodern **Davies Symphony Hall**, and the **Asian Art Museum**.

Farther up Market are two of the city's most storied neighborhoods. The **Castro District** has been at the forefront of American LGBTQ rights and culture since

A common waterfront sight, sea lions lounge on Pier 39.

Just a short drive from San Francisco, Napa Valley offers some of the world's best vineyards, like Castello di Amorosa.

the early 1960s and remains an enclave for the community. **Castro Street History Walk** is a self-guided tour of plaques that illuminate the district's story from Spanish through modern times, while the **Castro Theatre** (opened in 1922) remains one of the city's top independent movie houses.

The Castro borders on the fabled **Mission District**, a long-time Latin American barrio that has transformed into a multicultural hub for food, art, and music since the turn of the century. The neighborhood derives its name from **Mission Dolores**, the city's oldest building, founded by Franciscan friars in the fall of 1776.

The Mission is also renowned for **Paxton Gate** and other funky one-off shops and offbeat entertainment spots like the **Alamo Drafthouse** bar and cinema. Although it was once declared one of the city's best places "to score a used couch or refrigerator," **Mission Dolores Park** is picnic-friendly once again thanks to a $20 million makeover.

Haight-Ashbury is a cliché these days rather than a genuine counter-culture haunt. But **Golden Gate Park** remains true to itself as the city's favorite outdoor escape. The Bay Area's premier art collection is housed inside the **De Young Museum**, which was torn down

and totally rebuilt in the early 21st century. Its masterpieces of American and modern art are now housed inside a neoteric copper shell crowned by the **Hamon Observation Tower** with views across the park.

The science and natural history exhibits at the **California Academy of Sciences** also occupy a new Golden Gate Park home: a half-billion-dollar structure beneath a "living roof" that nourishes 1.7 million native California plants. Among the museum's multiple attractions are the **Steinhart Aquarium**, the **Morrison Planetarium**, and the indoor **Osher Rainforest**. ■

San Juan
Puerto Rico, U.S.A.

Puerto Rico's capital city may fly the Stars and Stripes, but it definitely marches to the beat of a Latin drummer—an exotic tropical city that relishes its homegrown culture, cuisine, language, and music.

THE BIG PICTURE

Founded: 1508

Population: 2.3 million

Size: 77 square miles
(199 sq km)

Language: Spanish and English

Currency: U.S. dollar

Time Zone: GMT -4 (AST)

Global Cost of Living Rank: 89

Also Known As: Caparra, Puerto Rico de San Juan Bautista
(Spanish colonial)

Prior to searching for the fountain of youth, Spanish conquistador Ponce de León started a settlement on the north shore of Puerto Rico that would grow into the city of San Juan. Two massive citadels kept the enemy at bay until 1898, when U.S. forces captured the city during the Spanish-American War.

Those bastions—**Castillo San Felipe del Morro** and **Castillo San Cristóbal**—are now centerpieces of a national historic site where visitors can explore a veritable maze of tunnels, towers, moats, gun emplacements, and even a dungeon. Artifacts and historical displays reflect the 500-year history of the largest fortification the Spanish built in the New World. The lofty walls offer panoramic views of San Juan, and the grassy **Esplanade** beside **El Morro** is a popular kite-flying and picnic spot. **Paseo del Morro,** a national recreational trail, runs along the bay shore beneath the citadel walls.

Old San Juan and its blue cobblestone streets sprawl across the rest of the peninsula, the multicolored colonial buildings filled with trendy bars, restaurants, boutique hotels, and museums. Directly opposite El Morro are the Ballajá Barracks, built by the Spanish and occupied by the U.S. Army until after World War II. The current tenant is the excellent **Museum of the Americas**, which showcases the Indigenous, colonial, and African-inspired art and culture of the Western Hemisphere.

Casa Blanca, Ponce de León's whitewashed mansion—and the oldest European home in the Americas—seems little changed from when his family lived there. The immortality-seeking Spaniard is buried in nearby **Cathedral of San Juan Bautista** (completed in 1540). The old town's other famous residence is **La Fortaleza**, the official home of Puerto Rico's governor, built atop another 16th-century bastion and only viewable through the imposing front gate. Right around the corner from the governor's mansion is the tiny but extravagant **Capilla del Cristo** (finished in 1780) with its gold-and-silver altar.

The old town reaches fever pitch during the annual **San Sebastián Street Festival**, four days of carnival parades, music, dancing, food, and

On January 20, Calle San Sebastián swarms with festivalgoers.

El Morro rises on the rocky coastline of San Juan, overlooking aquamarine waters and the city skyline along the shore.

craft stalls that revolve around the saint's feast day on January 20. The party runs all night, especially in the bars along **Calle San Sebastián**.

San Juan's year-round party scene can be found in **Condado**, the ritzy beach area that's sometimes compared to Miami Beach. Flanked by scores of bars, restaurants, and high-rise hotels, **Ashford Avenue** runs the length of the beach strip. Away from the beach, tranquil **Condado Lagoon** is popular for paddleboarding and kayaking, with a chance that you might spot one of the resident manatees. Those who prefer sun, sea, and sand with fewer people should venture farther east, to the uncrowded beaches at **Isla Verde**, **Carolina**, and **La Pocita**.

Along the south shore of Condado Lagoon, old barrios like **Miramar** and **Santurce** have refashioned themselves in recent years into hip urban enclaves with gourmet restaurants and cultural outlets. Foremost among the latter are the **Puerto Rico Museum of Contemporary Art** and the **Luis A. Ferré Performing Arts Center** with its year-round slate of theater, ballet, opera, and concerts. ■

LOCAL FLAVOR

• **Cocina Abierta:** The seasonal menu at "Open Kitchen" might include Caribbean bouillabaisse with conch and lobster, Cambodian-style pork tenderloin with coconut milk and grilled pineapple, or Bolivian rabbit and peanut stew with Andean potatoes. *Calle Caribe 55, Condado;* cocinaabierta.net.

• **Casita Miramar:** Antique furnishings, breezy balcony tables, and great local food make this Miramar restaurant a favorite with visitors and locals alike. *605 Avenida Miramar;* facebook.com/casitamiramarpr.

• **Deaverdura:** A simple setting offers incredible food in the heart of the old town. The Puerto Rican sampler includes a great entree to local favorites like *pasteles* (turnovers), *tostones* (fried plantains), and plantains. *Calle Sol 200 (at Calle De La Cruz), Old San Juan;* facebook.com/deaverdura.

• **Sage Steak Loft:** Created by celebrity chef Mario Pagán, the menu at this chic little eatery features fusion dishes like grass-fed rib-eye steak with cassava gnocchi, ahi tuna balls stuffed with crab salad, and shrimp cocktails with tomato sorbet and aioli. *Olive Hotel, Calle Aguadilla 55, Condado, (787) 728-3535.*

Caribbean Cities

Kingston, Jamaica; Port of Spain, Trinidad and Tobago; Georgetown, Guyana

A string of English-speaking cities across the Caribbean combine colonial relics and vibrant modern cultures where some of the world's best music (and cricket) were born.

A statue of legendary reggae icon Bob Marley welcomes visitors to his museum in Kingston.

KINGSTON

Population: 590,000
Size: 40 square miles (104 sq km)

Jamaica's capital offers insight into what makes the island tick.

Reggae fans pay homage at the **Bob Marley Museum** in the singer's former home on Hope Road. The **Peter Tosh Museum** highlights the work and influence of another reggae superstar, while the **Trench Town Culture Yard** preserves the tenement where Marley and other struggling musicians once lived.

Devon House was built in 1881 as home of George Stiebel, Jamaica's first Black millionaire. It's been lovingly restored into a house museum with places to eat and drink, including **Scoops Unlimited** ice cream.

Hope Botanical Gardens & Zoo showcases native Caribbean flora and fauna, while **Emancipation Park** celebrates the end of slavery. **Mona Reservoir** is where locals head for a peaceful jog.

Located at the mouth of Kingston Harbor, **Port Royal** was founded by the Spanish in 1494 and served as a base for notorious buccaneers until its destruction by an earthquake in 1692. Modern-day Port Royal features partially rebuilt **Fort Charles** and the half-sunken **Giddy House**, as well as casual fish fry joints like **Gloria's Top Spot**. Skippered motorized canoes depart for **Lime Cay** and other sandbars for a day of fun in the sun.

Downtown Kingston is home to colonial-era relics like **Ward Theatre** (opened in 1911), **Coke Memorial Methodist Church** (finished in 1840), and sand-floored **Shaare Shalom Synagogue** (built in 1912). In addition to the **National Gallery of Jamaica**, downtown is garnished with **"Yard Art."**

Kingston Craft Market on the waterfront sells locally made souvenirs. The West Indies plays matches at the legendary Sabina Park stadium (opened in 1895).

PORT OF SPAIN

Population: 525,000
Size: 110 square miles (285 sq km)

Port of Spain is one of the top spots to experience pre-Lent carnival. Soca, steelpan, and calypso provide the soundtrack, with the center stage at Queen's Park Savannah smack dab in the city center.

The 260-acre (105 ha) Savannah is bounded by the Royal Botanic Gardens and Emperor Valley Zoo, the National Museum & Art Gallery, and the National Academy for Performing Arts (NAPA). Along the park's western edge are the Magnificent Seven, historic structures like Queen's Royal College and the Scottish Baronial–style Stollmeyer's Castle.

The Central Bank Money Museum boasts exhibits that trace the evolution of currency from cowrie shells to hard cash. Trinidad's Parliament convenes in the beaux arts–style Red House (rebuilt in 1907) on Woodford Square.

Hilltop Fort George (completed in 1804) offers the best views across the city and Gulf of Paria from fortifications lined with original 19th-century cannons. Waterfront Park promenade is good for a sunset stroll followed by some liming ("hanging out") in the bars, restaurants, and street corners along Ariapita Avenue.

Beyond the urban area are the renowned Asa Wright Nature Centre in the mountains of north-central Trinidad and Caroni Bird Sanctuary, the mangrove-lined waterway with boat tours to view

A great kiskadee perches on a branch in Georgetown.

the nesting place of the scarlet ibis (Trinidad's national bird).

GEORGETOWN

Population: 200,000
Size: 30 square miles (78 sq km)

Georgetown is the only South American city where English is the official language, and Guyanese Creole is closely related to the English patois spoken in Jamaica and Trinidad.

But the city's similarities to its Caribbean island cousins run much deeper than just language. Cricket rather than soccer is the game that

residents flock to watch at Guyana National Stadium. Reggae, calypso, and soca music play to liming every Sunday along the Seawall at the end of Main Street.

Although new buildings are sprouting up all around Georgetown, many of the city's landmarks date from colonial days, including Parliament (completed in 1834) and the Supreme Court (1887). Housed inside another classic 19th-century building, the Walter Roth Museum of Anthropology showcases Guyana's Amerindian peoples.

St. George's Cathedral (consecrated in 1894) was the world's tallest wooden building (143 feet/ 44 m) until quite recently, while St. Andrew's Kirk is even older (1818). However, Stabroek Market (1881) along the banks of the Demerara River was designed and built by Americans.

Bourda Market is even bigger, a warren of street stalls, tiny shops, rum bars, and Asian eateries along Robb Street just east of downtown. The Guyana Shop at Robb and Alexander is a great place to buy local delicacies like farine, andiroba, jeera, mauby bark, and married man pork. ∎

The Whitehall Office of the Trinidad prime minister was newly renovated in 2019.

Latin America

Guadalajara, Mexico (p. 116), broke the Guinness World Record for largest folk dance with 886 dancers.

Mexico City
Mexico

North America's largest city—and among the 10 biggest on planet Earth—Mexico City has dazzled visitors since Aztec times as an urban area with an unquenchable lust for life and an aura that anything is possible on any given day.

B ig, bold, and boisterous, Mexico City is a hard place to wrap your brain around. Oftentimes it seems more intimidating than inviting. And the Mexican federal capital definitely has its downside: poverty, pollution, and crime.

But in the same breath, it might be the most stimulating city in the Western Hemisphere. As author Junot Díaz points out, "Every single famous person in Latin American history and art and politics seems to have found their way to Mexico City."

It's also a city of dreams, where anything might be possible. A place where a kid who grew up on the streets (Alejandro Iñárritu) transformed himself into a world-conquering film director who's won multiple Academy Awards. Where the illegitimate son of a peasant woman (Lorenzo de Tena) escaped the tribulations of his childhood by gazing at the stars and emerging as a groundbreaking astronomer. Where a teenager paralyzed by a bus accident (Frida Kahlo) overcame her disability to become one of the most beloved artists of the 20th century.

If there's a neighborhood that exudes the city's lust for life more than any other, it's **Coyoacán** on the south side of the city center. Five hundred years ago it was a village on

THE BIG PICTURE

Founded: 1521

Population: 20 million

Size: 915 square miles (2,370 sq km)

Language: Spanish

Currency: Mexican peso

Time Zone: GMT -5 and -6 (CDT and CDMX)

Global Cost of Living Rank: 152

Also Known As: CDMX, Distrito Federal, Ciudad de los Palacios (City of Palaces)

the shore of Lake Texcoco where conquistador Hernán Cortés sojourned while the Spaniards were looting the Aztec capital. By the early 20th century it was evolving into a gathering place for artists and intellectuals.

Although it attracts far more Instagrammers than iconoclasts these days, Coyoacán retains its bohemian vibe in relics like **Casa Azul** (Blue House), where Frida Kahlo lived and painted, and the **Museo Nacional de Culturas Populares** with its emphasis on Mexican folk art and Indigenous traditions, as well as in the laid-back cafés around the **Plaza Hidalgo** and adjoining **Jardín Centenario**.

The Coyoacan barrio also has had its darker moments: The **Museo Case de Leon Trotsky** is where the Russian revolutionary took refuge and was famously assassinated (via ice pick) in 1940, and the **Casa de la Maniche** is the early 16th-century hideaway of Cortés and the Indigenous woman who was both his translator and lover. Many Mexicans consider La Malinche ("The Captain's Woman") an enemy collaborator and traitor, despite her notoriety.

Ancient artifacts are on display at the Museo Nacional de Antropología.

Construction began on the Palacio de Bellas Artes in 1904; it now hosts music, dance, theater, and opera events.

The artsy vibe spills over into the nearby **San Ángel**, where Kahlo and her artist husband, the legendary muralist Diego Rivera, shared a multihome complex that's now the **Museo Casa Estudio**. The barrio's well-respected **Museo de Arte Carrillo Gil** (MACG) exhibits works by Rivera, Orozco, Siqueiros, and other prominent Mexican artists, while **Museo de El Carmen** revolves around the religious art of Mexico's past.

The Aztec metropolis that Cortés conquered and then meticulously destroyed to create a new European-style city was called **Tenochtitlan**. Stone from the great pyramids and palaces was used to construct the cathedral and other Spanish colonial buildings.

But the infamous conquistador never figured on modern underground utilities. Because in 1978, electric company workers discovered the **Coyolxāuhqui Stone**, a huge disk bearing the likeness of an Aztec goddess, which led to the discovery of the ruined **Templo Mayor**, the paramount religious shrine of ancient

ARTBEAT

- **Best Movies:** *Los Olvidados* (1950), *Amores Perros (2000), Frida* (2002), *Man on Fire* (2004), *Güeros* (2014), *Roma* (2018).

- **Best Books:** *Tristessa* by Jack Kerouac, *Battles in the Desert & Other Stories* by José Emilio Pacheco, *Caramelo* by Sandra Cisneros, *The Skin of the Sky* by Elena Poniatowska, *Malinche: A Novel* by Laura Esquivel, *La Perdida* by Jessica Abel, *The Savage Detectives* by Roberto Bolaño, *Faces in the Crowd* by Valeria Luiselli.

- **Best Art:** The landscapes of José Velasco, the self-portraits of Frida Kahlo, the photography of Manuel Bravo, the "street art" of Carlos Almaraz, and the murals of Diego Rivera, David Siqueiros, and José Orozco.

- **Best Music:** *Ballet Folklorico De Mexico* by Amalia Hernandez, *Hecho En México* by Alejandro Fernández, *Musica Mexicana: Chavez, Ponce & Revueltas* by Carlos Chávez and others, *En Éxtasis* by Thalía, *Exiliados en la Bahía* by Maná.

Tenochtitlan. The famous stone and hundreds of other relics—as well as a scale model of the temple's original appearance—are displayed inside **Museo del Templo Mayor**.

The big dig lies at the heart of the **Centro Histórico** district that served as the nucleus of Spanish colonial and post-independence Mexico City. Many of the touchstones of Mexican history have unfolded in the vast **Zócalo** plaza, where a towering flagpole flies an enormous Mexican tricolor.

The powers that be in the colonial era—church and state—staked claims around the Zócalo with grandiose structures like the **Catedral Metropolitana**, the **Antiguo Palacio del Ayuntamiento** (old city hall), and the **Palacio Nacional de Mexico**, built as the official residence of the Spanish viceroy but now occupied by Mexico's president. The latter flaunts the famous **Campana de Dolores** bell that announced the country's rebellion against Spain and Diego Rivera's fabulous **"Historia de Mexico"** murals. **El Laberinto**, the city's biggest independent bookstore, is located just a block off the plaza.

The old town's coolest street is **Avenida Madero**, a pedestrian zone connecting the Zócalo and the

Mexico City has a robust food scene, and churros are a must-try.

Alameda. Along the avenue are some of Mexico's most cherished buildings. **Palacio de Iturbide** (completed in 1785) is named for the War of Independence hero who became the first Emperor of Mexico. The **Casa de los Azulejos** (built in the 1790s) is clad in thousands of blue-and-white glazed tiles. The jewelry house called **La Esmeralda** (opened in 1892) is now the **Museo del Estanquillo**, which encompasses photography, cinema, books, music, and other forms of Mexican popular culture.

At the western end of the street is something completely different: the 44-story **Torre Latinoamericana**, the tallest building in Latin America when it topped out in 1956. And yes, there's an observation deck on the top floor. It looms high above another Mexico City landmark, the **Palacio de Bellas Artes**, a wondrous blend of art nouveau and art deco that harbors the **Museo Nacional de Arquitectura** and a grand **Sala Principal**, where opera, dance, symphony, recitals, tributes, and other events take place.

LAY YOUR HEAD

- **Las Alcobas:** Set along the chic Avenida Masaryk, the Alcobas provides a perfect location for delving into the Polanco's upscale shopping, eating, and art scene; restaurants, bar, fitness center, spa; from $298; lasalcobas.com.

- **Umbral:** Located just a block off the Zócalo, this 59-room boutique blends edgy decor and historic architecture with a roof terrace where guests can

sun, swim, or sip cocktails; restaurant, bar, pool, fitness center, art gallery, hanging garden, mini cinema, book and vinyl record library; from $148; hotelumbral.com.

- **Gran Hotel Ciudad de Mexico:** Old World elegance in the Centro Histórico, this art deco treasure features one of the world's four largest stained-glass ceilings, as well as guest rooms that mix period decor

and modern amenities; restaurant, bar, gym; from $140; granhoteldela ciudaddemexico.com.mx.

- **NH Collection Mexico City Reforma:** Sleek, contemporary, high-rise hotel near the Glorieta de los Insurgentes in the heart of Zona Rosa; restaurant, bar, outdoor pool, gym; from $72; nh-hotels.com.

Considered the oldest public park in the Western Hemisphere, the **Alameda Central** was laid out in 1592 and still functions as a leafy escape from the helter-skelter city center. It's surrounded by half a dozen museums, including the huge **Museo Nacional De Arte** (MUNAL), **Museo de Arte Popular**, and heartrending **Museo Memoria y Tolerancia**, dedicated to tolerance, nonviolence, human rights, and the memory of those who died championing these noble causes.

Just north of the Alameda are **Plaza Garibaldi**, where mariachi bands gather each evening to serenade the crowd, and the **Ballet Folklórico de México**, an exuberant performance of Mesoamerican dance, music, and costumes. A few blocks

south of the Alameda is the chromatic **Mercado De Artesanías La Ciudadela** where hundreds of artists and artisans make and sell their wares.

Beyond the Alameda is **Paseo de la Reforma**, a broad and perpetually busy avenue that cuts a nine-mile (14 km) path across the middle of Mexico City. Traffic circles along the famous avenue are graced by monuments to figures both real and imagined, from Columbus and Aztec ruler Cuauhtémoc to the Greek goddess Diana and the golden **Ángel de la Independencia**, where large crowds of *chilangos* (Mexico City residents) gather on Independence Day and New Year's Eve.

Reforma passes through **Zona Rosa**—basically the city's central business district—full of high-rise

banks, corporate headquarters, hotels, upscale boutiques, restaurants, and after-dark hangouts.

Zona Rosa nudges up against **Chapultepec Park**, a huge green space that takes its name from a rocky outcrop sacred to pre-Columbian people. A chapel erected by Franciscan missionaries over an Aztec altar was replaced in the 1700s by **Chapultepec Castle**. Renowned as the "Halls of Montezuma" that the U.S. Marines stormed in 1847 during the Mexican-American War, the castle's grandiloquent halls now host the **Museo Nacional de Historia**.

The park boasts several outstanding art museums—the **Museo Rufino Tamayo** and **Museo de Arte Moderno**—as well as the unrivaled

The history of the Catedral Metropolitana spans three centuries, between 1573 and 1813.

Museo Nacional de Antropología (MNA) with its priceless artifacts from every era of Mexican civilization. Spread across 10 buildings, the MNA is one of the world's foremost museums of any kind. The park also offers the **Chapultepec Zoo** and various hiking, biking, and equestrian trails through wooded areas. And on the first Saturday of every month, **Lago de Chapultepec** screens **"Lanchacinema"** outdoor movies that that can be watched from a rented paddleboat.

The forest of skyscrapers along the park's northern edge is **Polanco**, one of the city's hippest districts. The neighborhood is littered with upscale hotels, restaurants, and shopping, especially along swank **Avenida Presidente Masaryk**.

Museo Jumex exhibits innovative and experimental modern art, while the adjacent **Museo Soumaya**—housed inside an amorphous structure covered in 12,000 glimmering aluminum tiles—specializes in Mesoamerican art and European masters. **Acuario Inbursa** harbors Mexico's largest ensemble of undersea creatures. On the other hand, **Autocinema Coyote** is a good old-fashioned drive-in theater.

In 1531, a villager claimed that the Virgin Mary had appeared to him atop the **Cerro del Tepeyac** along the north shore of the lake that still filled much of the Valley of Mexico at that time. The wooded hill is still there—protected within the confines of **Parque Nacional El Tepeyac**—but nowadays it takes a backseat to the **Basílica de Santa María de Guadalupe**. The basilica is actually two churches: a hulking, colonial-style shrine originally

Traditional flat-bottom boats called *trajineras* are moored awaiting passengers.

"Fuente de los Coyotes" was sculpted in 1967 and sits in Plaza Hidalgo in Coyoacán.

finished in 1709 and a circular modern church with a huge copper roof that opened in 1976.

One of the few remnants of the waterways that once covered much of the Valley of Mexico is **Xochimilco** on the city's south side. The area is crisscrossed by 110 miles (170 km) of canals the Aztecs constructed and maintained throughout colonial days and post-independence. Brightly decorated *trajineras* take visitors on scenic cruises past *chinampas* (floating gardens) still used for agriculture, although one owner transformed his property into the creepy **Isla de las Muñecas** (Island of the Dolls). ∎

URBAN RELIC

Teotihuacan

Although the Aztec capital was hugely impressive, it paled in comparison to another ancient Mesoamerican city that was abandoned by the time the conquistadors arrived.

Located around 30 miles (48 km) north of Mexico City, **Teotihuacan** reached its height between A.D. 300 and 600, when it was the largest urban area in the Americas (an estimated 200,000 population) and location of several of the largest and most impressive structures of pre-Columbian times.

Unlike Tenochtitlan, which the Spanish eagerly dismantled, much of Teotihuacan survives into modern times. A broad cobblestone passage now called the Avenue of the Dead is flanked by the immense Pyramid of the Sun (world's third largest), the elegant Pyramid of the Moon, Temple of the Feathered Serpent, Palace of the Jaguars, and more than 2,000 residential compounds where ordinary citizens once lived.

Declared a UNESCO World Heritage site in 1987, Teotihuacan's archaeological site features two good museums, including one that showcases ancient murals discovered amid the ruins. Two unique ways to discover the ancient city are private tours with an archaeologist and hot-air balloon excursions.

Oaxaca
Mexico

Huddled in a broad valley in southern Mexico, Oaxaca City offers an artful blend of Indigenous and colonial traditions, exuberant fiestas, and offbeat culinary treats.

THE BIG PICTURE

Founded: 1532

Population: 660,000

Size: 33 square miles (85 sq km)

Language: Spanish

Currency: Mexican peso

Time Zone: GMT -5 or -6 (CST or CDT)

Global Cost of Living Rank: Unlisted

Also Known As: Huaxyacac (Aztec); Santa María Oaxaca and Antequera (Spanish colonial)

The Zapotec, Mixtec, Aztec, and Spanish all left their mark on Oaxaca City. And as the birthplace of Benito Juárez, it was also a cradle of Mexican democracy. Nowadays it's one of the most delightful cities in Latin America, especially the area around **Zócalo** square with its sidewalk cafés, street performers, and matchless people-watching.

Stretching north from the Zócalo, the pedestrian-friendly **Andador Turístico Macedonio Alcalá** is flanked by the 18th-century **Cathedral of Our Lady of the Assumption,** the **Museum of Contemporary Art** in the colonial-era **Casa de Cortes**, and the **Santo Domingo de Guzmán** monastery complex. The latter includes a meticulously restored baroque church and the superb **Museum of Oaxacan Cultures**, where gold and silver treasures from Monte Álban are displayed.

Oaxaca's unofficial "art avenue" is the **Calle de Manuel Garcia Vigil**, which parallels the Andador one block to the west. Starting from the **Alameda de León** public park beside the cathedral, Vigil runs a broad gamut of artistic expression, from the **Museum of Oaxacan Painters** and **Alvarez Bravo Photographic Center** to handicraft shopping at **La Casa de las Artesanías de Oaxaca**.

Although the area north of the Zócalo caters largely to visitors, the neighborhood south of the plaza is where locals shop for everyday items. Sprawling across an entire block, **Mercado Benito Juárez** offers a chromatic array of fruit, vegetables, herbs, spices, and other foodstuffs, as well as locally produced chocolate and mezcal. Two blocks farther south, the **Mercado de Artesanías** specializes in local handicrafts: rugs, blankets, hats, bags, tablecloths, and more.

For a bird's-eye view of Oaxaca, hike or taxi to the summit of **Cerro del Fortín**. The modern amphitheater at the base of the hill is ground zero for the annual **Guelaguetza** festival of traditional dance, music, clothing, and food in July.

Another great time to visit is during the three-day **Dia de los Muertos** (Day of the Dead) celebration at the end of October, when brass bands and costumed

Dancers carry baskets on their heads at the Guelaguetza festival.

Consecrated in 1733, the Cathedral of Our Lady of the Assumption can be found on Avenida de la Independencia.

dancers roam the old town, locals feast and play music throughout the night at local cemeteries, and stalls sell *pulque* (cactus moonshine), *pan de muerto* (bread of the dead), sugar skulls, and other ghoulish artifacts.

Oaxaca's hinterland is also worth investigating for both its natural and cultural wonders. The city is surrounded by villages where many of the region's handicrafts are created: **Atzompa** and **Ocotlán** (green-glazed pottery), **Arrazola** (wood carvings of creatures), **Teotitlán del Valle** (textiles), and **Coyotepec** (black clay pottery).

An hour's drive east of the city, the **Mitla ruins** flaunt unusual geometric designs. Nearby **Hierve el Agua** is renowned for its hot springs and "frozen waterfall" rock formations. In the highlands above Mitla, the **Pueblos Mancomunados** maintain a network of hiking, biking, and horseback trails through the **Sierra Norte**. ■

URBAN RELIC

Monte Albán

Perched on a mountaintop 1,300 feet (396 m) above Oaxaca City, **Monte Albán** is one of Mexico's largest and most important archaeological sites. The vast stone city was founded around 500 B.C. and eventually grew into the political hub of the Zapotec civilization with an estimated population of more than 40,000.

Monte Albán's enormous Grand Plaza is surrounded by ceremonial and residential structures, including two large pyramid-like platforms, several palaces, and a large ball court. The site's famed *danzantes*—human images found around the plaza—were originally thought to be Zapotec dancers, but researchers now believe they portray contorted, tortured prisoners of war.

Quito
Ecuador

Not long after the program was launched in the 1970s, UNESCO declared the entire city of Quito a World Heritage site, citing Ecuador's capital as "the best-preserved, least altered historic center in Latin America." But that's far from being the only thing that makes the city special.

THE BIG PICTURE

Founded: 1534

Population: 2.5 million

Size: 207 square miles (536 sq km)

Language: Spanish

Currency: U.S. dollar

Time Zone: GM -5 (ECT)

Global Cost of Living Rank: 161

Also Known As: Kitu (Quechua), Carita de Dios (God's Face), Luz de América (Light of America), Ciudad de los Cielos (City of the Heavens)

Poised at a truly breathtaking elevation of 9,350 feet (2,850 m), Quito is one of the few cities that verges on both the Andes *and* Amazon.

Its western suburbs have gradually crawled up the slopes of **Cerro Pichincha**, an active stratovolcano that sits amid a chain of lava-spewing mountains that runs down the middle of Ecuador. **Cayambe Coca Ecological Reserve**—which preserves a huge tract of cloud forest and montane watershed along the upper edge of the Amazon Basin—starts just 20 miles (32 km) east of the city center.

It's also a geographical oddity. One of the world's skinniest metro areas, Quito stretches roughly 35 miles (56 km) from north to south, but only around five miles (8 km) from side to side. And given its latitude (0°13'47" S), the city is located closer to the Equator than any other national capital.

Built on the foundations of an ancient Inca city, Quito was founded by the Spanish in 1534, a mere three years after Francisco Pizarro and his conquistadors "discovered" Ecuador. Although a 1917 earthquake heavily damaged the city, the **Centro Histórico** is incredibly well preserved. The old town boasts more than 200 structures built during the 16th and 17th centuries—churches, convents, monasteries, and mansions—that blend Iberian, Moorish, and Indigenous styles. Meanwhile, the area's cobblestone streets and plazas still adhere to the grid patterns laid out by Dionisio Alcedo y Herrera in 1734, a huge factor in determining Quito's early World Heritage status.

Located in the heart of the old town, the palm-shaded **Plaza Grande** is affectionately called "Quito's living room" because its enduring role as a meeting place for residents and visitors alike. The power brokers of colonial days surrounded the square with edifices that expressed their clout: the **Palacio de Carondelet** that now accommodates the president of Ecuador (free guided tours in English), the ecclesiastical **Palacio Arzobispal**, the **Alcaldía** city hall, and the immense **Catedral Metropolitana** with its distinctive green-and-gold dome.

Quito's best churches are actually elsewhere in the old town. Inspired

A puppeteer puts on a musical performance in Quito's colonial center.

The monument of the winged Virgin stands tall above the city of Quito, a uniquely long stretch of city rising into the Andes.

by the architecture of France's Gothic cathedrals, the **Basílica del Voto Nacional** isn't nearly as old as it looks, finally consecrated in 1988 after a century of on-and-off construction. However, the **Iglesia de la Compañía de Jesús** is most definitely vintage, an ornate baroque creation founded by the Jesuit order in 1605. Even older is the **Iglesia y Convento de San Francisco**. Started in 1537, it grew into a complex that eventually encompassed three churches, 13 cloisters, a vast art collection, and historic library—one of the largest clusters of colonial architecture in the Western Hemisphere.

Several of Quito's 60-plus museums stand out from the crowd. Among the best collections in the old town are the **Museo de la Ciudad**, which depicts everyday life in Quito from Indigenous through modern times, and **Casa del**

Alabado, which showcases pre-Columbian art and artifacts in one of the city's oldest colonial homes.

If there's one place outside of the historic center that everyone in Quito should visit, it's the recently renovated **Casa de la Cultura**

Ecuatoriana (CCE). The giant glass-enclosed complex beside **Parque El Ejido** features the **Museo de Arte Moderno**, **Museo de Instrumentos Musicales**, and **Museo Etnográfico**, as well as movies in the **Cinemateca Nacional**, and live dance, music, and

LAY YOUR HEAD

• **Plaza Grande Quito:** Located near the Plaza de la Independencia in the heart of the old town, this historic five-star hotel occupies the site of the first Spanish house in Quito; restaurants, bar, spa, wine cellar; flamenco shows, horse-drawn carriage city tours; from $211; hotelplaza grandeingles.com.

• **Hotel San Francisco de Quito:** Lovingly restored 17-century colonial house with 30 guest rooms near the Calle la Ronda nightlife zone; restaurant, atrium reading area/

courtyard, roof terrace; from $69; sanfranciscodequito.com.ec.

• **Le Parc Hotel:** Upscale high-rise in Quito's financial district with 30 contemporary designer suites and a comfy rooftop lounge; restaurant, bar, spa, gym; from $157; leparc.com.ec.

• **Hotel Ibis Quito:** Modern budget hotel in Barrio La Pradera near downtown Quito and Carolina Park; restaurant, bar, fitness center; from $52; all.accor.com.

Shops and restaurants line the streets of Quito, with the Basílica del Voto Nacional in the distance.

theater performances in **El Agora**. Yet the high point of the CCE is undoubtedly the **Museo Nacional**, which offers a broad overview of Ecuadorian culture, art, and history that culminates in a **Sala de Oro** (Gold Room) where Inca and other pre-Columbian treasures are on display.

High above the city center, an old military hospital has been converted into the **Centro de Arte Contemporáneo**. Farther out, the former home of renowned 20th-century sculptor and painter **Oswaldo Guayasamín** is now a museum devoted to his life's work. The nearby **Capilla del Hombre** ("Chapel of Man"), considered his masterwork, wasn't completed until three years after his death in 1999.

Northeast of the old town, **Barrio La Floresta** was named for its abundance of wildflowers. **La Floresta Mercado Agroecológico**, the city's first all-organic market, flaunts food stalls serving local favorites. La Floresta also boasts a cinema that plays foreign films, and as night falls, the area's club scene and microbreweries come to life.

With its edgy street art, galleries,

LOCAL FLAVOR

- **ZFood Pescaderia:** Dedicated to responsible fishing and fair trade, this popular seafood restaurant offers sandwiches, poke bowls, tacos, filets, and super-fresh ceviche; *Avenida la Coruna 30-135;* zfoodquito.com.

- **Mercado Central:** Market stalls serving authentic Ecuadorian street food including homemade soups, sausages, and seafood dishes like *corvina con arroz y papas* (sea bass with rice and potatoes) and *ceviche de camarón* (shrimp ceviche); *Avenida Pichincha at Calle Manabi.*

- **Chez Jerome:** Contemporary and sophisticated French and South American cuisine complemented by elegant decor and a good wine selection; *Calle Wymper N30-96 at Avenida Coruña;* chezjerome restaurante.com.

- **Zazu:** This Relais & Châteaux eatery pairs Ecuadorian traditions with new gastronomic techniques and flavors in a seven-course taster menu and à la carte seafood and grilled meats; *Mariano Aguilera 331;* zazuquito.com.

and hipster bars, the nearby **Barrio Guápulo** exudes an artsy bohemian vibe on the slopes of a hill that offers excellent views of snowcapped **Cotopaxi** volcano. Founded in 1620 and one of the nation's oldest churches, **Iglesia de Nuestra Señora de Guápulo** is renowned for its ornately carved wooden pulpit.

Barrio Mariscal, Quito's culinary hotbed, hosts live music and cultural performances in the **Plaza Foch**, as well as the annual **Fiestas de Quito** in early December. It's also home to three traditional chocolate makers: **Republica del Cacao** experiments with hot peppers, coffee nibs, and banana chips while **Pacari** distinguishes its products by growing region—for example, cacao from Los Rios creates a fruity-tasting chocolate. **Kallari Café** is the official distributor of the Kallari cooperative, which provides fair-trade compensation to the Indigenous people who grow their cacao. Among their flavors are ginger and salt, chili and cinnamon, and vanilla chocolate.

Quiteños escape their crowded, often hectic city at large, diverse green spaces like the **Parque Metropolitano Guangüiltagua**, which offers hiking and biking paths, as well as places to picnic and even camp overnight amid its expansive 1,400 acres (567 ha). Astride the modern business district, the popular **Parque La Carolina** renders a small zoo, children's museum, lake with paddleboat rentals, and the flora-infused **Jardín Botánico**.

Rising to 13,000 feet (3,962 m) above sea level, the **TelefériQo** cable car offers panoramic views of Quito and the Andes on a 20-minute flight up the side of **Pichincha volcano**. Reaching **Cruz Loma** station at the top, visitors can grab a snack in the mountainside café, saddle up for a horseback ride across the heights, or set off on a trail leading to the summit of **Rucu Pichincha**, the volcano's second highest peak (15,300 feet/4,663 m).

Although it's a bit of a tourist trap, the **Ciudad Mitad del Mundo** offers a monumental tower with a viewing platform, a line marking zero degrees latitude, and a kitschy museum that celebrates the fact that Quito is the closest capital city to the Equator. Located just beyond the northern edge of the metro area, the "Middle of the World" village also features restaurants and souvenir shops. ∎

Founded in 1534, Iglesia de San Francisco is Ecuador's oldest church.

Salvador
Brazil

One of the oldest European cities in the Western Hemisphere—and one of the closest to Africa—Salvador's cultural smorgasbord spawned a melodious heritage expressed in everything from the annual pre-Lent carnival to earthy music clubs and the city's grandiose churches.

THE BIG PICTURE

Founded: 1549

Population: 3.3 million

Size: 165 square miles (427 sq km)

Language: Portuguese

Currency: Brazilian real

Time Zone: GMT -3 (BST)

Global Cost of Living Rank: Unlisted

Also Known As: São Salvador da Bahia de Todos os Santos, Bahia

"The samba is so cool you never want it to end," sings Brasil '66 in their signature tune "Mais Que Nada." And the city where it truly never ends is Salvador, the 17th-century birthplace of the iconic music and dance tradition.

Although samba spread to the rest of the nation and around the world, it's forever linked to the tropical coastal city in northeast Brazil. So much so that the local *samba de roda* dance circle was added to UNESCO's Intangible Cultural Heritage list in 2005.

Samba is the best-known manifestation of the Afro-Brazilian culture that suffuses Salvador and surrounding Bahia state. After years of hiding in the shadows of the nation's European customs, age-old traditions like percussion, capoeira martial arts, and Candomblé religious rites have emerged again.

Carnival and other festivals are the best times to experience the city's Afro-Brazilian side. But for those who arrive at other times, there are plenty of ways to get into the groove. **Rhythm of Bahia** and other outfits offer half-day samba and capoeira workshops. Nightlife areas like **Rio Vermelho**, **Pelourinho**, and the edgy **Beco dos Artistas** are flush with clubs and bars with live bands playing a variety of Brazilian music.

Balé Folclórico da Bahia performs five nights a week at the **Teatro Miguel Santana while the Teatro Castro Alves** hosts concerts by contemporary headliners like Gilberto Gil, Adriana Calcanhotto, and Sérgio Mendes of Brasil '66 fame. Meanwhile, the 18th-century **Igreja e Convento de São Francisco**—one of the world's most lavishly decorated churches—gets in on the action on Tuesday evenings by organizing live music in the cobblestone plaza out front.

For the more intellectually inclined, the **Museu Afro-Brasileiro** showcases a wide range of locally produced handicrafts and fine art, while the **Casa do Carnaval** delves into the origins of the city's most flamboyant festival via an English-language audio guide.

Salvador's other claim to fame is the pastel hues that adorn many a facade in the aforementioned **Pelourinho** neighborhood. The

Salvador de Bahia sits on the coast of the Atlantic.

Enjoy the colorful costumes, lively music, and dance performances of Carnival, held every February ahead of Lent.

mosaic-tiled **Terreiro de Jesus** plaza is flanked by four ancient churches including the 17th-century **Catedral Basílica de Salvador**, whose austere facade masks an elaborate, gold-splashed interior.

Three blocks north of the plaza, the **Casa de Jorge Amado** celebrates the life and work of the celebrated Brazilian author, who set *Dona Flor and Her Two Husbands* in Pelourinho. Heading south from the plaza, the cobblestone streets lead past the **Museu da Misericórdia** and its colonial-era relics to the **Elevador Lacerda,** a historic art deco lift that descends to the waterfront. Along the bay is **Mercado Modelo**, once a slave market and customs house but filled nowadays with vendors selling handmade crafts.

From there you can follow the waterfront north to the dazzling **Igreja de Nosso Senhor do Bonfim** church, or south around the end of the Salvador peninsula to various fortifications that once protected the bay and beaches that line the city's Atlantic coast. ∎

LAY YOUR HEAD

• **Aram Yamí:** Romantic boutique digs with bay views tucked inside a colonial building in the old town; breakfast, swimming pool; from $144; aramyamihotel.com.

• **Hotel Fasano Salvador:** Read all about it in this posh five-star abode in an art deco–style high-rise that was once an office of a leading newspaper; restaurant, rooftop pool, spa, fitness center; from $264; fasano.com.br.

• **Fera Palace:** Salvador's equivalent of New York City's Flatiron

Building has been transformed into a chic boutique hotel with incredible bay views and elegant modern decor; restaurant, rooftop pool, and bar, fitness room; from $120; ferapalacehotel.com.br.

• **Vila dos Orixás:** Go full tropical at this swank beach hotel in Morro de São Paulo, an island resort two hours by ferry south of Salvador; restaurants, bar, swimming pool, beach, spa services, bike rental; from $117; hotelviladosorixas.com.

Buenos Aires
Argentina

With a passion matched by few urbanites around the globe, the citizens of Buenos Aires have created a larger-than-life existence on display every day in the city's dance, music, food—and an extraordinary talent for kicking a black-and-white ball.

THE BIG PICTURE

Founded: 1536

Population: 15.1 million

Size: 1,240 square miles (3,212 sq km)

Language: Spanish

Currency: Argentine peso

Time Zone: GMT -3 (AST)

Global Cost of Living Rank: 168

Also Known As: B.A., Paris of the Americas, Queen of the Plata

Buenos Aires presents a conundrum for anyone trying to define the globe's great cities. It doesn't flaunt any instantly recognizable landmarks, doesn't boast epic museums or theme parks, and there isn't a single World Heritage site.

Yet there's something very alluring about the Argentine capital. An elegance that comes across in the city's beloved tango, in the fiery passion of its soccer teams, the flavors exuded by local taste treats like maté tea and *bife de lomo* steak, and the way that Porteños—both men and women from the city—carry themselves. It's not far-fetched to claim that everyday life is the main attraction of Buenos Aires.

Much like Los Angeles, "B.A." (as it's so often called) boasts more than a single hub. Recoleta, Palermo, La Boca, Belgrano, and many of the countless suburbs move in largely self-contained worlds, almost like they're separate cities rather than integral parts of a large metropolis.

If there's anything like a central focus, it's the ancient and often tumultuous **Plaza de Mayo**. Ancient because the square traces its origins to 1608, when the Jesuits first arrived on the scene. Tumultuous because it's often been the place where Argentines have gathered to celebrate, protest, and even grieve.

Oldest among the plaza's many landmarks is the whitewashed **Cabildo**, an early 18th-century government building that now houses a museum dedicated to the May Revolution of 1810 that gained Argentina its independence from Spain. The imposing **Catedral Metropolitana**—which looks far more like an ancient Roman temple than a typical Latin American Catholic shrine—safeguards the tomb of Argentine liberator **José de San Martín** but is better known these days as the church that Cardinal Jorge Mario Bergoglio presided over before becoming Pope Francis.

However, the plaza's best known structure is the **Casa Rosada**. Originally opened as a post office in the 1870s, the pink palace soon transitioned into the home of Argentina's president (and tyrant, depending on the era). Of the many events that have transpired there, the most famous was a 1951 rally during which Eva Perón addressed an estimated two million people from the

Street performers stage an elegant tango for passersby in Plaza de Mayo.

The Obelisco de Buenos Aires is a national historic monument honoring the 400th anniversary of the city's founding.

building's main balcony. Explore artifacts from Argentina's turbulent political history in the **Museo Casa Rosada** or catch the free English-language guided tours on Saturday.

A block beyond the cathedral, the ornate rounded facade of the **Banco de Boston** building anchors the southern end of a **Calle Florida** pedestrian zone. Flanked by seemingly every sort of shop and fast-food eatery, the street runs more than half a mile (0.8 km) across the **Microcentro** neighborhood to **Plaza General San Martín**, a former slave market where the great liberator bivouacked his troops during the revolution. All those people scurrying across the plaza are probably on their way to **Estación**

Retiro and trains to the far corners of Argentina, even an old express to Patagonia.

Running along the western edge of B.A.'s Microcentro is broad

Avenida 9 de Julio and its eye-catching **Obelisco**, a towering tribute to the city's 400th anniversary in 1936. Inspired by the **Champs-Élysées** in Paris, the

---

LOCAL FLAVOR

● **Steaks by Luis:** Traditional *asado* (Argentine barbecue) with all the rituals and fixings—including a five-course set menu paired with wine—served at big family-style tables; *St. Jeronimo Salguero 1410, Palermo;* steaksbyluis.com.

● **La Morada:** The menu at this busy and reasonably priced restaurant sticks to delicious empanadas, tortas, and salads; *Avenida Hipolito Yrigoyen 778, near Plaza de Mayo;* lamorada.com.ar.

● **El Gato Negro:** Established in 1929 as a spice store, the atmospheric "Black Cat" offers coffees, tea, snacks, full meals upstairs, and an aromatic selection of spices; *Avenida Corrientes 1669, near the Obelisk;* donvictoriano.com.ar.

● **Es Ruiz:** Amazing selection of gourmet pastries, coffees, and snacks from master chef Eduardo Ruiz; *Calle José Terry 300 at Valle, Caballito;* esruiz.com.ar.

avenue is often cited as the world's widest street—460 feet (140 m) across.

Obsessed with everything French, the city's late-19th-century movers and shakers also created the **Teatro Colón**. Although modeled after the Paris Opera House, Latin America's grandest theater surpasses its archetype in size and sound. Considered one of the world's five best concert venues in terms of its acoustics, the theater offers a year-round slate of opera, ballet, and symphony. If you cannot catch a show, consider a behind-the-scenes guided tour.

The Buenos Aires waterfront boasts a passageway of a much different sort—**Rio Dársena Sur**—an artificial harbor that once ushered

cargo and immigrant steamers from the **Rio de la Plata** into a thriving docks area called **Puerto Madero**. By the 1920s the old port was obsolete and fell into disrepair.

However, a massive urban renewal effort in the early 21st century transformed Puerto Madero into a venue for restaurants, hotels, yachts, and cultural outlets like the **Museo del Humor** in a lovely art deco building that once housed a sailor-filled beer hall. The historic Argentine Navy warships **A.R.A.** *Uruguay* (launched in 1874) and the **A.R.A.** *Presidente Sarmiento* (launched in 1898) are preserved as floating museums.

A block away, the grandiose central post office (opened in 1928) has

found new life as the **Centro Cultural Kirchner**, an entire city block of exhibition galleries and performance spaces, including the **Ballena Azul** ("Blue Whale") concert hall.

Yet the area's most intriguing structure is something ultramodern—the **Puente De La Mujer** (Women's Bridge)—a futuristic pedestrian span that rotates 90 degrees to facilitate maritime traffic. Architect Santiago Calatrava says a couple dancing the tango inspired the design.

Promenades run down both sides of the **Rio Dársena**, but the area's best walks meander through the **Reserva Ecológica Costanera Sur.** The park has an incredible origin story: Decades worth of debris from

Take a guided tour of Teatro Colón, named one of the best opera houses in the world by National Geographic.

constructing the city's subway, motorways, and docklands was dumped into the Río de la Plata. Nurtured by the river's nourishing silt, plants began to sprout spontaneously, and this in turn attracted copious birdlife to a place that was declared a nature reserve in 2005.

Another great stroll is **Avenida Alvear** between **Plazoleta Pellegrini** and **Plaza San Martín de Tours**. Lined with designer shops and five-star hotels, the avenue expedites window-shopping and people-watching in the heart of **Recoleta**, the city's most fashionable neighborhood. Be sure to pop into the **Alvear Palace Hotel** (opened in 1932) for a peek at the opulent lobby, high tea in **L'Orangerie**, or a cocktail at the open-air **Roof Bar**.

Just past the **Eros statue** at the upper end of Avenida Alvear is the marvelous **Centro Cultural Recoleta**, an 18th-century Franciscan church and convent transformed into a vibrant youth-oriented nexus for visual and performing arts. Next door is the entrance to **Recoleta Cemetery**, one of the world's most renowned burial grounds. **Eva Perón's tomb** attracts a steady stream of pilgrims, but plenty of other notables have been laid to rest within the cemetery's elaborate tombs.

Having explored Buenos Aires on foot, it's time to sink into a comfortable seat and relish the abiding passions of nearly every Porteño: soccer and tango.

The Argentine national team and Club Atlético River Plate of the country's Primera División play their matches in colossal **River Plate Stadium** in **Belgrano**. Boca Juniors, the city's other highly regarded professional team, takes the field at the

Pay homage to Eva Perón and other luminaries at the Recoleta Cemetery.

legendary **La Bombonera** stadium in the **La Boca** neighborhood.

Matches at La Bombonera are easily combined with a stroll past the multicolored houses of **Caminito** (named for a popular 1920s tango song). Decorated with cartoonlike effigies of Diego Maradona (who played for Boca Juniors), Che Guevara, and other renowned Argentines, the street is kitschy but fun. There's more street tango in **Plaza Dorrego** in the **San Telmo** neighborhood renowned for its Sunday **Feria de San Telmo**.

Although tango is primarily known as a dance, Argentines also cherish tango music and lyrics. Buenos Aires is flush with tourist-oriented tango performances. Among the best are the **Piazzolla Tango** in the sumptuous **Galería Güemes** off Calle Florida and **Rojo Tango** in Puerto Madero. Less glitzy but far more authentic are neighborhood tango salons like **Maldita Milonga** in San Telmo and **Milonga Parakultural** in Palermo. ∎

Rio de Janeiro
Brazil

Soccer, sand, and samba may grab most of the attention, but Brazil's exotic seaside metropolis offers plenty of other diversions, from jungle trails and tropical gardens to amazing street art and space-age architecture.

THE BIG PICTURE

Founded: 1565

Population: 12 million

Size: 740 square miles (1,917 sq km)

Language: Portuguese

Currency: Brazilian real

Time Zone: GMT -3 (BRT)

Global Cost of Living Rank: 191

Also Known As: Cidade Maravilhosa (Marvellous City)

Appalled by shameless apparel and decadent behavior, the mayor of Rio de Janeiro issued a proclamation in 1917 warning beachgoers to wear "proper clothing with the necessary decency and composure." Citizens were not to transit streets without a bathrobe or suit jacket covering their swimming costumes. And furthermore, they should refrain from shouting or using loud voices unless truly in distress, or behaving "in an offensive way to public morals and decorum."

He was referring specifically to a trendy new beach area called **Copacabana**, frequented at that time by women in bathing suits that dared to show their bare arms and men getting a little rambunctious at milk bars operated by coastal dairy farms.

All that changed just a few years later when the **Copacabana Palace** (opened in 1923) and other posh digs began attracting movie stars, musicians, and other luminaries from around the globe. By the time the skimpy Brazilian bikini debuted on Copacabana in the 1970s, South America's most famous beach really was threatening public decorum. But that just made it even more popular.

Stretching 2.5 miles (4 km) from stem to stern, Copacabana is broken into six sections by numbered *postos* (lifeguard towers) and flanked by a **Calçadão De Copacabana** boardwalk renowned for its black-and-white wave pattern. There are countless places to grab a snack or sip a *caipirinha* (Brazil's national drink). Copacabana often hosts professional beach volleyball and beach soccer events, as well as live music (Rod Stewart drew an estimated three million fans—the largest concert crowd in world history—to a 1994 beach concert.)

For a dose of local history, explore the old bastions at either end of Copacabana. **Forte Duque de Caxias** (completed in 1779) offers historical exhibits, marmosets scampering through the forest that engulfs the fort, and great views of Sugarloaf Mountain on the other side. **Forte de Copacabana** (completed in 1914) houses the Brazilian **Army Historical Museum**, a collection of vintage artillery pieces, and the outdoor **Café 18 do Forte** with its panoramic beach views.

The rocky **Pedra do Arpoador** peninsula divides Copacabana from

Visit the Selarón Staircase in downtown Lapa for a colorful tile display.

Christ the Redeemer, atop Corcovado Mountain, looks over the coastal downtown area of Rio de Janeiro.

its sandy accomplice: **Ipanema**. The beach originally gained fame as an early 20th-century bohemian hangout for writers, artists, and other creative types. It wasn't until the bossa nova song "Girl from Ipanema" soared to the top of the charts in the mid-1960s that the beach achieved global celebrity. Renamed **Garôta de Ipanema** after the song, the casual café where Tom Jobim and Vinícius de Moraes wrote the tune is still there, just a block up from the beach.

Although the overall beach scene is similar to its neighbor, Ipanema is far more conducive to board and bodysurfing, especially the **Arpoador** end, one of Rio's best surf breaks. Other urban tribes have also staked claims to Ipanema: samba musicians around **Posto 7**, beach volleyball players at **Posto 10**, and an LGBTQ crowd that rallies around the rainbow flags at **Posto 8**. The **Leblon** neighborhood at the west end of the beach strip is ground zero for

101

nightlife, with dozens of restaurants, bars, and nightclubs, especially along **Rua Dias Ferreira**.

These legendary beaches are backed by a hinterland with attractions like the varied art and music events of the plush **Parque Lage** estate, photo exhibits and film screenings at the **Instituto Moreira Salles**, and thoroughbred horse racing at the **Jockey Club Brasileiro**. Best of all is the **Jardim Botânico**, a floral wonderland founded in 1808 that's now home to more than 8,000 plant species, as well as resident capuchin monkeys, marmosets, crab-eating raccoons, and more than 100 species of tropical birds.

Beyond its sandy strands, Rio's foremost natural landmarks are the granite peaks that rise all around, some of them climbing literally straight up from the sea. Most iconic of all is **Pão de Açúcar**, or Sugarloaf as it's called in English. Towering nearly 1,300 feet (396 m) above the entrance to **Guanabara Bay**, the twin summits are accessed on modern **Bondinho cable cars** that depart every 20 minutes from **Praia Vermelha**.

With its restaurants, shops, and amphitheater for open-air concerts

Beautiful murals line the walls on pedestrian-friendly Boulevard Olimpico.

Seemingly out of *Star Trek,* the futuristic Museu do Amanhã hovers at the pier of Praça Mauá.

and carnival events, **Morro da Urca** (the lower summit) often feels like a theme park. It's not until you reach the very top that Mother Nature comes into its own. Jungle trails meander through indigenous Atlantic coastal forest to sheer overlooks of the ocean, bay, and the aforementioned beaches. You can also reach the summit on foot, via an ultra-steep trail that takes about two hours of hiking. **Rio Hiking** offers guided walks along the precipitous route.

The world's largest art deco statue tops another granite peak—

a 98-foot (30 m) figure of **Cristo Redentor** (Christ the Redeemer) that crowns **Corcovado** mountain. Voted one of the New Seven Wonders of the World in a global poll, the sublime figure became an instant icon upon its dedication in 1931. The **Trem do Corcovado**, an electric cog railway, carries around 300,000 people to the summit each year for a view that really does take your breath away at times. The peak is far and away the best place to snap an aerial photo of Rio, especially at sunset when the bay glows molten gold.

LOCAL FLAVOR

• **Aprazível:** Funky house and garden restaurant with dreamy views and new takes on Brazilian favorites like *cabrito* (goat in red wine), *duo mineiro* (pork and manioc), and *palmito fresco assado* (roasted palm hearts); *Rua Aprazível 62, Santa Teresa;* aprazivel.com.br.

• **Carretão Ipanema:** Carnivores

flock to this classic Brazilian steakhouse for all-you-can-eat barbecued beef, chicken, and salmon; *Rua Visc. de Pirajá 112, Ipanema;* carretaipanema.com.br/en.

• **Restaurante Shirley:** Marinara paella is the specialty on a menu flavored with other seafood and Spanish dishes in an eatery that's been around

for more than 65 years; *Rua Gustavo Sampaio 610/A, Copacabana.*

• **Scenarium:** Cocktails and other libations complemented by art exhibits, poetry readings, samba bands, and other entertainment in an old warehouse in the hipster Lapa district; *Rua do Lavradio, 20, Centro;* rioscenarium.art.br.

The Sugarloaf cable cars whiz tourists to the summit of Sugarloaf Mountain, with Botafogo Bay behind them.

Cristo Redentor graces the eastern end of **Tijuca Forest National Park**, a mountainous preserve that safeguards one of the world's largest urban forests. The area was once covered in coffee and sugar plantations. Starting in 1861, forester Manuel Gomes Archer led a campaign to plant more than 100,000 native trees that resurrected the Tijuca Forest. The 11-mile (18 km) **Peaks Circuit** across 10 summits is the most formidable of the park's many hiking routes.

Rio owes its very existence to a third natural phenomenon: **Guanabara Bay**. Brazil's second largest harbor stretches 19 miles (31 km) inland from a single, narrow outlet to the Atlantic beneath Sugarloaf. At first

assuming they had discovered a river, Portuguese mariners christened it the "River of January" when they sailed into Guanabara on January 1, 1502.

The Portuguese colony first took root around the base of Sugarloaf, but later moved farther into the bay where the bustling **Centro** business and government district now sits. Several of the city's oldest colonial buildings huddle around the **Praça Quinze de Novembro**, including the **Paço Imperial** (built in 1738), where Brazil's governors and kings once resided, and the **Igreja Nossa Senhora do Monte do Carmo** church (consecrated in 1761) with its elaborate rococo interior.

Cinelândia square (Praça Floriano) is surrounded by grand belle

epoque–era buildings like the **Biblioteca Nacional do Brasil** (opened in 1810), **Museu Nacional de Belas Artes** with its artistic treasures from around the world (opened in 1937), and the marvelous **Theatro Municipal** (1905), which is still the place to see opera, ballet, and symphony in Rio. Another gorgeous structure from that same era now houses the eclectic **Centro Cultural Banco do Brasil** (built in 1906).

The belle epoque buildings offer an eye-catching contrast to the nearby **Catedral Metropolitana**, a starkly modern, pyramid-shaped church renowned for its soaring 210-foot (64 m) stained-glass windows. Nearby **Estação de Bondes** is the place to catch the antique mustard yellow **Bondes trams** that rumble across the monumental **Arcos de Lapa** aqueduct with its 42 arches.

Once upon a time, the tram continued uphill to **Santa Teresa**, an old but avant-garde neighborhood where many of the bygone mansions are now filled with bars, restaurants, and art spaces with inspiring views. Make your way back to the city center via the remarkable **Escadaria Selarón**, a flight of 215 steps transformed over a quarter century into a brilliant outdoor canvas by artist Jorge Selarón who lived in a tiny house beside the stairs until his death in 2013.

On the other side of Centro, the once derelict dock area has morphed in recent years from handling passengers and cargo to showcasing Brazilian nature, science, and human creativity. One of the old piers is now occupied by the **Museu do Amanhã** science museum in a huge gleaming, futuristic building. Rising on the other side of the **Praça Mauá** plaza is another astonishing structure: **Museu de Arte do Rio** (MAR), housed in adjoining buildings linked by a pedestrian bridge

Colorful umbrellas and street vendors brighten Copacabana Beach.

beneath the museum's giant wave-shaped roof.

Boulevard Olímpico—the waterfront walkway created for the 2016 Summer Games in Rio—retains many of its maritime warehouses. But nowadays many of them are filled with restaurants, bars, and entertainment, and covered in giant street murals painted by prominent Brazilian artists. The promenade stretches about a mile (1.6 km) between **Praça Mauá** and the new **AquaRio**, the largest aquarium in

South America with more than 5,000 sea creatures in 28 tanks, as well as a surf museum and walk-through shark and ray habitat. Almost hidden among the tangle of warehouses are the remains of the **Cais do Valongo**, a stone wharf where as many as a million Africans disembarked from slave ships in the early 1800s.

Sixteen lanes across, **Avenida Presidente Vargas** cuts a wide path across the city center in places where two other great Brazilian institutions unfold: samba and soccer. Designed by renowned architect Oscar Niemeyer, the immense **Sambódromo** is a purpose-built stadium for Rio's dazzling **Carnival** parades. Larger still is 78,000-seat **Maracanã Stadium**, base for two of the city's professional soccer teams (Fluminense FC and CR Flamengo) and venue for many international games. There are two ways to catch a match at Maracanã: purchase a ticket and transport package beforehand or buy single tickets at the stadium box office two or three hours before a league game. ∎

BACKGROUND CHECK

• The French rather than the Portuguese established the first European settlement on Guanabara Bay, a 1555 settlement called France Antarctique that endured for just 12 years.

• Rio de Janeiro was the capital of colonial and independent Brazil from 1763 to 1960, when Brasilia was completed. During the Napoleonic Wars of the early 19th century, it was also the capital of the Portuguese Empire.

• Rather than a Brazilian Amerindian word, Copacabana Beach is named for a Bolivian town on the shore of Lake Titicaca that derives its name from Kotakawana, the ancient Andean god of fertility.

• Over the years, Rio's residents have called themselves the Fluminense (river people), Guanabarinos (after the state), and most recently Cariocas, an old Amerindian term that may have been the name of a precontact tribe or village.

Santiago
Chile

Emerging from the dark age of the military junta that ruled Chile for so long, Santiago is quickly catching up with Rio and Buenos Aires as one of South America's most exciting cities in terms of culture, charisma, and culinary flair.

THE BIG PICTURE

Founded: 1541

Population: 6.4 million

Size: 440 square miles (1,140 sq km)

Languages: Spanish

Currency: Chilean peso

Time Zone: GMT -3 and -4 (CLST and CLT)

Global Cost of Living Rank: 108

Also Known As: Santiago del Nuevo Extremo (Spanish colonial)

Author Isabel Allende described Santiago as feeling "like the end of the world" while she was growing up there during the ultraconservative days of dictatorship. Like many Santiaguinos she eventually fled overseas. It wasn't until democracy was restored in 1990 that many of those exiles returned to the Chilean capital, helping to stoke an urban renaissance that embraces just about everything new, from gourmet foods and fashion to modern architecture and alternative lifestyles.

Unlike the time-warped, worn-out metropolis of the past, the Santiago that dashed into the 21st century is confident and incredibly creative, ready to take its place among the great cities of Latin America. It's definitely no longer at the end of the world, rather at the forefront of many global trends.

Nowhere is this more evident than in a high-rise business, shopping, and residential district that has sprouted over the past two decades. Geographically, it's the western end of the upscale **Las Condes** district, but locals have taken to calling it **"Sanhattan"** since the advent of the local skyscraper boom. A blue monolith called the **Gran Torre Santiago** (aka Costanera Center) is the highest of them all, soaring 984 feet (300 m) and the tallest building in South America since its completion in 2013. Perched at the summit, the **Sky Costanera** observation deck offers a condor's-eye view of Santiago and the snow-capped Andes.

Down at street level, busy **Avenida Apoquindo** cuts a wide path across the middle of Las Condes to the other architectural icons like the huge glass box that wraps around the **Teatro Municipal Las Condes** (music, dance, theater) and another eye-catching skyscraper called the **Edificio Génesis.** But there's also a reminder of the days when this was still rural: The 19th-century **El Rosario** farmhouse, with its whitewashed walls and red-tiled roof, hosts cultural performances, workshops, and art exhibits organized by the district cultural center.

The **Barrio el Golf** and **Vitacura** neighborhoods north of Avenida Apoquindo harbor some of the city's best hotels and restaurants, as well as **Los Leones Golf Club** that gives the former its sporting name. There

See penguins and other critters at the Chilean National Zoo, founded in 1925.

A monument of Pedro de Valdivia stands at the center of Plaza de las Armas, which also boasts the Metropolitan Cathedral.

are also two small but highly intriguing museums: **Museo de la Moda** offers a riveting overview of fashion from the baroque and rococo periods through the outrageous pop art garb of the psychedelic sixties. Although dedicated to modern Latin American art, the **Museo Ralli** also boasts works by European and American masters like Chagall, Calder, and Dalí.

From the Costanera end of Avenida Apoquindo, you can hike or bike along the south bank of the **Rio Mapocho** along paths that meander through riverside parks to older neighborhoods in the central city.

The most astonishing of these is **Lastarria**, a neighborhood that, like much of Santiago, fell into neglect and disrepair during the junta years.

But its phoenixlike rise over the past three decades is nothing short of phenomenal. As fast as the old buildings could be restored, new restaurants, bars, museums, and galleries occupied them.

Lastarria's cultural cornerstones include the **Chilean Museum of Pre-Columbian Art**, which exhibits weaving, clothing, woodwork, ceramics, and metal objects created over the past 3,000 years by Chile's various Indigenous cultures. Following the junta's downfall, Pinochet's

---

LOCAL FLAVOR

• **Fuente Mardoqueo:** In a nation that loves sandwiches, this old-time Las Condes eatery makes some of the best—*lomito bávaro, churrasco,* and *atún* served with traditional condiments and draft beer; *El Bosque Nte. 145, Las Condes.*

• **Peumayén Ancestral Food:** As the name implies, this distinctive restaurant offers modern takes on the traditional food and drink of the Mapuche, Rapa Nui, and other Indigenous peoples; *Constitución 136, Bellavista.*

• **Boragó:** Consistently ranked among the world's 50 best restaurants, this riverside eatery revolves around tasting menus with offbeat seasonal and local ingredients harvested from all around Chile; *Avenida S.J. Escrivá de Balaguer 5970, Vitacura.*

LAY YOUR HEAD

• **The Aubrey:** Chic boutique hotel set on the grounds of an elegant 1920s mansion in Bellavista; rooms feature hardwood floors, oversize soak tubs, and alfresco seating areas; restaurant, piano bar, spa, heated outdoor pool; from $172; theaubrey.com.

• **Ritz Carlton Santiago:** Set amid the skyscrapers of Sanhattan, the Ritz radiates luxury with its gourmet cuisine, outstanding wine cellar, and rooftop swimming pool beneath a palm-filled sky deck; restaurants,

bars, spa; from $362; ritzcarlton.com /en/hotels/santiago.

• **Casa Bueras:** This 1927 mansion in Barrio Lastarria provides an apt venue for an elegant hotel in the artsy Lastarria and is within walking distance of the neighborhood's many museums, galleries, and nightlife; restaurant, bar, swimming pool, garden; from $169; lastarriahotel.com.

• **Hotel Casa Real:** Located in the Maipu wine region on the southern edge of Santiago, the "royal house" is

located on the grounds of Viña Santa Rita winery; restaurant, tasting room, museum, replica Roman baths, the world's second largest bougainvillea; from $382; santarita.com/en/ casa-real-hotel.

• **Noi Puma Lodge:** Around 2.5 hours from the city center, on the edge of Río Cipreses National Reserve, this wilderness retreat rotates winter sports (like heli-skiing and snowshoeing) with summer outdoor adventure in the Andes; restaurant, bar, spa; from $154; noihotels.com.

massive Defense Ministry was transformed into the **Centro Cultural Gabriela Mistral** (GAM), an eclectic venue for art, music, dance, theater, and comedy.

Parque Forestal on the barrio's riverside hosts two of Chile's biggest and most important museums: the **Museo Nacional de Bellas Artes** and the **Museo de Arte Contemporaneo**. The popular green space is also renowned for its user-friendly

lawns, and handsome Oriental plane trees *(Platanus orientalis)*.

The heavily wooded hill on Lastarria's west side is **Cerro Santa Lucia,** a rocky outcrop that was sacred to the area's Amerindian inhabitants. The Spaniards didn't bother to fortify the hill until the early 1800s, when Chile's independence movement was reaching fever pitch. One of those bastions is the redbrick and rather medieval-

looking **Castillo Hidalgo** on the summit, reached via a network of trails that meander through a park that also renders a small **Japanese Garden** and the extravagant **Neptune Fountain**.

Named after the Chilean general who helped expel the Spanish in the 1820s, **Avenida Libertador Bernardo O'Higgins** connects Lastarria and Cerro Santa Lucia with the city's **Centro Historico**. The spot where conquistador Pedro de Valdivia founded Santiago in 1541, the historic center was a bastion of Chile's political and religious power brokers for nearly 600 years.

Designed by one of Valdivia's minions, the leafy **Plaza de Armas** is surrounded by Spanish colonial gems like the **Metropolitan Cathedral** (completed in 1799) with its **Museo de Arte Sagrado**, and a former colonial courthouse and one-time national capital building called the **Palacio de la Real Audiencia de Santiago** (completed in 1807). A few blocks north of the plaza, the magically restored **Mercado Central** (opened in 1872) and **Estación Mapocho** train depot (opened 1913) offer a medley of gourmet

A funicular runs to the top of San Cristóbal Hill, 984 feet (300 m) above Santiago.

The Neptune Fountain was built in the early 1700s in a neoclassical design.

eateries, handicraft shops, exhibitions, and performing arts.

In the 1840s, Chile's government relocated to the nearby **Palacio de La Moneda**, a structure that derives its name (Palace of the Coin) from the fact that it was originally the Spanish colonial mint. Tucked beneath the capitol plaza are the **Centro Cultural Palacio La Moneda**, a modern art venue with exhibition galleries and the **Cineteca Nacional** movie house, while across the street is the **Crypt of Bernardo O'Higgins**.

Farther west, **Barrio Brasil** and **Barrio Yungay** have also undergone dramatic transformations in the past 30 years, with restaurants, bars, and boutiques that cater to students

from the area's various educational institutions. One of its newer landmarks is the **Museo de la Memoria y los Derechos Humanos**, which honors the disappeared and details the human rights abuses of the Pinochet era.

On the north side of the river is bohemian **Bellavista**. Santiago's intellectual heart for much of the 20th century, the barrio was home to Nobel Prize–winning poet, diplomat, and politician Pablo Neruda, and many other prominent writers, artists, and thinkers who gathered in local cantinas and cafés. Guided tours of **La Chascona**, Neruda's eccentric hillside house, expose the life and times of Chile's greatest writer. One of the city's best nightlife

areas, Bellavista boasts numerous clubs, cafés, and theater spaces.

Rising high above the barrio is **Cerro San Cristóbal**, a 2,800-foot (853 m) peak that crowns the western end of **Parque Metropolitano**, Santiago's largest park. The vintage 1925 **Funicular de Santiago** rises from Bellavista to the summit, where a towering statue of the **Virgin of the Immaculate Conception** gazes down on the city. Visitors can ride the **Teleférico** cable car along the ridge or hike the **Sendero de las Grandes Travesías**, a six-mile (9.7 km) trail that leads to the **National Zoo**, several botanical gardens, two public pools, an adventure playground, and numerous viewpoints. ■

Montevideo
Uruguay

Perched on the edge of the Atlantic, Uruguay's capital offers a mixed bag of big city and beaches, fine arts and *fútbol* that often feels like a throwback of long-ago Latin America.

THE BIG PICTURE

Founded: 1724

Population: 1.9 million

Size: 633 square miles (1,640 sq km)

Language: Spanish

Currency: Uruguayan peso

Time Zone: GMT -3 (UYT)

Global Cost of Living Rank: 132

Also Known As: N/A

Lounging in an outdoor café around a palm-shaded plaza, watching couples trip the light fantastic in an old tango club, or watching La Celeste (The Sky Blue) play at Estadio Centenario alongside 60,000 local soccer fans, one can easily imagine that it's 1930 in Montevideo rather than the present day.

The Uruguayan metropolis has that effect on people, especially if you've hopped the ferry from helter-skelter Buenos Aires over on the other side of the estuary. The residents are also something of an anachronism, thoroughly laid-back, utterly cultured, and passionate about just nearly everything they do in life.

Life swirls around the **Ciudad Vieja** (Old City), which crowds a peninsula between the muddy Rio de la Plata and busy Montevideo Harbor. Although founded by early 18th-century Spanish settlers, most of the buildings date from the belle epoque and art deco eras when Montevideo was one of the world's fastest-growing (and hippest) cities.

The Ciudad Vieja is punctuated by three plazas connected by the **Peatonal Sarandí** pedestrian street.

Many of the city's museums, cafés, and entertainment outlets are arrayed around these open spaces.

Closest to the ferry terminal, **Plaza Zabala** is named for the Spaniard who founded Montevideo. Overlooking the square is the elegant **Palacio Taranco**, home of the **Museum of Decorative Arts**, and nearby is the revamped **Mercado del Puerto**, a former train station now filled with gourmet restaurants and food shops.

Four blocks along Sarandi is **Plaza Constitución**, the heart of the Spanish colonial town and the venue for two of the city's finest colonial structures (both of them opened in 1804): the baroque **Metropolitan Cathedral** and the **Cabildo Museum** with its historical and cultural exhibits, concerts, poetry readings, and puppet shows. Just off the square is the poignant **Museo Andes 1972**, which honors the Uruguay rugby players who died and survived the *Alive* airplane disaster.

Farther east is the rectangular **Plaza Independencia**, the city's paramount square. The plaza is anchored at one end by the **Puerta de la Ciudadela**, one of the few relics of the Old City fortifications. Looming over the eastern end of the

A statue of national hero José Artigas marks the center of Plaza Indepencia.

Take in a sunset and skyline views from the sandy shores located just steps away from the neighborhood of Pocitos.

square is the **Palacio Salvo**, an outlandish blend of Gothic and art deco that debuted in 1928 as the city's tallest building.

The **Museo del Tango** on the palace's ground floor illuminates a music and dance form just as important in Uruguay as in neighboring Argentina. The mausoleum of Uruguay independence hero **José Gervasio Artigas** dominates the middle of the plaza. And tucked over in one corner is the wonderful **Teatro Solís**, where opera, symphony, and ballet take the stage.

Beyond the plaza is busy **Avenida 18 de Julio** and bustling downtown Montevideo. And at the far eastern end of the avenue is leafy **Parque Palermo** and **Estadio Centenario**, built for the inaugural 1930 World Cup and home of the sky blue

Uruguay national team, commonly called La Celeste. The stadium's **Museo del Fútbol** pays tribute to the "beautiful game."

La Rambla boulevard hugs much of the Montevideo waterfront, with the portion east of the Old City boasting a seaside promenade that

curls around the Punta Carretas with its **historic lighthouse** and then onward to another bizarre structure called the **Castillo Pittamiglio** and **Playa de los Pocitos** beach. **Punta del Este**—Uruguay's chicest beach retreat—is a two-hour drive east of Montevideo. ∎

GATHERINGS

• **Iemanjá:** Tracing its roots to animism brought by colonial-era enslaved Africans, this February 2 event draws thousands to Playa Ramírez for chanting, drumming, and supplications to the Yoruba goddess of the sea.

• **Carnival:** Montevideo lets its hair down the week before Lent (February or March) with a series of music, dance, and costume-suffused

carnival events that culminates with the Las Llamadas parade through the Palermo and Barrio Sur neighborhoods.

• **La Milonga Callejera:** Plaza Liber Seregni provides a venue for these Wednesday night "street dance" sessions during the Southern Hemisphere's warm summer months. Everyone is welcome and group tango lessons are gratis.

Cusco
Peru

Like its distinctive Escuela Cuzqueña artwork, this Andes city offers a fusion of Amerindian and Iberian traditions, as well as a place where people from around the globe orient themselves to modern Peru and vast mountains that lie beyond the city limits.

THE BIG PICTURE

Founded: ca A.D. 1100

Population: 430,000

Size: 149 square miles (386 sq km)

Language: Spanish

Currency: Peruvian sol

Time Zone: GMT -5 (PET)

Global Cost of Living Rank: Unlisted

Also Known As: Qosqo, Tawantinsuyu (Inca), Ciudad Imperial (Imperial City), Ombligo del Mundo (Navel of the World)

From Manila and Havana to San Juan and San Antonio, Texas, many of the cities of the Spanish Empire were built around a Plaza de Armas. In addition to functioning as a military parade ground (hence the name), the square served as a power center of church and state.

Many of these plazas are now sideshows to the huge cities that grew up around them. But one case where that's definitely not true is the **Plaza de Armas** of Cusco in the Peruvian Andes, which retains its essential role in just about every aspect of local life.

The **Catedral del Cuzco**, the **Iglesia De La Compañia De Jesús**, and the **Iglesia del Triunfo** dominate the square, their baroque domes and steeples rising high above all other structures in the **Centro Histórico**.

The interiors of all three are decorated with masterpieces of the **Escuela Cuzqueña** (Cusco School), a brightly colored and highly dramatic style of religious art that blends aspects of Andean and Iberian traditions. Among the more evocative examples are **"The Wedding of Martín García Oñas de**

Loyola with Doña Ñusta Beatriz Clara Qoya"** in the Jesuit church (which depicts the marriage of a Spanish conquistador and Inca princess) and **"The Last Supper"** in the cathedral (where guinea pig is the main dish).

Spanish governors ruled from the nearby **Cabildo del Cusco**, its last reconstruction undertaken in the 1840s after Peruvian independence and still home of the city's mayor and its government, as well as the **Museo Municipal de Arte Contemporáneo**.

Many of Cusco's other landmarks huddle within a block of the plaza, including the **Museo Inka** in the Casa del Almirante (Admiral's House) and the **Museo Machu Picchu** in the Casa de las Concha (Shell House), where more than 360 artifacts unearthed by Hiram Bingham during his early 20th-century rediscovery of the fabled Inca city are on display.

Although founded shortly after the Spanish conquest, the convent and church of **La Merced** was rebuilt after the earthquake of 1650. Among its ecclesiastical treasures is the **Mercedarian monstrance**,

Andean women craft traditional knit belts at the Chinchero market.

Plaza de Armas is the central hub of Cusco, a famous landmark surrounded by Inca ruins and churches.

made with 49 pounds (22 kg) of gold, silver, pearls, and gems. In addition, the basilica is one of the churches that receives a Holy Monday visit from the **Señor de los Temblores** (Lord of the Earthquakes), a crucifix that portrays Cusco's patron saint and is believed to have spared the city other natural disasters.

Yet the Plaza de Armas is far more than vintage buildings and ancient relics. It's also a hub of modern life, a heady blend of tourists and locals that's alive from early morning until late at night beneath its plentiful streetlights.

Arrayed around the square are agencies that book guided **Inca Trail** treks and **Peruvian Amazon** adventures, and even a **North Face** store to buy equipment for those journeys. Burger joints and coffee shops compete with local restaurants that serve ceviche and other Peruvian delights. **King of Maps** hawks artful reproductions while tattoo parlors around the square are more than glad to supplement your body art.

The historic colonial-era balconies along the **Portal de Comercio** are prized for sunset drinks and people-watching, while **Inka Team** offers a chance to dance the night away.

URBAN RELIC

Machu Picchu

Four hours by train or bus from Cusco, the 15th-century Inca citadel at **Machu Picchu** is poised at 8,000 feet (2,438 m) in the Andes on a ridge that overlooks the rainforest-filled Urubamba River Valley.

One of the world's most recognizable landmarks, the "lost city" was rediscovered by American explorer Hiram Bingham in 1911. Adding to its mystique is the fact that its actual use has never been discerned. The World Heritage site comprises stone terraces and more than 200 structures likely used as religious sites, workshops, and residences.

In typical Inca fashion, Machu Picchu was constructed with sophisticated building techniques including astronomical alignments and the fusing of huge rocks without mortar. Due to its remote and almost inaccessible location, the Spanish never discovered the citadel. Best guess is that the Inca occupied the site for around 80 years before they mysteriously vanished.

Plaza de Armas is also a place to imagine what it must have been like before the conquest because it was created directly over the **Huacaypata**, the main square of pre-Columbian Cusco when it served as capital of the Inca realm. The cathedral was built atop the foundations of **Kiswarkancha**, a sacred temple precinct that included the palace of Inca ruler Viracocha. The square was surrounded by other stone palaces—each Inca ruler erected his own royal residence—and important buildings like the **Ajlla Wasi** (House of the Virgins of the Sun) and the **Suntur Wasi** (House of God), where royal emblems and weapons were safeguarded.

All of this disappeared during the first decades of Spanish rule. As writer Christopher Isherwood observed after visiting Cusco in the 1940s, "What remains with you is the sense of a great outrage . . . one of the most beautiful monuments to bigotry and sheer stupid brutality in the whole world."

Although it's hard to tell nowadays, the Inca city was designed in the shape of a giant leaping puma (mountain lion), its metaphorical heart and other vital organs wrapped around Huacaypata plaza and its head on the hilltop that shelters the great fortress of **Sacsayhuamán.** The ultimate expression of Inca stonework, many of the blocks weigh 90 to 120 tons (82 to 109 metric tons), the massive walls constructed without mortar and their zigzag pattern thought to represent the puma's teeth.

The Spaniards cannibalized many of the blocks to construct the churches, convents, and casas of

Machu Picchu, the renowned Inca site, is five hours from Cusco, where visitors often stop to acclimate to the higher elevations.

GATHERINGS

- **Carnival:** Cusco celebrates the run-up to Lent with a week-long bash in February or March that reaches a joyous crescendo on Sunday with a parade at the Plaza de Armas, as well as water balloon and foam fights, and *yunza* trees laden with decorations and gifts.

- **Semana Santa:** Easter week in May or April means religious processions in the streets of Cusco, and vendors hawking traditional food and drinks in the plazas.

- **Inti Raymi:** This modern resurrection of the Inca religious rite pays homage to the sun god Inti on the winter solstice (June 24). In ancient times, as many as 200 llamas were sacrificed on Inti Raymi; nowadays it's just one.

- **Santurantikuy:** Locals and tourists alike gather in the Plaza de Armas on Christmas Eve (December 24) for a colorful "Saints for Sale" arts and crafts fair that specializes in items for Nativity scenes and other religious displays.

---

colonial Cusco, and they buried the remaining structures so that Inca rebels couldn't use the bastion as a base for retaking the city. Sacsayhuamán wasn't excavated until the 1930s. In addition to the massive walls, archaeologists uncovered the **Muyuq Marka**, the foundation for a huge round tower, as well as the remains of other Inca-era buildings destroyed by the Spanish. Today the ruins provide an evocative site for **Inti Raymi,** a revival of the Inca celebration of the winter solstice on June 24.

Farther up the hill is a series of natural rockslides called the **Rodadero** and the modern **Planetarium Cusco**, which offers both a window onto the

contemporary cosmos and an interpretation of Inca astronomy, including their own unique constellations.

Inca stonework is also on display at the **Iglesia de Santo Domingo**, or rather below it, because the 17th-century convent and chapel sit atop the remains of the **Qorikancha**. Dedicated to the Inca sun god, the "Golden Temple" was considered the most sacred of all Inca shrines. As such, the walls were literally paneled with sheets of gold, the temple precinct filled with gold and silver amulets and other ritual objects, all of it stolen, melted down, and shipped back to Spain by the invaders.

Another great Inca wall—along **Calle Hatunrumiyoc** in the city's **San Blas** neighborhood—features the much photographed **Piedra de los 12 Ángulos** (12-Angled Stone). Centered around the adobe **Iglesia de San Blas**, the barrio has long catered to Cusco's artists and bohemian underbelly. Nowadays it's more known for small restaurants and boutique hotels, as well as offbeat sights like the Museo de la Coca, which explores the traditional place of the coca plant in Andean culture.

Cusco lends itself to day trips into the surrounding **Sacred Valley**. Far and away the most popular—although it really deserves multiple days—is the train or bus trip to **Machu Picchu**. But there's plenty more in the countryside.

Out west are multicolored **Laguna Humantay**, perpetually snowcapped and glacier-covered **Salkantay** peak (highest in the Cusco region), and the perfectly round Inca agricultural terraces at **Moray.** Heading northeast, the highway skirts the Inca ruins at **Q'enco** and **Puka Pukara** on its way to **Pisac** with its lively market, colonial relics, and a miniature version of Machu Picchu. ■

Guadalajara
Mexico

Many of the traditions that the outside world considers quintessentially Mexican were born and raised in Guadalajara, the big city in western Mexico that offers a whole different take on south-of-the-border experiences.

THE BIG PICTURE

Founded: 1542

Population: 4.9 million

Size: 310 square miles (803 sq km)

Language: Spanish

Currency: Mexican peso

Time Zone: GMT -5 and -6 (CDT and CST)

Global Cost of Living Rank: Unlisted

Also Known As: City of Roses, Pearl of the West

Anyone who savors tequila knows it comes from the spiky blue agave plants that thrive in the hot, humid climate of central Jalisco state. But Mexico's second largest city is also the cradle of mariachi music, *charreada* rodeos with cowboys (and increasingly cowgirls) who wear huge silver buckles and wide-brimmed sombreros, and the exuberant *jarabe tapatío* or Mexican hat dance that once served as a local courtship ritual.

Tapping into those roots and experiencing these traditions in the place where they began is what makes a sojourn in Guadalajara so out of the ordinary compared to other Mexican cities.

Declared an Intangible Cultural Heritage of Humanity by UNESCO, mariachi music originated in the town of **Cocula** in the Jalisco countryside. But it came of age during the 19th century in the plazas and cantinas of Guadalajara. Nowadays it seems like just about every eatery features strolling musicians. But two restaurants are especially well known for mariachis: **El Parián** in the Tlaquepaque neighborhood and **Casa Bariachi** in the Arcos Vallarta area. A more visceral experience awaits at the **Plaza de los Mariachis**, where musicians gather after dark in their black, silver-studded regalia to play songs on request (for a tip, of course).

Jarabe tapatío is a little harder to catch on a regular basis. It's always part of huge annual events like the **Festival Cultural de Mayo** in the spring and **Fiestas de Octubre** in the fall. Otherwise, plan on attending the Sunday afternoon performance of the **Ballet Folclórico** from the University of Guadalajara at the historic **Teatro Degollado** (opened in 1866) in the old town.

Mexican rodeo, another UNESCO-designated Intangible Cultural Heritage, is a way of life in Guadalajara. **Charrería** traces its roots to the stock-raising culture that began shortly after the Spanish conquest when *charros* (cowboys) from different haciendas competed against one another to determine who had the best roping and riding skills. By the early 20th century,

A worker harvests blue agave for tequila on a plantation in the Amatitán Valley.

Street performers in colorful costumes honor Día de Muertos at Guadalajara's Plaza de Armas.

many of these traditions were starting to fade away. The National Charros Association was founded to keep charrería alive and it was named Mexico's national sport in 1933.

Charros de Jalisco, Mexico's premier competitive rodeo group, practices every Sunday at noon at the **Lienzo Charros de Jalisco** arena in Barrio La Aurora. Spectators are welcome to watch the three-hour practice for a small entry fee. Among the city's annual rodeo events are the **International Mariachi & Charrería Festival** in late August, the **Children's Charro Championship** in late April, and the **Feria de Escaramuzas** women's rodeo in early May.

Tequila gets its name from **Tequila**, the town just a 45-minute drive west of Guadalajara. Although Mexico's pre-Columbian peoples consumed agave-based beverages, it was the Spanish conquistadors who distilled the first tequila in the 16th century. Most of the distilleries (nearly 200 total) and blue agave fields fall within the Agave Landscape and Ancient Industrial Facilities of Tequila UNESCO World Heritage site.

The **Museo Nacional del Tequila** offers a great orientation to Mexico's national drink before you head for the factories. The most renowned is

---

HIDDEN TREASURES

• **Panteón de Belén:** Late-night guided tours are the best to explore a creepy 19th-century cemetery that features ghost sightings, vampire stories, and tales of buried treasure; last tour starts at 11:30 p.m.

• **Barranca de Oblatos-Huentitán:** The Río Santiago cuts a deep ravine through the wooded highlands north of Guadalajara, the focus of a large nature reserve with hiking trails, panoramic viewpoints, the ruins of Casa Colorada hacienda, and the old Arcediano suspension bridge (built 1893).

• **Museo de Arte Huichol Wixárica:** Ceremonial masks, animal figures, and other objects fashioned from yarn and beads are the focus of this Zapopan museum dedicated to the colorful creations of the Indigenous Huichol people.

Destileria La Rojeña, home base of Jose Cuervo, with a factory tour and tasting experience that promise a sip of *blanco, reposado,* and *añejo,* or premium margaritas if you're not into shots. Other Tequila-area adventures range from horseback riding through the agave fields and helping with the agave harvest on **Rancho El Chiquihuitillo** to *talabartería* (saddle-making) shops and a broad range of art and artifacts at the new **Centro Cultural Juan Beckmann Gallardo**.

Some of those artifacts are pre-Columbian treasures from **Guachimontones**, an ancient Teuchitlán religious complex on the south side of **Volcán de Tequila** from the agave country. Occupied as early as the third century B.C., the site features an unusual round pyramid and other grass-covered ruins beside a modern interpretive center.

Some of the other traditions you come across in Guadalajara may not be homegrown but they're just as alluring. Like *lucha libre*—Mexican professional wrestling—with its colorful masks, outrageous antics, and rowdy fans. **Arena Coliseo** near the Plaza de los Mariachis is the place to catch the sweat fest.

Tapatíos (residents of Guadalajara) also adore baseball and soccer. Named for the local cowboys, the Charros de Jalisco play their Mexican winter league games in the modern **Estadio de Béisbol** while the old and venerable Chivas soccer team (founded in 1906) takes the field against other Liga MX teams at **Estadio Akron**. Both stadiums are in **Zapopan** on the city's northwest side.

Tlaquepaque is the place to head for traditional Mexican arts and crafts. The cobblestone lanes around **Jardín Hidalgo** plaza are flush with handicraft stalls, galleries, and specialty shops like **Nuestros Dulces,** an artisanal sweet shop that sells chocolate, marzipan, *rompope* (fermented eggnog), *cajeta de leche* (caramelized goat milk), and other goodies.

Farther down Calle Matamoros, **Taller Nuñez Panduro** offers clay figurine classes as part of a family business that specializes in Nativity

Explore colorful shops, cafés, and restaurants along Calle Independencia.

sets and miniatures of Mexican presidents and celebrities. **Galería Sergio Bustamante** offers animal and humanoid sculptures, jewelry, and other works by Guadalajara's best known living artist.

Guadalajara's biggest street market is the **Tianguis Artesanal de Tonalá**, which unfolds every Thursday and Sunday when as many as 4,000 vendors offer their wares in **Tonalá** on the city's eastern edge. Handicrafts are all the rage, but there's plenty of food and drink too. Combine the market with a visit to the nearby **Museo Nacional de la Cerámica**.

Like so many Latin American urban areas, Guadalajara grew up around a Spanish colonial nucleus now called the **Centro Histórico**. Located in the heart of the old town, the vast **Plaza de la Liberación** is surrounded by the **Palacio de Gobierno** with its priceless José Clemente Orozco murals, the aforementioned **Teatro Degollado**, and the enormous **Catedral de Guadalajara**, a blend of Spanish Gothic, baroque, and Renaissance design with an interesting crypt and museum of sacred objects. The **Museo Regional de Guadalajara** and **Rotonda de los Jaliscienses Ilustres** honoring celebrated Jalisco

Delicious mole is freshly prepared and ready for sale at a market stall.

artists and political leaders also overlook the plaza.

The restaurant-flanked **Paseo Degollado** offers vehicle-free passage to **Plaza Tapatía** and what most folks consider the city's crown jewel: the **Hospicio Cabañas**. Founded in the 1790s as a Catholic orphanage, hospital, and rehab center, the labyrinthine structure (yet another UNESCO World Heritage site) includes the domed **Capilla Mayor**, a chapel adorned with Orozco's incredible **"Hombre de Fuego (Man of Fire)"** and other art deco–era murals.

West of the cathedral is the **Zona Rosa**, an energetic blend of old and new with tree-lined streets that stretch between **Parque Revolución** and the **Glorieta La Minerva** roundabout. The area's Victorian mansions and mid-century modern buildings harbor many of Guadalajara's top shops, galleries, and restaurants, as well as stylish hotels like the **Villa Ganz** and **Bellwort**. ∎

LOCAL FLAVOR

• **Casa Luna:** Modern Mexican cuisine and rum-spiked *cazuela* cocktails (aka Guadalajara sangria) in a courtyard restaurant that could easily double as a folk-art museum; *Calle Independencia 211, Tlaquepaque.*

• **Hueso:** Decorated with more than 10,000 whitewashed animal bones, and specializing in the cuisine of Jalisco and Sonora, this uber-trendy

eatery is a feast for both the eyes and the taste buds; *Calle Efraín González Luna 2061, Obrera;* huesorestaurant.com.

• **Tortas Ahogadas José El de la Bicicleta:** It's a mouthful, both the name of this tiny neighborhood restaurant and Guadalajara's beloved *torta ahogada* ("drowned sandwich")—a pork roll floating in chili

sauce; *Calle Gante 150, Las Conchas.*

• **Allium:** Huitlacoche risotto with smoked oyster and shiitake mushrooms, poached shrimp ceviche in a tomatillo and coriander seed emulsion, and braised lamb in a tomato and maguey worm salsa flavor the menu at this sleek, modern restaurant; *Calle Manuel López Cotilla 1752, Zona Rosa;* allium.com.mx.

Lima
Peru

One of the Western Hemisphere's largest cities, the sprawling Peruvian capital offers an eclectic urban experience that runs all the way from gourmet seafood and world-class surfing to pre-Columbian ruins and Spanish colonial relics.

THE BIG PICTURE

Founded: 1535

Population: 11.5 million

Size: 345 square miles (894 sq km)

Language: Spanish

Currency: Peruvian sol

Time Zone: GMT -5 (PET)

Global Cost of Living Rank: 150

Also Known As: Ciudad de los Reyes (City of the Kings), La Perla del Pacífico (The Pearl of the Pacific)

A lot of travelers breeze through Lima on their way to Cusco and Machu Picchu, but that's a huge mistake because the Peruvian capital has transformed itself since the start of the 21st century into one of South America's most tantalizing cities.

First (and possibly foremost) is the food. The cuisine was always scrumptious. But now visitors at world-class restaurants can sample a bounty that blends dishes, ingredients, and cooking methods from the Amazon, Andes, and coastal desert.

As much as Egypt or Greece, Peru is celebrated for its ancient civilizations and awesome archaeological sites. Lima is the place to view many of the incomparable treasures discovered at those sites including the pre-Columbian ceramics of the **Museo Larco**, the ancient textiles of the **Fundación Museo Amano**, the eclectic relics of the **National Museum of Archaeology, Anthropology, and History**, and the golden artifacts of the **Museo Oro del Perú**, which seems to have recovered from the scandalous 2001 discovery that many of its artifacts were reproductions rather than originals.

After vanquishing the Inca,

Spanish conquistador Francisco Pizarro founded Lima along the Rímac River. As the capital of the Viceroyalty of Peru, it grew into a thriving cultural, commercial, and political hub second to none in the Western Hemisphere. That 16th-century city has passed into modern times as the **Historic Center of Lima** (Cercado de Lima), a UNESCO World Heritage site that safeguards scores of historic structures and blocks around the broad **Plaza de Armas**.

It took more than a century (1535–1649) for the Spanish to complete **Lima Cathedral** and its 14 chapels, one of which boasts Pizarro's tomb. On the other side of the plaza, the stately **Palacio de Gobierno** isn't nearly as old as it looks, built in the 1930s as the official residence of Peru's president.

Side streets lead to other impressive Spanish-era buildings like the exquisite **Iglesia de Santo Domingo** church and convent with its distinctive pink facade, tranquil cloisters, and the remains of New World saints like San Martín de Porres and Santa Rosa de Lima. Not to be outdone by the Dominicans, the Franciscan friars erected the equally impressive **San Francisco de Assisi** church and

See ancient Peruvian artwork—such as ceremonial vessels—at the Museo Larco.

Footmen march in the changing of the guard ceremony outside the Government Palace of Peru at the Plaza de Armas.

convent, renowned for its magnificent library and creepy skeleton-filled catacombs where around 25,000 Spanish-era residents are buried.

Lima's old town also offers more than a dozen museums. Foodies should sample the **Casa de la Gastronomía Peruana**, dedicated to the various culinary cultures that flavor Peru. The **Museo Nacional Afroperuano** is dedicated to the culture and history of a largely coastal ethnic group descended from the colonial-era enslaved population. On the other hand, the **Casa de Aliaga** showcases the culture and lifestyle of Lima's Spanish aristocracy in a mansion built in 1535 by one of Pizarro's captains.

Miraflores may boast more great eating places than any other neighborhood in South America, with fresh-off-the-boat seafood (especially ceviche) as the biggest taste treat.

The seaside burb also boasts the city's best shopping: silver and antique shops along **Avenida La Paz** and handicrafts from all around Peru at the **Centro Artesanal Indios**, **Gran Chimú Market**, and **Artesanías Miraflores** (all three along Avenida Petit Thouars).

The cliff-top **Malecón** drive and walkway stretches 3.4 miles (5.5 km) down the Miraflores coast with numerous parks, viewpoints, and public artworks along the way. **Parque del Amor** with its folksy mosaics and amorous sculpture is the best place to snap a selfie, and

---

URBAN RELIC

Chan Chan

Prior to Spanish conquest, the metropolis of the Peruvian coast was **Chan Chan**, located near present-day Trujillo about 350 miles (560 km) north of Lima.

Capital of the Chimú Empire between 900 and 1470 when the Inca captured the city, Chan Chan was South America's largest urban area in pre-Columbian times with a population that may have reached

60,000. Though built in a desert climate, the elite capital flourished thanks to irrigated canals and wells.

Nowadays the walled adobe city—and its many palaces, temples, tombs, and other structures—is a UNESCO World Heritage site sprawling across almost eight square miles (21 sq km) along the shore. Many of the walls are decorated with intricate geometric patterns and stylized images of birds, fish, turtles, and other animals.

LOCAL FLAVOR

• **Astrid y Gastón:** One of South America's ultimate taste treats, a 16-course El Reencuentro ("Reunion") menu takes diners on a culinary journey across the length of Peru; *Avenida Paz Soldán 290, San Isidro;* en.astridygaston.com.

• **Brujas de Cachiche:** "Double, double toil and tasty" could very well be the motto for a restaurant inspired by the legendary witches *(brujas)* of the Peruvian desert. The menu features regional dishes like *cabrito a la norteña* (northern-style goat stew) and Arequipa-style *rocoto relleno* (stuffed peppers); *Calle Bolognesi 472, Miraflores.*

• **La Mar Cebichería:** It's difficult to choose from a world-renowned menu that renders tuna, corvina, sea urchin, calamari, clam, scallop, and other types of ceviche; *Avenida Mariscal La Mar 770, Miraflores;* lamarcebicheria .com/en/Lima.

• **ámaZ:** As the name suggests, this offbeat but utterly delicious eatery specializes in blending ingredients from the Amazon rainforest into classic Peruvian dishes like *lomo saltado,* ceviche, and arroz con chorizo; *Avenida la Paz 1079, Miraflores;* amaz.com.pe.

there are plenty of ways to pass the time, from clay tennis courts and tandem paragliding to the bowling alley, cinema, and food court at the modern **Larcomar** shopping mall.

The promenade's north end is anchored by the **Lugar de la Memoria** (Place of Memory), a museum and monument that honors those who perished during Peru's domestic strife of the 1980s and '90s. At the southern end, the glass-box **Museo de Arte Contemporaneo** (MAC) exhibits modern Peruvian and international artists. Below the palisades are the black-sand beaches of the **Costa Verde**, renowned for its radical surf breaks.

Farther south along the shore is funky Barranco. Home to Nobel Prize–winning writer Mario Vargas Llosa, beloved singer Chabuca Granda, and many other artists, the bohemian burg cultivates a vibrant cultural life via restaurants, theaters, and traditional music bars *(peñas)* where Lima's hipsters hang. The historic wooden **Puente de los Suspiros** (Bridge of Sighs) leaps across the **Bajada de Baños** ravine and a path that leads down to the beach.

Named after the man who collected

An Andean girl dresses in traditional clothing for a procession around Plaza de Armas.

almost everything inside, the **Museo Pedro de Osma** displays Peruvian art and artifacts from ancient through colonial days in a gorgeous beaux arts mansion where Osma once lived. Barranco's **Museo MATE** couldn't be more different, with its modern black-and-white work of renowned Peruvian fashion photographer Mario Testino.

The **Pan-American Highway** leads along the shore to **Pachacámac**, a ruined mud-brick city that flourished from around A.D. 200 to the Spanish conquest. Farther south along the Pan-Am is the **Paracas** and **Ballestas Islands** National Reserves. Often called the "Peruvian Galápagos," the parks are home to a vast array of wildlife—whales, dolphins, sharks, seals, flamingos, condors, and many more—that thrive in the adjoining marine and desert environments. **Guided boat trips** and **four-wheel-drive safaris** are the best way to discover the critters and incredible landscapes.

You can also venture north on a day trip to **Lomas de Lachay National Reserve** with its unique coastal vegetation and onward to the secluded hilltop ruins of **Caral** in the Supe Valley. Considered the oldest city in the entire Western Hemisphere, Caral was an urban center of the Norte Chico civilization between the 26th and 20th centuries B.C. ■

LAY YOUR HEAD

• **Belmond Miraflores Park:** Steps away from the coastal hiking/biking path and black-sand Playa Redondo, this modern high-rise seems more like a tropical island resort than a big-city hotel; restaurants, bar, spa, swimming pool, fitness center; from $293; belmond.com.

• **Hotel B:** An elegant belle epoque summer mansion provides an atmospheric venue for an art-savvy boutique hotel with its own celebrity chef restaurant and Relais & Châteaux membership; restaurant, bar, beach access; from $200; hotelb.pe.

• **Antigua Miraflores:** Lima's first boutique hotel is still one of its best, a 1920s mansion converted into romantic, modern accommodations without losing any of its bygone charm; restaurant, bar, garden, cooking classes, airport shuttle; from $132; antiguamiraflores.com.

• **El Golf:** Two blocks from the Lima Golf Club, this comfy but reasonably priced boutique hotel provides a quiet escape in the upscale San Isidro district; restaurant, bar, fitness center, swimming pool; from $85; elgolfhotelboutiqueperu.com/en.

Latin American Cities

Antigua, Guatemala; León, Nicaragua; Paramaribo, Suriname

Smaller cities in Central and South America that have managed to preserve their historic cores form a cluster of UNESCO World Heritage sites where it's easy to imagine what local life was like a century ago or more.

Stroll the arches of the Palacio de los Capitanes Generales in Antigua.

ANTIGUA

Population: 46,000
Size: 108.5 square miles (281 sq km)

Located 25 miles (40 km) west of Guatemala City, Antigua was declared a UNESCO World Heritage site in 1979, owing to its pristine Spanish baroque architecture and bygone ambience.

Founded as the seat of Spanish power in Central America, the city was largely destroyed by an earthquake in 1773. Wandering the cobblestone streets today, visitors encounter a mix of ruins and structures that were rebuilt post-disaster. With its bird's-eye view of the city, **Cerro de la Cruz** is a great place to get your bearings.

The saffron yellow **Arco de Santa Catalina** offers an elegant 17th-century frame for photographing **Volcán de Agua** in one direction and the restored **Iglesia de La Merced** in the other. **Iglesia de San Francisco el Grande**, the city's oldest church, wasn't restored until the 1960s.

Arrayed around **Parque Antigua** in the town center are the ruined **Catedral San José**, **Museo de Arte Colonial**, and partially restored **Palacio de los Capitanes Generales**, seat of the Spanish viceroys for more than 200 years.

Mercado Central offers an artisan section and a food court that specializes in Guatemalan street foods. But the **Mercado de Artesanias** beside the ruined **Iglesia El Carmen** is the place to buy embroidered *huipils* (blouses).

New Sensation dance studio on Avenida Norte offers salsa lessons, while **Dyslexia Libros** treats patrons to a free beer (with each book purchased) at the neighboring **Café No Sé**. Tours of small coffee farms on the volcanic slopes around Antigua can be arranged through **De La Gente**.

LEÓN

Population: 170,000
Size: 316.6 square miles (820 sq km)

Located in northwest Nicaragua, León is actually two cities (old and new) that offer their own unique takes on the UNESCO World Heritage sites experience.

León Viejo lies in the shadow of smoldering **Momotombo** volcano along the shore of **Lago de Xolotlán**.

A one-of-a-kind experience: boarding down the slopes of a volcano

A volcano-induced earthquake convinced the occupants to abandon their colonial city in 1610. León Viejo was largely forgotten until the 1960s, when excavations revealed the city's onetime magnificence.

The clincher to obtaining World Heritage listing was the fact that it's the only early 16th-century settlement in the Americas whose original city plan has never been altered. The small **visitors center** offers an orientation to ruins that include the foundations of **Santa Maria Cathedral**, **La Merced** church and convent, the conquistador **Hernando de Soto's** house, and the **Fortaleza** citadel.

The residents moved 20 miles (32 km) west to "new" **León**, which grew into a thriving regional center and Nicaragua's university town. Eventually they got around to constructing **León Cathedral**, the largest church in Central America and a World Heritage site in its own right.

León also boasts three great museums: the cultural and historical artifacts of the **Museo de Leyendas y**

Tradiciones, contemporary Central American creations at the **Museo de Arte Fundación Ortiz-Gurdián**, and tales of Nicaragua's long civil war at the **Museo Histórico de la Revolución**.

Beyond the city limits, the black-sand slopes of **Cerro Negro** volcano are ideal for sandboarding, while **Playa Las Peñitas** offers a black-sand beach.

PARAMARIBO

Population: 240,000
Size: 70 square miles (181 sq km)

An inner city rich in Dutch colonial buildings prompted UNESCO to declare Suriname's capital a World Heritage site in 2002.

Fort Zeelandia, an amazingly well-preserved citadel, renders panoramic river views and the Surinaams Museum with exhibits on the history, art, and ethnic groups of the country. Outside the fort's main gateway, tree-shaded **Zeelandiaweg** is lined by beautiful heritage homes.

The 18th-century **Presidential Palace** isn't open to the public, but right behind is the wonderful **Palmentuin**, a colonial-era palm garden with walkways, monuments, and a new **Wakapasi** promenade with stalls selling Amerindian foods, crafts, and holistic cosmetics.

Henck Arron Street runs up from the waterfront to **Saint Peter & Paul Cathedral Basilica**, one of the largest wooden structures in the Western Hemisphere. A few streets over, **Neveh Shalom synagogue** and **S.I.V. Mosque** sit side by side along Keizerstraat Street. Erected in the early 18th century by Sephardim fleeing the Inquisition in Europe, the synagogue features a sand-covered floor.

The savory **Saoenah (Javanese) Market** on Anamoe Street is packed with locals in traditional Indonesian garb buying batik fabric or munching barbecued satay slathered in peanut sauce. The intersection of Tourtonnelaan and Verlengde Mahonylaan is ground zero for a **Chinatown** area that jumbles restaurants and general stores. ∎

Fort Zeelandia was built to protect the interests of the Dutch West India Company.

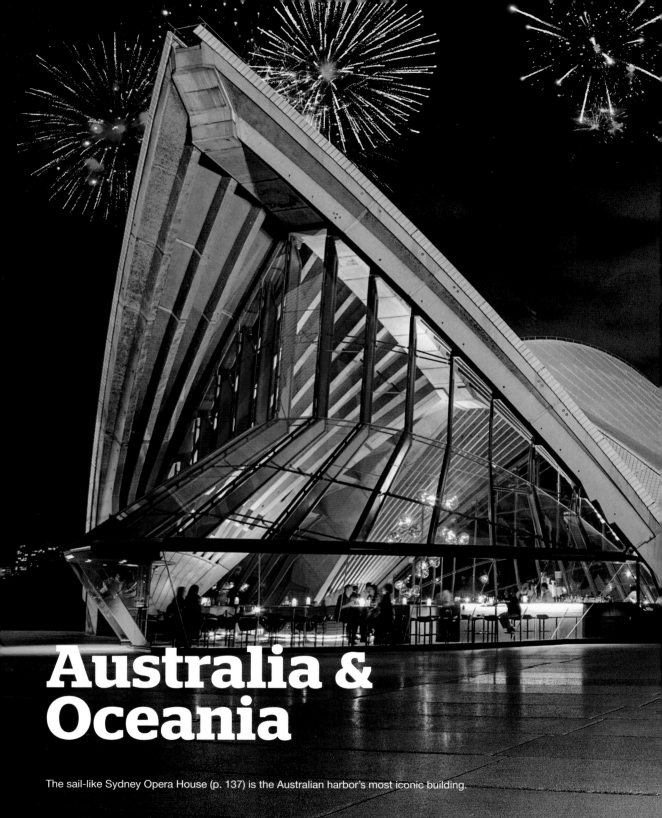

Australia &
Oceania

The sail-like Sydney Opera House (p. 137) is the Australian harbor's most iconic building.

Honolulu
Hawaii, U.S.A.

The mid-Pacific metropolis reveals itself as a far more complex city than picture postcards would have us believe: a multifaceted urban area with an array of neighborhoods and a backstory filled with royals, robber barons, and imperial ambitions.

THE BIG PICTURE

Founded: 1804

Population: 865,000

Size: 170 square miles (440 sq km)

Languages: English

Currency: U.S. dollar

Time Zone: GMT -10 (HST)

Global Cost of Living Rank: 43

Also Known As: The Big Pineapple, Crossroads of the Pacific, Sheltered Bay, Kou (ancient Hawaiian)

"Nothing had prepared me for Honolulu," wrote British author Somerset Maugham after a visit in 1916. "You had expected something wholly beautiful and you get an impression which is infinitely more complicated than any that beauty can give you."

More than a century later, the Hawaiian metropolis still feels that way. Captured by the camera lens and painter's eye, the city is visually immortalized as palm-fringed Waikiki Beach with Diamond Head looming in the background.

Although the landscape certainly exists—and is even more stunning in person than pictures—there's so much more to Honolulu.

Visitors are often surprised by its size: nearly a million people are squeezed along the south shore of Oahu. After Auckland, it's the second biggest city in Polynesia, which translates into highways with rush-hour traffic, noisy construction zones, homeless encampments, and all the other issues that plague large urban areas.

On the flip side of the coin—what makes Honolulu such an appealing destination—are incredible beaches, a blissful tropical island climate, a laid-back lifestyle, and a cultural melting pot matched by few cities on planet Earth.

Honolulu's reign as the capital of the Hawaiian Islands stretches back to 1804, when King Kamehameha I moved his court there from the Big Island. Those regal days are reflected in the **ʻIolani Palace,** home to the last Hawaiian monarch (Queen Liliʻuokalani). After American planters overthrew the monarch in 1893 and the U.S. annexed Hawaii, the Victorian-era structure served as the territorial capitol building. The palace is now surrounded by other historic structures like the **Aliʻiolani Hale** courthouse with its golden King Kamehameha statue, the art deco–style Honolulu Hale municipal building, and the modern Hawaii state capitol building, completed a decade after Hawaii became the 50th state in 1959.

The city's other blast from the long-ago past is the **Bishop Museum.** Established in 1889 by Charles Bishop, the American husband of a prominent Hawaiian princess, the collection includes the

View ruins of Pearl Harbor from the bridge of the U.S.S. *Arizona* Memorial.

Take in sweeping views of the ocean and the Honolulu skyline from the summit of Diamond Head.

royal heirlooms of the **Kāhili Room**, as well as natural history specimens and cultural artifacts from all around Polynesia. The museum campus also features the family-oriented **Science Adventure Center** with a faux volcano that explodes every seven minutes, **Na Ulu Kaiwiʻula Native Hawaiian Garden**, and the **Hawaii Sports Hall of Fame**.

History buffs can continue their regal tour at other sights around Honolulu like **Queen Emma's Summer Palace** and the **Royal Mausoleum** or learn more about Hawaii's rich cultural landscape at the **Mānoa Heritage Center** and **Kūkaʻōʻō Heiau** archaeological site in Manoa Valley.

Although Honolulu attracted its fair share of wandering scribes during royal days—Mark Twain, Herman Melville, and Robert Louis Stevenson, just to name a few—tourism didn't take off until the early 20th century with the opening of the **Moana Surfrider**, the first hotel on **Waikiki Beach**. Even if you're not staying at the Moana, visit the hotel's beach bar banyan trees and savor one of the Pacific's most romantic sunsets.

GATHERINGS

- **Honolulu Festival:** The city's ethnic diversity is the focus of this March event that promotes harmony between all Pacific peoples through dance, music, food, art, and crafts; honolulufestival.com.

- **Lantern Floating Hawaii:** Ala Moana Beach is the venue for this moving Memorial Day ceremony that honors Hawaiians who have died while serving their country; lanternfloatinghawaii.com.

- **Nā Hula Festival:** Founded in 1940, the city's longest-running, noncompetitive dance event takes the stage at the Waikiki Shell in Kapiʻolani Park in early August.

- **Aloha Festivals:** Honolulu's biggest block party. A floral parade through Waikiki and Ala Moana, a royal court investiture, ukulele concerts, and hula dancing are the mainstays of this August celebration; alohafestivals.com.

- **Hawaii International Rugby 7s:** Teams from around the world square off at this December tournament at Kapiʻolani Park.

Watch surfers take on some of the world's best waves at the Ala Moana Bowls surf spot on the south shore of Oahu.

Waikiki is all about surf, sand, and sun, with manifold ways to enjoy them. Aloha Beach Services and other outfitters organize surfing lessons and outrigger canoe rides, and paddleboard, umbrella, and beach chair rentals. There's plenty of shopping (both upscale and not) along busy **Kalakaua Avenue**. And at the south end of the beach strip, **Queen Kapiʻolani Regional Park** harbors the **Honolulu Zoo**, open-air concerts at the **Waikiki Shell** bandstand, **Waikiki Aquarium**, and the **War Memorial Natatorium**, a unique saltwater pool that doubles as a World War 1 remembrance.

Although **Diamond Head** rises nearby, there is no path from beach to summit. The one-mile (1.6 km) **Summit Trail** starts from the state park visitors center inside the crater, accessed via Diamond Head Road on the volcano's north side. The road curls around to the seaside **Kaʻalāwai** neighborhood and the outstanding **Shangri La Museum of Islamic Art, Culture & Design**.

The U.S. Navy didn't waste any time moving into **Pearl Harbor**, establishing its first facility in 1899, just a year after American annexation. By the 1920s, it was a hub of U.S. naval power in the Pacific and a defense against the aggressive expansion of imperial Japan. The most poignant reminder of December 7, 1941, is the **U.S.S. *Arizona* Memorial**, which can only be reached by U.S. Navy shuttle boats from the **Pearl Harbor Visitor Center** on the mainland.

Ford Island, the main focus of the Japanese attack in 1941, offers several other wartime attractions, including the **Pearl Harbor Aviation Museum** and battleship **U.S.S. *Missouri***, where the Japanese surrendered to Allied forces at the end of World War II. More than 30,000 American service personnel who perished in that conflict (and three other wars) are buried at the **National Memorial Cemetery of the Pacific** inside Punchbowl Crater high above Honolulu.

More than 80 percent of Honolulu's residents are ethnic minorities, in particular Asians and Pacific Islanders, a statistic that easily makes it the nation's most diverse city. The first Asian immigrants arrived in the mid-1800s to work the sugarcane fields, and nowadays they account for more than half the city's population.

A revitalized **Chinatown** stretches inland between Honolulu Harbor and the green-roofed **Kuan Yin Temple** (the city's oldest place of Buddhist worship) in **Foster Botanical Garden**. The neighborhood reaches fever pitch at the intersection of Kekaulike and King Streets, a cluster of fresh produce stalls, meat and fish markets, noodle factories, and seafood eateries in vintage buildings like **Oahu Market** (opened 1904).

Farther inland, **Chinatown Cultural Plaza** offers another warren of Asian shops and eateries, as well as daily sunrise luk tung kuen movement and thrice-weekly tai chi and kempo sessions in the courtyard, karaoke, and Chinese musical instrument classes. On the edge of the plaza, a statue honors **Dr. Sun Yat-sen**, who studied in Honolulu's 'Iolani School before overthrowing China's last imperial dynasty in 1911.

Chinatown segues into the adjacent **Arts District**, which revolves around the restored **Hawaii Theatre** (opened in 1922) with its eclectic slate of plays and musical performances. Around the block, the **ARTS at Marks Garage** offers rotating exhibits and more than 150 lectures, workshops, movies, and live performances each year. The Arts District also shelters the city's oldest

drinking establishment: **Murphy's Bar**. The establishment opened in 1890 as the Royal Hotel but with a history that stretches back to a royal retail spirit license granted in the 1860s.

Honolulu is the ideal base for exploring the rest of **Oahu**. Road-trippers can undertake a coastal circumnavigation that's easily accomplished in one day or a two-day adventure using **Pali Highway** (Route 61) or **Burns Freeway** (H3) to cut across the island's rugged volcanic spine.

The **North Shore** offers diverse attractions, from ATV adventures, ziplines, and horseback riding at **Kualoa Ranch** (where much of the original *Jurassic Park* was filmed) to the traditional music, dance, foods, and handicrafts of the **Polynesian Cultural Center**. Radical waves are the lure at **Waimea Bay** and its **Banzai Pipeline**, a holy grail of the surfing world, while **He'eia State Park** boasts the island's best kayaking and snorkeling in a pristine bay that includes **Coconut Island** of Gilligan's Island fame (now a marine research station). ■

The Bernice Pauahi Bishop Museum houses traditional Hawaiian artifacts.

Melbourne
Victoria, Australia

Australia's second biggest urban area has refashioned itself from an Anglo-centric outpost of the British Empire into a brilliant 21st-century multicultural mosaic with a high-rise cityscape that towers over Victorian landmarks of old.

THE BIG PICTURE

Founded: 1835

Population: 4.4 million

Size: 1,044 square miles (2,704 sq km)

Language: English

Currency: Australian dollar

Time Zone: GMT + 10 and +11 (AEST and AEDT)

Global Cost of Living Rank: 59

Also Known As: Dootigala (Aboriginal), Batmania (early pioneer), Bleak City, City by the Bay, Europe of Australia

Once upon a time, Melbourne was the epitome of the ideal British colonial city, an antipodean version of the mother country when it came to lifestyle and culture, its high regard for cricket, lawn tennis, and equestrian sport, and even the fact that it's rainier than most Australian cities.

"If Queen Victoria were still alive," British politician Jonathan Aitken wrote after a 1971 visit, "not only would she approve of Melbourne, she would probably feel more at home there than in any other city of her former Empire."

Melbourne was already starting to change by that time. Postwar immigration from Greece, Italy, and Eastern Europe—together with an end to the infamous "White Australia" policy that had strictly controlled immigration before the 1970s—transformed the Aussie metropolis into a vibrant multicultural urban area, ironically in the same manner as contemporary London, Manchester, and other British cities.

Perhaps Queen Victoria would also be proud of the fact that Melbourne's multiculturalism reaches fever pitch at one of the landmarks that bears her name: **Queen Victoria Market**. With more than 600 stalls, it's the largest open-air market in the Southern Hemisphere, and one of the most diverse, a lineup that ranges from **Shen Chinese Medicine Centre** and **Made in Japan** kitchenware to **Gözleme Turkish Cafe** and **Hellenic Deli**.

The Melbourne mix also flourishes in the city's many ethnic neighborhoods. Italian shops and cafés flank lower Lygon Street in **Carlton**. Vietnamese eateries flavor the **Footscray** neighborhood on the west side. Little Bourke Street and the **Chinese Museum** anchor the city's old and venerable **Chinatown**. The intersection of Lonsdale and Russell in downtown Melbourne is the hub of a historic **Greek Precinct** and the nearby **Hellenic Museum** in the old **Royal Mint Building.** The city's global intake is the focus of the excellent **Immigration Museum** on the spot where settlers from Tasmania founded Melbourne in 1835.

Although Melbourne's demographic mix was evolving in the late 20th century, the city's movers and shakers were radically altering the cityscape with bold new landmarks,

Find cheese and other fresh fare at the Queen Victoria Market.

Take a break from sightseeing in Federation Square, surrounded by St. Paul's Cathedral and the SBS Building.

attractions, and gathering places that expedited Melbourne's transition into a 21st-century metropolis and a place that consistently ranks among the world's most livable cities.

Unveiled in 2002 to mark a century of Australian nationhood, **Federation Square** offers a mishmash of museums, performance spaces, and eating places. The **Ian Potter Centre: NGV** safeguards the world's largest and finest collection of Aboriginal art, while the **Australian Centre for the Moving Image** (ACMI) showcases outstanding examples of film, television, video, and digital media. The square also hosts Australia's **Special Broadcasting Service** (SBS), a radio, television, and online network with

programming in more than 70 languages.

Federation Square's quirky modern design offers a stark contrast to nearby colonial-era buildings— the neo-Gothic bulk of **St. Paul's Cathedral** (completed in 1891),

the **Duke of Wellington** pub (established in 1853 and now the city's oldest licensed drinking establishment), the flamboyant Moorish Revival exterior of the **Forum Theatre** (opened in 1929), and the art nouveau–style **Flinders Street**

BACKGROUND CHECK

• Melbourne was founded by Anglo-Scottish Tasmanian ranchers seeking fresh pastures on the Australian mainland.

• With more than 36% of residents born overseas, Greater Melbourne is among the world's top 10 cities in immigrant population.

• Two of the earliest dramatic movies—*Soldiers of the Cross* (1900)

and *The Story of the Kelly Gang* (1906)—were filmed in Melbourne.

• With more than 165 miles (266 km) of double track and 1,763 stops, Melbourne operates the world's largest urban tram network.

• Nearly every year, Melbourne competes with Vienna, Zurich, and nearby Auckland for the title of World's Most Livable City.

Station (opened in 1910) with its trademark domes and clock tower.

Across the river, urban renewal has transformed the **Southbank** from the docks, factories, and warehouses of old into riverside restaurants, art spaces, and high-rise apartments. Highest of them all is the new **Australia 108** apartment building; at 1,039 feet (317 m), it's the nation's second tallest building. Nearby **Eureka Tower** is topped by the 88th-floor **Eureka Skydeck** with panoramic views across the city. Much closer to the ground, the **National Gallery of Victoria** safeguards more than 70,000 works of art spanning the entire globe and multiple eras.

Two other inner-city areas that have undergone radical makeovers are the west side **Docklands** and **Yarra's Edge**. Flashy yachts fill a

GATHERINGS

• **Australian Open:** This prestigious Grand Slam tennis tournament (founded in 1905) unfolds over the last two weeks of January at Melbourne Park; ausopen.com.

• **Moomba:** Australia's longest-running free community festival features music, rides, food, fireworks, a grand parade, and water activities like the Birdman Rally in riverside Birrarung Marr park over Australia's long Labor Day weekend in March; moomba.melbourne.vic.gov.au.

• **Royal Melbourne Show:** The outback comes to the city for the Aussie equivalent of a state fair, a shindig that includes animal and craft competitions, woodchopping contests, car racing, and good old country cooking over 10 days in September; royalshow.com.au.

• **Melbourne Cup:** The first Tuesday of November is the traditional date of Australia's biggest thoroughbred horse race, first run in 1861; flemington.com.au.

• **Italian Festa:** Music, comedy, circus, magic, food, and aperitivos are all part of the fun at this November fete around Lygon Street and Argyle Square in Carlton; melbourneitalianfesta.com.au.

basin where steamers once unloaded cargo from around the world, while the wharves are now stacked with restaurants, bars, and modern residential buildings. Among other Docklands attractions are the massive **Melbourne Star** Ferris wheel and **Marvel Stadium**, a 53,000-seat venue for Aussie rules football (AFL), cricket, rugby, and soccer.

St. Kilda Beach is one of the most famous in Melbourne, a prime spot for sunbathing and a dip in the water.

Melbourne may not be nearly as British as bygone times, but there are still plenty of relics of the days when the Union Jack rather than the Southern Cross waved above the city. Established in 1846, the **Royal Botanic Gardens Victoria** blends Victorian-era landscaping with distinctly Australia experiences like the **Bush Food Experience** and **Aboriginal Plant Knowledge Tour**.

Perhaps the finest Victorian structure in all of Australia, the **Royal Exhibition Building** in **Carlton Gardens** is a stunning relic of the Melbourne International Exhibition of 1880–81. Nowadays it hosts a wide variety of events from dog and bridal shows to the city's annual fashion fest and garden show. Carlton Gardens also hosts the **Melbourne Museum** with its manifold exhibits on the region's human and natural history.

It may not have "royal" in its name, but the **Melbourne Zoo** was born during Queen Victoria's reign. Founded in 1862, it is the world's third oldest zoo and the world's first carbon-neutral zoo. It's especially dedicated to the protection and propagation of endangered Australian animals like the Tasmanian devil, the orange-bellied parrot, the Lord Howe Island stick insect, and the giant burrowing frog.

Look up into the eucalyptus trees on Phillip Island, where you may spot a koala.

Native species are also the forte of the superb **Healesville Sanctuary**, the zoo's satellite campus in the **Yarra Valley** about an hour's drive east of downtown Melbourne. The sanctuary makes a great day trip from the city when combined with the valley's celebrated **wineries and art galleries**. More indigenous wildlife awaits on **Phillip Island**, a four-hour round-trip from the central city but well worth the drive for its bird-watching, fur seal colony, **Koala Reserve**, and nightly **Penguin Parade**.

Located close to the coast, Melbourne also has its sandy side. The seaside suburb of **St. Kilda** sprawls along a waterfront promenade splashed with beaches and palm-shaded lawns. Kitesurfing, paddleboarding, and fishing charters compete with an old-fashioned amusement area called **Luna Park** and the popular bars and restaurants along party-hearty **Fitzroy Street**. If you're more of a hang ten type, **Bells Beach** and its world-famous surf break are about 90 minutes down the coast along the route to **Great Otway National Park** and the **Great Ocean Road**. ∎

LOCAL FLAVOR

• **Pelligrini's Espresso Bar:** Founded in the 1950s, Melbourne's oldest Italian coffee bar still serves a pretty mean espresso, as well as canneloni, lasagne, spaghetti, and other delicacies; *66 Bourke Street;* pellegrinis.juisyfood.com.

• **Stalactites:** Another longtime favorite, this casual corner café in the Greek Precinct offers classic Hellenic treats like kalamari, giro, spanakopita, and moussaka; *177-183 Lonsdale Street;* stalactites.com.au.

• **Daughter in Law:** "Unauthentic Australian Indian" cuisine is the motto of a restaurant that prides itself on serving bona fide dishes from the subcontinent; à la carte options are available, but their $55 tasting menu is a true delicacy; *37 Little Bourke Street;* daughterinlaw.com.au.

• **The Fitzroy Pinnacle:** Tucked inside a wacky Victorian-era building, this neighborhood pub serves craft beer, pub grub, and live music; *251 St. Georges Road;* fitzroypinnacle .com.au.

Sydney
New South Wales, Australia

With an urban landscape and big-city lifestyles that revolve around nearly 100 beaches and one of the world's largest and most picturesque harbors, Australia's east-coast metropolis makes the most of its liquid bounty.

"I take, but I surrender" was Sydney's original motto, a phrase that was intended to reflect how the Anglo-Irish settlers who founded the city would also give back to their newfound home in the antipodes.

But the motto really should have been "Water, water everywhere." Because there isn't another global city—even those renowned for their canals—that has a more intimate relationship with the liquid that defines it. Sailing, surfing, and seafood are among the traditions that Sydney has cultivated on its numerous beaches and bays.

Stretching 11 miles (18 km) inland from the rocky **Sydney Heads** (North and South) that separate it from the open ocean, **Port Jackson** is one of our planet's largest and most spectacular harbors. Even more astounding is the fact that its deeply indented shoreline stretches for nearly 200 miles (322 km).

Scores of beaches are scattered along the city's eastern edge. World-famous strands like **Bondi** and **Manly** are where Australia's beloved beach culture reaches fever pitch, although expect them to be crowded any time the sun is out.

THE BIG PICTURE

Founded: 1788

Population: 4.5 million

Size: 841 square miles (2,178 sq km)

Language: English

Currency: Australian dollar

Time Zone: GMT +10 and +11 (AEST and AEDT)

Global Cost of Living Rank: 31

Also Known As: Harbour City, Emerald City

At the other end of the sandy spectrum are secluded shores like **Congwong Beach** at La Perouse Point, **Jibbon Beach** in the city's deep south, clothing-optional **Lady Bay Beach** at the tip of South Head, or **Store Beach** at North Head, where you might be lucky enough to stumble on the resident little penguins (smallest species of the aquatic birds).

A good portion of Sydney's waterways and corresponding shores are protected within the confines of six different national parks: **Ku-ring-gai Chase** in the **Pittwater** area on the city's north side; **Royal** and **Kamay Botany Bay** in the far south (including the spot where Captain James Cook landed in 1770 to stake Britain's claim to down under); **Lane Cove** and **Garigal** in the tidal creeks of the northwest; and **Sydney Harbour National Park** in the heart of the city.

Although Cook is credited with discovering (and naming) Port Jackson, the fabled captain never actually sailed into the bay. That didn't happen until 1788, when the First Fleet arrived with more than 1,000 convicts and instructions to establish a penal colony. The Union Jack was planted along the shore of a

Take the plunge in Bondi Beach's famous Olympic-size swimming pool.

For epic views, climb the Sydney Harbour Bridge at night to see the city light up during the Vivid Sydney light festival.

small inlet the Aboriginal people called Warrung ("Little Child") that the new arrivals would later christen **Circular Quay**.

No one could have imagined at the time that this modest little cove would grow into the epicenter of a great city and venue for two of the world's most recognizable structures, the gigantic steel frame of **Sydney Harbour Bridge** (opened in 1932) on one side, the elegant concrete sails of **Sydney Opera House** (unveiled in 1973) on the other. There are various ways to experience both of these architectural wonders. You can walk, bike, drive, or hop a train across the 3,770-foot (1,149 m) span. But the

best way to get up close and personal with the metal behemoth is the **BridgeClimb Sydney.** Not for the squeamish, this adrenaline-packed adventure includes a walk along catwalks beneath the bridge and then along one of the huge steel arches to the summit.

The Opera House experience is much more sublime (unless, of course, Wagner is playing). If you can't catch a performance, the venue offers a variety of **guided tours** (backstage, behind the scenes, wine, dine, and mobility access). If all else fails, grab a plate of oysters and an amber nectar in the **Opera Bar**.

The five finger piers at the head of Circular Quay dispense water taxis and the ubiquitous green-and-gold **Sydney Ferries** to the far reaches of Port Jackson. You can ferry across the water to hilltop **Taronga Zoo** with its emphasis on the captive breeding of the endangered wildlife of Australia and Southeast Asia. You can hop a water taxi to **Milsons Point**, home to the historic **North Sydney Olympic Pool**, a vintage carnival amusement area called **Luna Park**, and the bimonthly **Kirribilli flea market**. If time isn't pressing, consider a three-hour round-trip ferry ride up the **Parramatta River** to the western suburbs.

There are almost countless ways to experience Sydney's water world. One of the newest is camping on **Cockatoo Island**, a former penal settlement and shipyard that's also one of the major staging grounds of the **Biennale of Sydney** contemporary arts festival every two years.

In 2020, the Sydney Opera House honored bushfire emergency responders by projecting images on its sails.

LOCAL FLAVOR

- **The Gantry:** Modern Australian cuisine created with locally sourced meats, fish, and produce served harborside at Pier One; vegetarian and vegan options and a bar that's a destination in itself; *11 Hickson Road, Dawes Point;* thegantry.com.au.

- **Pavilion:** Panoramic views of Coogee Beach, Dolphins Point, and the Tasman Sea are the main attractions of this Victorian-era establishment spread across three floors, including a breezy rooftop bar and large game room with a giant Scrabble board, table tennis, and more; *169 Dolphin Street, Coogee;* merivale.com.

- **The Glenmore:** Serving Sydneysiders since 1921, the Glenmore offers shared dishes, snacks, full "knives and forks" meals, and fabulous city views from outdoor and indoor tables; *96 Cumberland Street, The Rocks;* theglenmore.com.au.

- **Sixpenny:** Fine dining with locally sourced ingredients—including a seasonal, ever changing seven-course tasting menu of Australian cuisine—on the outskirts of artsy Newton. *83 Percival Road, Stanmore;* sixpenny.com.au.

Sydney Harbour Kayaks offers a four-hour guided paddle tour of the tranquil **Middle Harbour** and the mangroves of Garigal National Park. **Sydney Seaplanes** offers scenic flights over Port Jackson and the coast from a dock in **Rose Bay**, where the big flying boats from London once landed.

Back in the day, Circular Quay served as the watery divide between the ruling class and commoners of early Sydney. **The Rocks** area on the west side was originally inhabited by convicts and later working-class people. The **Rocks Discovery Museum**, historic **Cadmans Cottage**, and the restored 19th-century row houses of **Susannah Place Living Museum** illuminate those days, as do the area's brilliant guided walking tours.

The establishment lived, worked, prayed, and played on the eastern side of the cove. The castle-like **Government House** (completed in 1847) is still home to the governor of New South Wales. The adjacent shoreline did time as farmland and the governor's private garden before its transformation into the lush **Royal Botanic Garden** and the sweeping lawns of **The Domain**.

As the ruling class settled in, they created the Gothic bulk of **St. Mary's Cathedral** (founded in 1821), the Georgian-style **Parliament House of New South Wales** (completed in 1816), and other heritage sites. **Hyde Park Barracks** (built 1811–19), nerve center of the early 19th-century convict colony, is now a museum devoted to the merciless penal system and the impact it had on the area's Indigenous peoples.

The barracks take their name from adjacent **Hyde Park**, established in 1810 as Australia's first public park. Used over the years for cricket matches, horse racing, and military drills (in case the convicts revolted), the park is renowned for its large ancient trees, art deco–style **Anzac Memorial** to those who fought in World War I, and the bullet-shaped **Yininmadyemi Memorial** that honors the Aboriginals and Torres Strait Islanders who served in Australia's armed forces.

Rising along the park's eastern edge is the recently renovated and

The Anzac Memorial in Hyde Park South pays tribute to the courage and sacrifice of all military servicemen and servicewomen.

expanded **Australian Museum**, a compilation of human culture and natural history that moved into its current premises in 1846. Another Victorian-era addition, the nearby **Art Gallery of New South Wales** (opened in 1874) presents a broad range of creativity from around the world, but is especially strong in Asian, Indigenous Australian, and Euro-Australian art.

Wedged between the privileged and the imprisoned were Sydney's early capitalists, a mix of merchants, traders, bankers, and other entrepreneurs who settled along the south shore of Circular Quay. Over many years, their modest shacks and shops would evolve into the central business district (CBD) and its jumble of modern skyscrapers.

Tallest of them all is the 1,014-foot (309 m) **Sydney Tower**, topped by an observation deck with views that run all the way from the Tasman Sea to the Blue Mountains on a clear day. Downtown's other darling couldn't be more different: A **Queen Victoria Building** festooned with domes, arches, and stained-glass windows feels more like a

---

LAY YOUR HEAD

- **Wildlife Retreat at Taronga:** Located *inside* Taronga Zoo, this stylish, eco-friendly retreat offers a fabulous bush setting; guests can sign up for guided wildlife tours and have complimentary access to the zoo; restaurant, bar, family-friendly activities; from $338; taronga.org.au/sydney-zoo/wildlife-retreat.

- **The Langham:** Situated in the Rocks between Circular Quay and Darling Harbour, the luxurious Langham raises the bar on Sydney hospitality; restaurants, bar, spa, indoor pool, fitness center, tennis court, Tesla charger; from $248; langhamhotels.com.

- **QT Sydney:** Tucked inside an art deco masterpiece beside the historic State Theatre in the city center, this 200-room boutique hotel oozes quirky glamour; restaurant, bar, spa; from $246; qthotels.com/sydney-cbd.

- **Darcy's Hotel**: Set along a laid-back, tree-lined street in suburban Homebush on Sydney's west side, this historic 11-room B&B is named for an early 19th-century settler who helped depose corrupt colonial governor William Bligh (of *Bounty* fame); breakfast, guest kitchen; from $70; darcyshotel.com.au.

medieval cathedral than a posh shopping arcade.

Like in many cities, the CBD is surrounded by satellite neighborhoods with their own (often offbeat) personalities. **Kings Cross** progressed from a mild-mannered residential area into a notorious red-light district and after-dark party zone. **Oxford Street** in funky **Darlinghurst** blends bohemian and gay restaurants, bars, and boutiques. **Haymarket** is the place to catch the *Indian Pacific* train across Australia or munch gourmet eats at **Paddy's Market** and the yummy Friday night street stalls in **Chinatown**.

Flavor is also the forte of the adjoining **Surry Hills** with its cocktail lounges, top-shelf restaurants, and lazy weekend brunches. Student life thrives in the **Newtown** neighborhood that borders on the **University of Sydney** campus. And so does creativity, expressed through Newton's copious street art as well as venues like art house **Dendy Cinemas**, comedy and music at the **Enmore Theatre**, and the sprawling **Carriageworks**, a 19th-century railway workshop converted into Australia's largest arts center.

The inner-city area that's changed the most in recent years is **Darling Harbour**, the gritty docks, warehouses, and railroad yards converted into a vast pleasure and leisure park. Among its dozens of attractions are the **Powerhouse Museum** of science and technology, the underwater **Sea Life Sydney** aquarium, and a submarine, destroyer, and other historic ships of **Australian National Maritime Museum**.

In the next cove over from Darling Harbour, the **Sydney Fish Market**—the world's third largest seafood emporium—offers oyster and sushi bars, wholesale outlets, gourmet shops, and a **Sydney Seafood School** with cooking classes and demonstrations. On the other side of **Anzac Bridge** (Sydney's most impressive span after the "Coat Hanger") is **Balmain**. The traditional working-class neighborhood has gentrified, especially the area along Darling Street. Yet Balmain retains many of its historic "hotels" (pubs), including the **Dry Dock** (established in 1857), **Riverview** (opened in 1888), and the **Exchange** (opened in 1885). ∎

Climb aboard a reproduction of Captain Cook's famous ship, the *Endeavour*, at the Australian National Maritime Museum.

Auckland
New Zealand

With one of the world's longest urban waterfronts, the "City of Sails" offers plenty of activities on and around its two great harbors. But New Zealand's big metropolis is good as gold for landlubbers, too.

THE BIG PICTURE

Founded: 1840

Population: 1.5 million

Size: 210 square miles (544 sq km)

Language: English

Currency: New Zealand dollar

Time Zone: GMT +12 and +13 (NZST and NZDT)

Global Cost of Living Rank: 70

Also Known As: Tāmaki Makaurau (Maori), City of Sails, Queen City, Big Little City

"Last, loneliest, loveliest, exquisite, apart," wrote Rudyard Kipling in 1897, comparing New Zealand's largest city to the other metropoli of the British Empire. His words could very well describe Auckland today, an urban area that still seems poised at the end of Earth.

Much as it did in empire days, Auckland continues to revolve around its waterways. The city center is sandwiched between two large bays—**Waitematā Harbour** in the north and **Manukau Harbour** in the south—separated by an isthmus that was settled by Maori people in the 14th century and colonial Brits in the 1840s. The long and deeply indented waterfront stretches for an amazing 2,300 miles (3,702 km), a figure that would place it among the world's top 30 longest coastlines if Auckland was an independent nation.

So it comes as no surprise that many of Auckland's major activities and attractions revolve around the water. **Queens Wharf**, **Princes Wharf**, and other finger piers that once welcomed clipper ships and steamers have morphed from handling cargo and colonists to modern-day pursuits like restaurants, hotels, and an eye-catching event venue called **The Cloud** (or "The Slug" as some critics have dubbed it).

Nearby **Viaduct Harbour** underwent an equally dramatic transition from fishing boats and rundown warehouses into a swish yacht basin and one of Auckland's nightlife havens. The neighborhood boasts numerous drinking spots (many with live music) as well as the **New Zealand Maritime Museum**, which offers harbor cruises on four of its historic vessels. The Viaduct is also the place to hop aboard **whale- and dolphin-watching cruises** or take the helm of a retired **America's Cup yacht** for a two-hour sail around the harbor.

The historic **Ferry Building** (opened in 1912) rises beside a modern ferry terminal with frequent boats across the harbor to Auckland's mild-mannered **North Shore** suburbs. **Devonport** is by far the most interesting of the ports, a jumble of cafés, art galleries, and antique shops, as well as the **Torpedo Bay Navy Museum** and views from the summit of an

See the Pataka Maori Court structure at the Auckland War Memorial Museum.

You can catch epic city views from the observation deck of the iconic Sky Tower.

extinct volcano called **Takarunga** (Mount Victoria).

Fullers 360 ferry company runs boats to Auckland's fascinating outlying islands, bits and bobs in the **Hauraki Gulf** that offer both natural history and hip human culture.

Protected as a nature area since the 1890s, **Rangitoto Island Scenic Reserve** nurtures more than 200 species of native flora on the slopes of the youngest volcano in the Auckland region. Visitors can explore the island via hiking trails or guided four-wheel-drive "road-train" excursions. In addition to New Zealand's oldest lighthouse (first lit in 1865), **Tiritiri Matangi Island** is a favorite bird-watching spot with

more than 70 species recorded. Big **Waiheke Island** offers vineyards, waterfront restaurants, and the best beaches in the Auckland region.

Ferries were the main way to reach the North Shore until 1959, when the **Auckland Harbour Bridge** was completed. Pedestrians

---

ARTBEAT

• **Best Movies:** *The Piano* (1993), *Once Were Warriors* (1994), *The Rainbow Warrior* (2006), *Orphans & Kingdoms* (2014).

• **Best Books:** *Whale Rider* by Witi Ihimaera, *Once Were Warriors* by Alan Duff, *In My Father's Den* by Maurice Gee, *A Field Guide to Auckland: Exploring the Region's Natural and Historic Heritage* by multiple authors, *The Penguin History of*

New Zealand by Michael King.

• **Best TV Shows:** *Outrageous Fortune* (2005–10), *Go Girls* (2009–13), *Hounds* (2012), *The Almighty Johnsons* (2011–13), *Harry* (2013), *Westside* (2015–present).

• **Best Music:** The 1970s and '80s alternative rock of Split Enz and anything by opera soprano Dame Kiri Te Kanawa.

Karekare Beach offers a sandy retreat just an hour from Auckland's city center.

and cyclists are still waiting for a proposed "SkyPath" across the bridge. However, **AJ Hackett Bungy** offers a bridge climb to the flag-topped summit of the span and a bungee jump from a pod fixed to the bottom of the bridge—during which you can dip your head and hands in the brackish bay water.

But it's not all about water. The skyscrapers of Auckland's central business district (CBD) rise behind the finger piers and ferry landings, reaching a peak in the 1,076-foot (328 m) **Sky Tower**. The tallest free-standing structure in the Southern Hemisphere offers panoramic views that extend across the entire city and surrounding waterways. But it's also a platform for more daredevil activities—like a **SkyWalk** along the metal catwalk that surrounds the

pergola and a **SkyJump** bungee plunge that features a 629-foot (192 m) free fall.

A venue for public art and spirited public demonstrations, **Aotea Square** is the cultural hub of the downtown area, flanked by **Aotea Centre** for performing arts and its modern **Kiri Te Kanawa Theatre** (symphony, opera, ballet, pop concerts), a glorious 1920s picture palace called the **Civic Theatre** (music, movies, plays), and **Auckland Town Hall** (opened in 1911). Rounding out the neighborhood's cultural offerings is the **Auckland Art Gallery Toi o Tāmaki**, New Zealand's top art museum, a collection that spans 15,000 works, including many noteworthy Maori and Pacific island artists.

Just west of the city center, the

funky **Ponsonby** neighborhood offers a blend of fashion boutiques, art galleries, handicraft stores, coffee shops, day spas, and an eclectic array of eateries with foods from around the globe.

That huge green space hovering on the outer edge of the CBD is **The Domain**. Set aside as a public park in the 1840s, shortly after the city was founded, its spacious lawns and shady trees sprawl across yet another dormant volcano (Pukekawa). Among its assorted attractions are outdoor summer concerts at the **Domain Rotunda**, the plants of a large Victorian-era greenhouse called the **Domain Wintergardens**, and the imposing **Auckland War Memorial Museum**. Opened in 1929, the museum covers a broad range of New Zealand natural and human history. Highlights include the Maori and military collections, the dinosaurs of the **Origins Gallery**, and daily Maori cultural performances.

The summit of **Mount Eden** (Maungawhau), another of the 50 ancient volcanoes that surround the city, provides more lofty views across Auckland, as well as a 150-foot-deep (46 m) grass-filled crater that visitors can hike around or descend into for a look-see or even a picnic. More breathtaking views await atop **One Tree Hill** (Maungakiekie) in **Cornwall Park**, which also harbors the **Native Arboretum** with sterling examples of indigenous trees and the **Stardome Observatory & Planetarium**.

Auckland makes an excellent base for exploring the middle section of the **North Island** on day trips. The dark sands and rugged surf of **Karekare Beach**—where scenes from *The Piano* were filmed—are about an hour's drive west of the city center. The drive meanders through **Waitākere Ranges Regional Park** with its many hiking trails.

The **Matakana wine region** lies about an hour due north of Auckland. **Brick Bay**, **Providence**, **Omaha Bay**, **Sawmill**, and

Take the plunge with a bungee jump off the Kawarau Bridge.

Ascension are among its foremost wineries. There's also culture—the **Brick Bay Sculpture Trail** and the whimsical art gardens of the **Sculptureum**—and the pristine nature of **Goat Island Marine Reserve** (scuba diving and snorkeling).

South of Auckland are two of the North Island's more unusual attractions: The **Waitomo Glowworm Caves** (five-hour round-trip) offer "blackwater rafting" along an underground river illuminated by millions of glowworms, and Tolkien fans can tour the **Hobbiton** movie set— where *The Lord of the Rings* and the *Hobbit* trilogies were filmed—on the outskirts of **Matamata** (four-hour round-trip). ∎

South Pacific Cities

Papeete, French Polynesia; Suva, Fiji; Nouméa, New Caledonia

Although village life still dominates the cultural landscape of the South Pacific, the region's island nations and territories nourish exotic capital cities that blend colonial relics with modern ways and means.

Find traditional crafts and food at the Papeete Market.

PAPEETE
Population: 137,000
Size: 115 square miles (298 sq km)

Many travelers breeze through Papeete on their way to beach resorts in Moorea or Bora Bora. Their loss: The territorial capital of French Polynesia—on the northeast shore of Tahiti—is one of the more intriguing cities in the South Pacific.

Lined with outdoor cafés, pearl jewelry shops, and **Les Roulottes** food carts, the **Boulevard Pomare** wraps around the city's waterfront, a great place to stroll, shop, or simply people-watch. The nearby old town revolves around the bustling **Municipal Market**, where fruit, fish, and flowers are complemented by handicraft stalls hawking hats, handbags, and traditional *pareo*

sarongs. The old town is also renowned for its traditional Polynesia **tattoo parlors**.

On the city's eastern outskirts is **Venus Point**, where Captain Cook recorded the transit of Venus across the sun in 1769 to calculate the distance between Earth and its star. It is now a gorgeous beach area with a black-sand strand, outrigger canoe racing, and a vintage French colonial lighthouse. Farther along is the **James Norman Hall Museum**, the seaside home where the American novelist penned *Mutiny on the Bounty* and many other works.

Out west (beyond the airport) are the **Museum of Tahiti and the Islands**, the ancient Polynesian temple complex at **Arahurahu Marae**, and Papeete's best beaches for watching the sunset. Across the water are the jagged volcanic peaks of **Moorea**, which makes for an easy day trip from the city via a 30- to 45-minute ferry ride or 15-minute flight.

Papeete is ground zero for various **aquatic and adventure activities** around Tahiti, from scuba diving and catamaran cruises to 4x4 rainforest safaris and waterfall hikes.

SUVA
Population: ca 200,000
Size: 790 square miles (2,046 sq km)

It's something of a miracle that Suva became a city, let alone the capital of Fiji. When Australian plantation owners settled the area in 1879, they found it mosquito-infested and inhospitable to sugar-cane cultivation. To recoup their investment, the Aussies hoodwinked the British into moving the colonial capital to the site.

Though present-day Suva boasts British colonial relics, the city is

Nouméa's Tjibaou Cultural Centre celebrates the Indigenous Kanak culture.

very much a product of post-independence (1970) Fiji, a vibrant melting pot and ever expanding metropolis that sprawls across a thumb-shaped peninsula on the south side of Viti Levu.

The modern high-rise towers of downtown Suva (aka Central) hover above the shops of old **Cummings Street**, the chromatic **Municipal Market**, and pungent **Fish Market** beside Nubukalou Creek.

Victoria Parade runs south along the waterfront past the **Suva City Carnegie Library** (opened 1909) to the playing fields of **Albert Park**, where rugby games are often in progress. Farther south are the tropical blooms of **Thurston Gardens**, the historic artifacts and cultural exhibits of the **Fiji Museum**, and the Georgian-style **Presidential Palace**, where a showy changing of the guard takes place the first week of each month.

The capital's "green lung" is **Colo-i-Suva Forest Park**, a 227-acre (92 ha) expanse of tropical woodland with hiking trails that lead to picnic areas and freshwater swimming holes. The park provides a habitat for butterflies, tree frogs, and more than half of Fiji's 57 native land bird species. A panoramic view of Suva and the south coast is the reward after the two-hour climb to the top of **Mount Korobaba**.

Suva's other outdoor playground are the beaches along Queen's Road west of the city, a strip that includes **Pacific Harbour** (suffused with diving and snorkeling, river rafting and tubing, boat tours and deep-sea fishing) and the natural coastal landscapes of **Sigatoka Sand Dunes National Park**.

NOUMÉA

Population: 180,000
Size: 634 square miles (1,642 sq km)

At first glance, New Caledonia's capital seems like a slice of the French Riviera that somehow washed ashore in the South Pacific. But on closer examination, it's clear the island's Indigenous Kanak culture is also woven into the urban fabric.

Nouméa is a thoroughly modern city that draws vacationers to its idyllic beaches, turquoise waters, and natural sightseeing opportunities.

Centered on the palm-shaded **Place des Cocotiers**, the old town maintains French colonial structures like the **City Museum**, **St. Joseph's Cathedral**, and **Bernheim Library**. Along the waterfront, the **Marché Municipal** is a mélange of fish and produce stalls, pastry shops, and souvenir outlets opposite the **Museum of New Caledonia**.

But the real action is on the shore, a necklace of bays and beaches that nearly encircles the city. **Baie des Citrons** ("Lemon Bay") transitions from swimming and sunbathing by day into a hip strip of bars, restaurants, and outdoor cafés at night. **Anse Vata** offers another fine stretch of sand. Perched on a peninsula between the beaches, the excellent **Aquarium des Lagons Nouvelle-Calédonie** harbors sea creatures like blacktip reef sharks, sea snakes, nautilus, and huge humphead wrasse.

One of the architectural gems of the South Pacific, **Tjibaou Cultural Centre** on the city's northeast coast is a cluster of conical pavilions inspired by traditional Kanak houses. The center presents a wide

Hiking trails lead to stunning waterfalls in Colo-i-Suva Rainforest National Park.

array of modern and traditional Oceanic art, live music and dance, and artistic workshops.

From the Nouméa waterfront, ferries, water taxis, and excursion boats whisk passengers to offshore attractions like sun-splashed beaches, coral reefs, and the outdoor art of the **Île aux Canards** (Island of Ducks), as well as historic **Amédée Lighthouse**, erected in 1865 and one of the tallest of its kind anywhere on the planet (184 feet/56 m). ∎

Africa

Markets in Cairo (p. 166) offer one-of-a-kind souvenirs, including beautiful lamps and lanterns.

Nairobi
Kenya

Born as a remote railroad way station, Nairobi cast aside its shortcomings to become East Africa's largest urban area and a city synonymous with safari adventure.

THE BIG PICTURE

Founded: 1899

Population: 4.5 million

Size: 269 square miles (697 sq km)

Languages: English, Swahili, and various tribal

Currency: Kenyan shilling

Time Zone: GMT +3 (EAT)

Global Cost of Living Rank: 145

Also Known As: Green City in the Sun, Safari Capital, Silicon Savannah

Nairobi may not be everyone's cup of tea—especially when it comes to the city's notorious traffic. But no one can deny that what the Kenyan capital lacks in overall charm, it more than compensates with the region's best nightlife, food scene, shopping, and hotels.

Many visitors give the city only a passing glance on their way from the international airport to the wildlife wonders of Tsavo or Masai Mara. But that's a mistake because just as much as the nation's legendary game parks, Nairobi offers an authentic African experience, albeit a modern one.

Downtown Nairobi revolves around busy **Kenyatta Avenue** and the 28-story **Kenyatta International Conference Centre**, where a rooftop viewing deck affords a 360-degree view of the city. The nearby **Nairobi Railway Museum** houses relics of the "Lunatic Express" that the British built between the Kenya coast and Uganda in the 1890s.

Another downtown landmark is the old **Norfolk Hotel**, opened in 1904 and one of the few colonial-era buildings remaining in the central city. Also worth a look-see is the refurbished **National Museum**, especially the Hominid Skull Room and the Cycle of Life exhibit on Kenya's various tribes and cultures.

North of downtown, the cosmopolitan **Westlands** area is Nairobi's nightlife nerve center, a cluster of bars, restaurants, and music clubs frequented by expats and the city's burgeoning middle class. The Woodvale Grove–Parklands Road–Mpaka Road triangle is especially dense with night spots.

Founded in 1946 when the city was much smaller, **Nairobi National Park** runs along the southern fringe of the metro area. Despite its diminutive size (and proximity to more than four million humans), the park harbors lions, elephants, rhinos, and other wildlife. Visitors can explore the park in their own vehicle or join a guided safari tour. Part of the park is devoted to the **David Sheldrick Wildlife Trust** elephant orphanage, a nonprofit that has rescued hundreds of baby pachyderms since the 1970s.

West of the national park, leafy suburbs like **Langata** and **Karen** have evolved from farmland and bush into upscale suburbs where many of Nairobi's wealthy reside. The latter is named after Karen Blixen, the Danish baroness who penned *Out of Africa* after living on a farm "at the foot of the Ngong

See the home of Danish author Karen Blixen, who famously penned *Out of Africa*.

Two zebras stand gracefully in the grasslands of Nairobi National Park, with the city skyline in the background.

Hills." Her house (built in 1916) is now the **Karen Blixen Museum**, filled with period furnishings and artifacts from the 1985 film version of her famous book.

The **Giraffe Center** in Langata does double duty as a place that educates people about the world's tallest animal as well as a rescue and captive breeding center for the highly endangered Rothschild giraffe. Like the nearby elephant orphanage, many of the animals have been reintroduced into the wild.

Nairobi's western suburbs are also preserving local culture. **Bomas of Kenya** is an open-air museum featuring Indigenous arts and crafts, dance and music, and traditional foods like *nyama choma* (roasted goat

meat), *mukimo* (mashed potatoes, beans, and corn), and *muthokoi* (maize, bean, and pea stew). On the other hand, the **Karen Village**

cultural complex offers a place for contemporary Kenyan artists and artisans to create, display, and sell their works. ■

URBAN RELIC

Stone Town, Zanzibar

"Earth, sea and sky all seem wrapped in a soft and sensuous repose," wrote British explorer Sir Richard Burton during an 1856 sojourn in **Stone Town**. And the old Zanzibar city still feels that way today, the epitome of an exotic, romantic, tropical port of call.

Named for the coral stone used to construct most of its buildings, Stone Town flourished as a spice, ivory, and slave market under the

Portuguese, British, and Omani sultans.

The city looks much as it did a hundred years ago, the waterfront lined with dhow sailing craft and historic buildings like the House of Wonders and the Old Fort (main venue for the annual Sauti za Busara music festival). The slave market is memorialized beside the Anglican Cathedral, while Mercury House honors singer Freddie Mercury, born in Stone Town in 1946.

Cape Town
South Africa

South Africa's window to the outside world for more than 500 years, Cape Town complements its phenomenal topography with an urban landscape that ranges from beaches populated by penguins, surfers, and other sunseekers to world-class museums, vibrant multicultural neighborhoods, and even its own wine country.

THE BIG PICTURE

Founded: 1652

Population: 4.2 million

Size: 315 square miles (816 sq km)

Language: English

Currency: South African rand

Time Zone: GMT +2 (SAST)

Global Cost of Living Rank: 178

Also Known As: Mother City, Kaapstad (Afrikaans), Camissa (Khoisan), Tavern of the Seas, Hoerikwaggo (Mountain in the Sea)

When in Cape Town, it's always best to start at the top—the summit of **Table Mountain**. Even more than maps or satellite images, the celebrated peak elucidates the city's layout and landmarks in a single glance. They're literally spread out beneath your feet—the bustling Victoria & Albert Waterfront, the ovular stadium that hosted the 2010 FIFA World Cup, crescent-shaped Camps Bay, notorious Robben Island in the hazy distance, and so much more.

The historic **Table Mountain Cableway**—launched in 1929 and revamped in the 1990s—begins its ascent from the visitors center on Tafelberg Road. It takes around five minutes for Africa's original high-wire act to reach the 3,558-foot (1,084 m) summit and its expansive views. Visitors can also hike to the top via several routes, including spectacular **Platteklip Gorge**, a trail that descends 2,195 feet (669 m) over a little more than a mile (1.6 km).

Longer hiking routes ramble through the wilds of **Table Mountain National Park**, which flows across ridgelines and valleys all the way to Cape of Good Hope. Trekkers can stay overnight in cottages, safari-style tented camps, and a historic washhouse along the way.

At the opposite end of the Cape Town spectrum—in both elevation and atmosphere—is the **Victoria & Albert Waterfront**. Starting in the late 1980s, the old docklands area was reworked into a residential, retail, hotel, and entertainment hub that lures more than 24 million visitors each year. Among its manifold attractions are the stunning **Zeitz Museum of Contemporary Art Africa**, the giant **Cape Wheel** observation ride, **Two Oceans Aquarium** with its penguins and predators, and the **Watershed marketplace** where more than 150 shops offer South African fashion and craft creations.

In addition to exhibits on South Africa's political history, **Nelson Mandela Gateway** at the waterfront is the starting point for boat trips and guided tours to **Robben Island**—where Mandela spent 18 years as a political prisoner of the apartheid regime. The Robben

Walk through Bo-Kaap, a formerly racially segregated area once known as the Malay Quarter.

Cape Town boasts many treasures, including Table Mountain, Cape Town Stadium, and a harbor along Table Bay.

Island tourist experience lasts around 3.5 hours, including the round-trip boat ride and a bus tour that's sometimes guided by a former prisoner.

Between the docklands and Table Mountain, an enormous natural amphitheater called the **City Bowl** harbors Cape Town's central business district and several of the city's more intriguing neighborhoods.

Erected in the mid-17th century by the Dutch East India Company, the **Castle of Good Hope** is South Africa's oldest surviving building. Although it's still an active military base, visitors are free to walk the ramparts or browse the castle's **Military Museum** and **William Fehr Collection** art gallery. **Good Hope Adventures** offers two-hour tours of

the tunnels and bunkers beneath the fortress.

Right up the road from the castle entrance is the old **City Hall** (completed in 1905), an elaborate

Edwardian-style structure now used for exhibitions and classical music concerts. The **Nelson Mandela statue** on the balcony marks the spot where he gave his first speech

after his release from Robben Island in 1990.

Downtown's biggest green space is the **Company's Garden**, established as a vegetable plot for early Dutch colonists and sailors but now a refreshing expanse of lawns, gardens, and wooded areas (including a pear tree planted in 1652). Flanking the garden are the **South African National Gallery** of art, the **Iziko South African Museum** of natural history, and the **South African Houses of Parliament**.

More vintage buildings huddle a block away at the busy intersection of Wale and Adderley Streets. In addition to its spiritual tasks, **St. George's Anglican Cathedral** (completed in 1901) hosts an inspirational social justice forum underground in what's called **The Crypt**. Across the street, **Iziko Slave Lodge** delves into the long history of slavery in South Africa in a structure used as a holding pen for chattels during the Dutch colonial era.

Nearby Long Street also retains many of its Victorian-era buildings, transformed in modern times into Africa-flavored restaurants, bars, and shopping spots like **Long Street**

Stairs climb to the lighthouse at Cape Point and its coastal views.

Antique Arcade, the homegrown fashion of **Mungo & Jemima**. Cutting-edge contemporary design colors uber-hip **Bree Street**, shops like **Skinny laMinx** with its unique household accessories, the high-end handbags of **Missibaba**, and the South African haute couture of **Alexandra Höjer Atelier**.

Poised at the west end of Wale Street is the **Bo-Kaap**, literally the city's most colorful neighborhood. Hundreds of brightly painted homes and commercial buildings saturate an area that's been home to Cape Malay people for more than 300 years. The **Iziko Bo-Kaap Museum** tells the story of their ancestors, Southeast Asian Muslims enslaved as servants for the Dutch colony. Scores of structures in the Bo-Kaap are national heritage sites, including the sea green **Boorhaanol Islam mosque** (opened in 1884) and a number of homes along Wale, Rose, and Dorp Streets.

Rising behind the Bo-Kaap is **Signal Hill**, a windswept ridge that defines the western edge of the City Bowl. A roadway and footpaths meander to the summit for sweeping views of the coast (especially at sunset) and paragliders who use the hill as their launch pad. Farther south along the ridge is **Lion's Head** peak and an even more demanding

hike to a summit with a view many Capetonians favor over Table Mountain.

Surf, sun, and sand castles rule Cape Town's epic west coast, which starts with the **Green Point** neighborhood that exemplifies the city's multiculturalism. **Cape Town Stadium**—which bears a striking resemblance to the ruff collars worn by 17th-century Dutchmen—has hosted World Cup matches and is now the city's professional rugby and soccer ground.

The annual **Cape Town Carnival** with its 2,000 costumed dancers and musicians takes to the streets of Green Point each March. Over the past 30 years, the **De Waterkant** area has evolved into the city's LGBTQ hub and epicenter of the **Cape Town Pride Festival** in February. Green Point is also renowned for its eclectic dining scene, restaurants like **Gold**, where a 14-course African tasting menu is complemented by traditional drummers and puppets.

Sea Point Promenade offers a three-mile (4.8 km) stroll along the coast between Green Point and **Bantry Bay**. Farther along is picture-perfect **Clifton**—often rated as one of the world's top 10 beaches—where Capetonian surf culture was born in the 1960s. Next up is

---

HIDDEN TREASURES

• **Blaauwberg Nature Reserve:** Blessed with panoramic views of Table Mountain and Robben Island, the reserve boasts a long, pristine beach and a 2.8-mile (4.5 km) loop trail through native coastal vegetation.

• **Camissa Township Tours:** Guided walking, Gospel, and soccer tours of Langa Township and African cooking

experiences in restaurants and private homes; gocamissa.co.za.

• **Norval Foundation:** Contemporary South African creativity is the forte for a stunning new private gallery and performance space in southern Cape Town that opened in 2018.

• **Coffee Beans Routes:** Jazz, art, beer, wine, bikes, and home cooking

are just a few of the urban "safaris" offered by this innovative local tour company; coffeebeansroutes.com.

• **Tygerberg Nature Reserve:** Around a 20-minute drive from the city center, Leopard Mountain offers hiking trails through 740 acres (300 ha) of wilderness that harbors more than 500 plant and 100 bird species.

Camps Bay, the de facto Malibu of Cape Town given how many celebrated actors, athletes, and authors live there. And don't forget to look up—a towering rock formation called the **Twelve Apostles** hangs off the side of Table Mountain.

Beyond the stony disciples is **Hout Bay** with its snug harbor and waterfront restaurants. Road-trippers can head inland from there along a highway that curls around the south side of Table Mountain to Constantia. Or continue down the **Cape Peninsula** on cliff-hugging **Chapman's Peak Drive**.

Around 50 miles (80 km) south of Green Point, the coast road finally comes to a majestic finale at the fabled **Cape of Good Hope**. Geographically speaking, it's definitely *not* the southernmost tip of Africa; that honor goes to **Cape**

ALTER EGO

Bloemfontein

The fragrant "City of Roses" **(Bloemfontein)** lies more than 600 miles (966 km) inland from Cape Town on the cusp of the Highveld and the semidesert Karoo.

Founded by Boer settlers who fled the cape in the 1840s to escape the British, the city was the capital of the independent Orange Free State for half a century.

"Bloem" remembers its past in sights like the **Fourth Raadsaal**

capitol building (completed in 1893), the **Anglo-Boer War Museum**, and the esteemed **National Museum** (established in 1877) with its dinosaur bones and faithful re-creation of a pioneer-era village street.

But the city is also stepping boldly into the 21st century with developments like the **Loch Logan Waterfront** shopping and entertainment district, as well as the **Oliewenhuis Museum** with its emphasis on modern (and often offbeat) South African art.

Agulhas three hours east of Cape Town. But for more than 500 years, Good Hope has been the historical and spiritual turning point around the bottom end of the continent.

Buffelsfontein Visitor Centre offers a good primer to the area's unusual fynbos vegetation and surprisingly rich animal life (ostrich, baboons, penguins, hyrax, and six antelope species). From the main parking lot, trails lead to **Cape Point** and its historic lighthouse (built in 1860), the **Cape Peninsula wooden walkway**, and white-sand **Dias Beach** beneath the cliffs. The

Take a tour of Boulders Beach where you can see a colony of African penguins.

Enjoy lush views on a stroll of the Tree Canopy Walkway at the Kirstenbosch National Botanical Garden.

Flying Dutchman Funicular provides an alternative means of transiting the last stretch of cape.

Main Road meanders up the **False Bay** side of the Cape Peninsula to **Simons Town** and the nation's largest navy base. Waterfront restaurants, fishing charters, and shark-diving outfits crowd around **Jubilee Square** and its compact harbor. But the town's main attraction is **Boulders Beach** and its colony of African penguins. The coast road continues to cutesy **Kalk Bay**, where the **SOSF Shark Education Centre** offers information and exhibits on one of the animal world's most fascinating creatures.

Reaching **Surfers Corner**, drivers can continue along the north shore of False Bay or turn inland for a rendezvous with Cape Town's **wine country** around charming **Constantia** village. Founded in 1685 by the colonial governor as the first wine estate, **Groot Constantia** preserves a classic Cape Dutch mansion as well as vineyards that produced wine sipped by the likes of Charles Darwin, Prussian emperor Frederick the Great, and Napoleon during his exile on St. Helena.

Snug against the side of Table Mountain is **Kirstenbosch National Botanical Garden**. Founded in 1913, the garden has always been dedicated to preserving South Africa's natural flora, from the Cape region's delicate fynbos vegetation to the arid-adapted plants of the Kalahari and Karoo. Among its many features are protea gardens, a tree canopy walkway, a braille trail, and the **Galileo Open Air Cinema**.

Completing the loop around the Cape Peninsula, Nelson Mandela Boulevard runs through **Woodstock**, the city's hippest neighborhood and one of the few in Cape Town that survived the segregation of the apartheid era. **Albert Road** is now lined with trendy restaurants, boutiques, and craft breweries, while a disused factory called the **Old Biscuit Mill** has morphed into an innovative space that blends fashion boutiques, art galleries, and upscale eating places like the **Test Kitchen**, considered one of the planet's top 50 restaurants. ◼

Johannesburg
South Africa

Many travelers pass straight through Johannesburg on their way to a safari. But South Africa's largest city deserves closer inspection, not just for its historical and cultural attractions, but also for a glimpse of an urban area still coming to terms with its turbulent past.

Egoli or "City of Gold" is what the Zulu people call Johannesburg, a reference to the shiny yellow metal that spurred the city's birth and growth. As much as half of all the gold on planet Earth (1.5 billion ounces/42,524 metric tons) has been mined in and around the metropolis, an area that remains one of the world's top gold producers.

The historic mining belt runs right through the heart of the city, separating the central business district and largely white northern suburbs from south-side communities like Soweto, where people of color have traditionally resided.

Although Indigenous people occupied the Gauteng region for thousands of years before the arrival of European settlers, a city didn't begin to rise in the area until the 1880s, when gold was discovered along an extenuated east-west plateau called the Witwatersrand. Since then, "Joburg" has

THE BIG PICTURE

Founded: 1886

Population: 9.3 million

Size: 1,000 square miles (2,590 sq km)

Language: English, Afrikaans, various tribal languages

Currency: South African rand

Time Zone: GMT +2 (SAST)

Global Cost of Living Rank: 184

Also Known As: Joburg, Jozi, Egoli (Zulu), Gauteng (Sotho-Tswana), Muḍi Mulila Ngoma (Venda), Joni (Tsonga)

grown into one of Africa's five largest cities.

Unlike Kimberley—which still revels in its diamond rush days—Johannesburg retains little of its golden past. Much of the pioneer-era town that arose around **Constitution Hill** disappeared during the 20th-century rush to create a modern, Western-style city. But a few vestiges of the past remain.

Erected in 1892, the **Old Fort** is the city's oldest remaining structure. Built by the Boers as a prison, the whitewashed citadel on Constitution Hill also served as a military bastion and POW camp during the Anglo-Boer conflict before reverting back to a penitentiary for hardened and political prisoners. Nelson Mandela and Mahatma Gandhi are among the famous names detained there. Nowadays the fort is part of a multibuilding museum that traces South Africa's journey from apartheid to democracy.

Across the street, South Africa's modern **Constitutional Court** was purposely built atop a demolished portion of the old prison as a symbol of the nation's human rights struggles. Visitors are free to attend

Taste local flavors from stalls at food markets throughout the city.

Students perform at Nelson Mandela Square to mark the day honoring the courageous freedom fighter.

court sessions or browse the works of prominent South African artists in the building's gallery.

The "Rand Lords" who made their fortunes on the gold and diamond rushes erected grand mansions in the **Parktown** neighborhood on the hill's north side. Among the more stylish are **Northwards** (built in 1904), **Dolobran** (built in 1906), **Hazeldene Hall** (built in 1902), and **The View**, a neo-Queen-Anne-style manse created in 1896 for Sir Thomas Cullinan, owner of a mine where the world's largest-ever diamond (3,106.75 carats) was discovered. **Microadventure Tours** offers guided peeks at several Parktown mansions.

The only inner-city mining shaft open to the public is **Gold Reef City**, which ceased mining operations in the 1970s and was later transformed into a sprawling entertainment zone

with a casino, a cinema complex, the **Lyric Theatre** performing arts venue, and a theme park with roller coasters and other thrill rides. There's also gold-rush history: the restored

---

URBAN RELIC

Great Zimbabwe

Around 500 miles (805 km) north of Johannesburg, a cluster of ruins marks the spot where an Iron Age city once stood.

Created by the ancestral Shona people in the 11th century, **Great Zimbabwe** was occupied by as many as 18,000 people until the 15th century, when it was inexplicably abandoned.

The stone buildings are divided into three groups: Hill Ruins, Valley Ruins, and the ellipsis-shaped Great Enclosure and its conical tower—the single largest ancient structure in all of sub-Saharan Africa.

According to legend, Great Zimbabwe was home to the Queen of Sheba. But there's no proof of that, for the city was already abandoned when the Portuguese came across the ruins in 1531

The Apartheid Museum offers an eye-opening display of South Africa's struggles with colonial domination and racial segregation.

Oosthuizen House and **Olthaver House** museums, the bulletproof **Bullion Coach** used to transport gold from Joburg to the seaports of Cape Town and Durban, and a heritage tour called **Jozi's Story of Gold** that includes a descent into the abandoned underground mine shaft.

Kromdraai Gold Mine on the northern outskirts of the city offers a more authentic gold mine experience, a three-hour tour (including an underground portion) of one of the first and most productive pits.

The tour is easily combined with a visit to the famous **Sterkfontein Caves**, which are both a geological wonder and the place where the remains of a 2.3-million-year-old human ancestor *(Australopithecus africanus)* was discovered in 1947.

The nearby **Maropeng Visitor Centre** features modern exhibits on the evolution of mankind over millions of years and the various anthropological digs that make up UNESCO's **Fossil Hominid Sites of South Africa World Heritage site**.

Joburg's claim to historical infamy is apartheid, a long struggle for civil rights commemorated at numerous places around town. The evocative **Apartheid Museum** at Gold Reef City chronicles that struggle with permanent and temporary exhibits, as well as a tribute to Nelson Mandela. The modest redbrick **Nelson Mandela Home**, where the liberation leader lived between 1946 and his arrest in 1962, is now a museum in Soweto township. **Mandela's cell** is

faithfully preserved inside the Old Fort prison, while **Liliesleaf Farm** in suburban Rivonia preserves the spot where Mandela hid out and plotted antiapartheid strategy with African National Congress (ANC) colleagues in the early 1960s.

Nothing evokes the apartheid era like **Soweto**, the massive township on the city's southeast side. Created in the 1930s when the South African government shifted its separation of the races into overdrive, Soweto would eventually house more than a million Black residents.

Guided tours are one of the more popular ways to visit the sprawling township. A number of local outfitters offer van, bus, and private car excursions, including tours that combine the Apartheid Museum and Soweto. But one of the more interesting ways to explore the township is private biking tours with **Cycle in Soweto**. They run anywhere from two to eight hours, and on Sundays one of the tours includes a Gospel church service.

Besides the aforementioned Mandela House, another township landmark is the **Hector Pieterson Memorial & Museum**, which honors schoolchildren killed by authorities during the 1976 Soweto uprising. **Walter Sisulu Square** features an open-air museum and brick monument that commemorates the spot where the antiapartheid Freedom Charter was written and read aloud to a large crowd in 1955.

After decades of neglect, Joburg's **city center** is undergoing something of a renaissance thanks to revitalized neighborhoods like **Maboneng**, which has morphed from urban decay to artsy enclave via an influx of students, artists, and entrepreneurs. Among its cornerstones are the **Bioscope** indie movie house and a collection of creative spaces and eateries called **Arts on Main**. On Sundays, the **Market on Main** offers a wide array of regional foods and local designer items. Peer down on the neighborhood with drink in hand from the plant-filled **Living Room** rooftop café.

Or snatch an even higher view from the **Top of Africa**, the 50th-floor observation deck at the summit of the **Carlton Centre**, the tallest building on the continent from 1974 to 2019 when the 55-story **Leonardo** tower was completed in suburban Sandton. Down at street level, the old and venerable **Rand Club** (founded in 1887) hosts high tea for those who wish for another glimpse of life as a gold or diamond mogul. On the other hand, **Mad**

Giant Brewery is a hangout for hipsters seeking South African–style craft beer. Those questing for something more cerebral should pop into **Collectors Treasury**, southern Africa's largest bookshop with over a million tomes, plus antique maps, photographs, and vinyl records.

Three museums in the downtown area also beg attention, including the **Museum Africa** (opened 1933) with its thousands of cultural artifacts from around the continent, the modern **Origins Centre** with its ancient rock art and fossil treasures, and the new **Wits Art Museum** on the campus of Witwatersrand University. Many of the Wits students hang in the bars and cafés of the adjoining **Braamfontein** neighborhood, which is also home to the weekend **Neighbourgoods Market**, renowned for its music, artisan vendors, and gourmet food stalls. ■

Take a tour of the Sterkfontein Caves, known as the Cradle of Humankind.

Addis Ababa
Ethiopia

Africa's fourth largest urban area offers a glimpse of the evolutionary forces that shaped the continent's prehistory and a preview of how many African cities are likely to evolve in the future.

THE BIG PICTURE

Founded: 1886

Population: 4.5 million

Size: 203 square miles (526 sq km)

Languages: Amharic, Oromo

Currency: Ethiopian birr

Time Zone: GMT +3 (EAT)

Global Cost of Living Rank: 194

Also Known As: Barara (15th–16th centuries); Finfinne (Oromo)

From its humble birth as a royal compound and Italian colonial outpost in the 1880s, Addis Ababa has matured into a budding economic superstar with glitzy high-rise hotels and office blocks, a massive new airport terminal, one of Africa's first light-rail commuter lines, and headquarters of the African Union.

But it's the old Addis that lures visitors into exploring the city before heading off into Ethiopia's remarkable hinterland—in particular, the **National Museum of Ethiopia**, which safeguards the world's most extensive collection of early hominid remains. Foremost among these is "**Lucy**"—a distant human kin that called Africa home approximately 3.2 million years ago.

Ethiopia's 80 ethnic groups are the focus of the **Ethnographic Museum** and its array of tribal artifacts, royal regalia, and priceless monastery murals. And given the fact that it's tucked inside the former royal palace, the museum also showcases the lavish private quarters of longtime emperor Haile Selassie. Ethiopia's dark recent history is the focus of the **Red Terror Martyrs' Memorial Museum**.

As a bastion of the Ethiopian Orthodox Church, the city flaunts several impressive churches including the 19th-century **St. George's Cathedral**, where Ethiopia's emperors were once crowned, and the flamboyant **Holy Trinity Cathedral**, which harbors the tombs of Haile Selassie and other members of the royal family, as well as the celebrated British social-political activist Sylvia Pankhurst.

Yet the city's foremost appeal is easy access into modern Ethiopian life and culture, including a lively dining scene. Thick *injera* bread, curry-like *wat* dishes, and sautéed *tib* meats form the heart of an Ethiopian cuisine that's become hip around the globe. Among the city's popular eateries are **Yod Abyssinia** near Bole International Airport, **Dashen Terara** in the Kazanchis neighborhood, and **Lucy Restaurant** beside the National Museum. And be sure to ask for a traditional Ethiopian coffee ceremony following your meal—an experience that's at once delicious and photogenic.

The Addis music scene is just as vibrant, with numerous places where you can catch live acts from the Horn of Africa. **Mama's Kitchen** is world renowned for its Ethio jazz. Great music clubs like the **Fendika Azmari Bet** huddle near Meskel Square.

The history of Lucy is just one display at the National Museum of Ethiopia.

Visit the colorful Entoto Maryam Church and Entoto Museum.

The mountainous heights around Addis render great city views and more historical insights. About a 30-minute drive north of the city center, 10,500-foot (3,200 m) **Mount Entoto** provides a lofty venue for **Emperor Menelik's palace**, monolithic **Kiddus Raguel Church**, and the chromatic **Entoto Maryam Church** and museum. Crowning the summit, **Entoto Natural Park** offers a number of scenic hiking trails through the forested highlands.

On the southwest side of Addis, **Menagesha Suba Forest Park** is Africa's oldest protected area. Established as a royal reserve in the 15th century by Emperor Zera Yacob, the park is laced with hiking and biking trails through old-growth woodland.

South of the capital, Highway B51 links three historical landmarks that can be combined into a great day trip from the city—**Melka Kunture** prehistoric site and museum, the large stone pillars of the UNESCO World Heritage site at **Tiya**, and the monolithic, 12th-century **Adadi Mariam Church**. ∎

URBAN RELIC

Lalibela

Founded by Emperor Lalibela in the 12th century, the high-altitude city of **Lalibela** in northern Ethiopia is celebrated for its rock-hewn churches and is often referred to as the red-rock shrine.

Sculpted from volcanic scoria, the 11 underground shrines were part of the king's attempt to create a "New Jerusalem" for Ethiopian Orthodox Christianity.

Shaped like a Greek cross, the exquisite Bete Giyorgis (Church of St. George) is the most photographed, while Bet Medhane Alem (House of the Savior of the World) is the largest. But each of them is unique and has its own stories to tell.

More churches and monasteries cling to the mountains around Lalibela—like Asheton Maryam, carved into a vertical cliff face and reached on foot via a precipitous trail.

Visiting the churches is especially evocative during daily rites or during Ledet (Christmas), Timkat (the Epiphany), and other Ethiopian Orthodox holidays.

Luxor & Aswan
Egypt

Located about 150 miles (240 km) from one another along the Nile, Luxor and Aswan are both gateways into the treasures of ancient days, as well as a touch of British colonial panache and vibrant modern Egypt.

THE BIG PICTURE

Founded: ca 2000 B.C.

Population: Luxor 500,000; Aswan 300,000

Size: Luxor 161 square miles (416 sq km); Aswan N/A

Language: Arabic

Currency: Egyptian pound

Time Zone: GMT +2 (EET)

Global Cost of Living Rank: Unlisted

Also Known As: Luxor—Thebes, Waset; Aswan—Syene, Swenett

LUXOR

Ancient Egypt reached the apex of its power and glory during the New Kingdom, between 3,000 and 3,500 years ago. For much of that era, Luxor was the empire's political, spiritual, and cultural capital. Nowadays the riverside city boasts what is arguably the world's greatest outdoor archaeological collection, dozens of global landmarks ranging from the Temple of Karnak to King Tut's tomb.

Perched along the Nile in the middle of town, the **Luxor Temple** offers a good starting point for any exploration. With its massive pylons, columns, and statuary artfully illuminated at night, the temple is especially evocative in the evening. Just up the **Corniche** waterfront road are the small but interesting **Mummification Museum** and the superb **Luxor Museum**, which displays many of the statues and other artifacts discovered at the area's various archaeological sites.

On the north side of town, the massive **Temple of Karnak** complex sprawls across nearly 500 acres (200 ha). Dedicated to a trio of ancient Egyptian gods—Amun-Re, Montu, and Mut—it was the most important of all the shrines scattered along the Nile, as well as a masterpiece of New Kingdom architecture.

Ferries and feluccas flit across the river to a **West Bank** area renowned for its sprawling funerary temples and once hidden tombs. Foremost among the memorial temples—owing to its unusual design and the fact that it's dedicated to a female pharaoh—is the **Temple of Hatshepsut**. But several others are just as impressive, including **Medinet Habu** (the temple of Ramses III) and the **Ramesseum** (temple of Ramses the Great), where a giant stone head of the pharaoh inspired the sonnet "Ozymandias" by Percy Bysshe Shelley.

Beyond the sandstone cliffs that line the West Bank is the **Valley of the Kings**, where the tombs of more than 60 pharaohs have been uncovered. Only a handful are open to the public at any one time, including the richly decorated **tombs of Pharaohs Seti I**, **Ramses VI**, the warrior king

See ancient Egyptian hieroglyphics on a tour of the Temple of Karnak complex.

Take a sail on the Nile in a felucca, a traditional Egyptian wooden boat with a canvas sail.

Tuthmosis III, and Boy King **Tutankhamun**, famously discovered by Howard Carter in 1922.

ASWAN

The river spangled with islands, the Sahara looming to the west, a waterfront lined with lateen-sailed feluccas—Aswan is easily Egypt's most picturesque city. For more than 4,000 years, it's been the country's southern gateway and a segue into a rich Nubian civilization that flourishes along the upper reaches of the Nile.

But the city is actually best known for something thoroughly modern: the **Aswan High Dam**. Completed in 1970, the barrier created **Lake Nasser**, which flooded the Nile for hundreds of miles upstream.

The inundation sparked a massive rescue effort that saved nearly two dozen ancient sites, including **Philae Temple** just behind the dam and the remarkable **Abu Simbel** temples with their massive statues of Ramses the Great, which can be visited as a day-long road trip or 45-minute flight from Aswan.

Aswan's Nubian roots are expressed in the marvelous **Nubian Museum**—winner of a prestigious Aga Khan Award for Architecture— and open-air **Solaih Nubian Restaurant** on Lake Nasser (you can watch the Philae sound-and-light show while dining there). ∎

Cairo
Egypt

The Arab world's largest metropolis may trace its roots to ancient Egypt, but its personality was largely molded by medieval sultans, an avidly progressive 19th-century ruler, and the Cairenes who are currently shaping it into a dynamic modern city.

THE BIG PICTURE

Founded: 969

Population: 17 million

Size: 740 square miles (1,917 sq km)

Language: Arabic

Currency: Egyptian pound

Time Zone: GMT +2 (EST)

Global Cost of Living Rank: 137

Also Known As: al-Qāhirah (Arabic), Kashromi (Coptic), Memphis (ancient Egyptian), Heliopolis (ancient Greek), Misr al-Qahira (ancient Arabic), Mother of the World, Paris of the Nile, City of 1,000 Minarets

For most first-time visitors to Cairo, it's best to get the big stuff out of the way first. Starting with the largest of them all, the **Pyramids of Giza**. Lone survivor of the original Seven Wonders of the Ancient World, the three massive structures were erected in the 26th century B.C. as grandiose tombs for the 4th-dynasty pharaohs Khufu, Khafre, and Menkaure.

"As long as Earth endures some vestige will remain of the pyramids," wrote Herman Melville in 1857. One glimpse and you know that's true. By any measure, they are global treasures. Yet perhaps their most impressive aspect is the fact that they were built more than 4,500 years ago by people with only rudimentary engineering skills and construction devices compared to what's available today.

The **Great Pyramid of Khufu**, largest and oldest of the three, rises 455 feet (138.8 m) into the Sahara sky and would loom even higher today if the smooth white limestone casing that once topped the edifice had not been pillaged to construct the medieval mosques and palaces of Cairo. Researchers estimate that it took around 100,000 workers laboring for years to move the 2.3 million stone blocks that make up the pyramid.

Although nothing to sneeze at, the other two Giza monoliths seem almost puny by comparison: the **Pyramid of Khafre** (448 feet/ 136.4 m), which retains some of its limestone cap, and the **Pyramid of Menkaure** (204 feet/62 m). The three pyramids are surrounded by a veritable city of the dead, with the *mastabas* (Egyptian tombs) and cut-rock tombs of queens and nobles, as well as the much more modest tombs of pyramid workers who died on the job. The **Khufu Boat Museum** harbors an ancient wooden sailing craft that had been buried in a pit beside the Great Pyramid and discovered in 1954.

Climbing the pyramids has been banned for decades. But visitors can discover the ancient tombs and surrounding desert in numerous other ways: They can venture inside several of the **burial shafts** that burrow deep inside the pyramids. They can watch nightly, hour-long **sound-and-light shows** from seats arrayed along the eastern edge of the

The Cairo Tower stands at the center of the capital's skyline.

Cairo is a jumping-off point for seeing Egyptian artifacts, including the Sphinx at the Giza Pyramids necropolis.

archaeological park. The best overall view is from an observation area just west of the pyramids, where visitors can also arrange desert **horseback and camel rides**.

Nearly as famous as the pyramids, the enigmatic **Sphinx** crouches on the southeast edge of the complex. Each of its paws larger than a bus, the colossal man-lion statue was chiseled from local limestone rather than constructed with blocks moved to the site. Given its location in front of Khafre pyramid, the face most likely bears his image.

The other really big thing on the Giza Plateau is the new **Grand Egyptian Museum** (GEM). More than a decade in the making at a cost estimated at around $1 billion, it's the world's largest museum dedicated to a single culture. More than 50,000 artifacts elucidate ancient Egyptian history from Stone Age

LOCAL FLAVOR

• **Abdel Rahim Koueider:** Popular confectionary chain serving a wide range of traditional Egyptian sweets, including *kanafa* (noodle-like pastry soaked in syrup, stuffed with creamy cheese, and topped with nuts) that's to die for; *eight branches in Cairo including downtown at 42 Talaat Harb Street;* ar-koueider.com.

• **Naguib Mahfouz Cafe:** Named after the Nobel Prize–winning author, this atmospheric café in the Khan el Khalili market serves tea, coffee, and snacks in the neighborhood where Mahfouz set many of his stories; *5 El Badistan Lane.*

• **Kebabgy Oriental Grill:** Middle Eastern–style barbecue is the forte of this romantic alfresco restaurant overlooking the Nile in the Sofitel Cairo; *3 El Thawra Council Street;* sofitel-cairo-nile-elgezirah.com.

• **Felfela:** Opened in 1959, this historic restaurant in downtown Cairo specializes in *ta'ameya,* delicious deep-fried falafel balls made with broad beans that traces its roots to pharaonic times; *15 Hoda Shaarawy Street near Tahrir Square;* felfelaegypt .com.

• **Abou el Sid:** Traditional Egyptian dishes like baba ghanoush, *koshari,* and *molokhiya* flavor the menu at this Zamalek eatery that seems like something out of the movie *Casablanca; 157 26th of July Corridor, Street;* abouelsid.com.

through Greco-Roman times, many of the priceless objects relocated from the old Egyptian Museum in downtown Cairo. The main event is the **King Tut Collection**, nearly every item from the Boy King's tomb displayed in one place for the first time.

The Giza Pyramids are easily combined with a visit to their prototype—the **Step Pyramid of Saqqara**—around a 45-minute drive south of Giza in the Nile Valley. Erected as the last resting place of Pharaoh Djoser about a century before the others, the structure reveals how Egyptian tomb design took a great leap forward by placing ever smaller mastabas atop one another to achieve a pyramidal shape. The nearby **Imhotep Museum** offers an orientation film and artifacts recovered from the sprawling **Saqqara necropolis**. Closer to the river, **Mit Rahina** village offers a small archaeological park with the ruins of **Memphis**, capital of the Old Kingdom during the reign of the pharaohs who built the pyramids.

Memphis was the ancient ancestor of a city founded in A.D. 969 by Fatimid invaders from the Arabian Peninsula. The name Cairo is derived from al-Qāhirah ("The Victorious"), the capital's original name. More than a thousand years later, it's far and away the largest urban area in the Middle East, and until recently passed by Lagos, the most populous metropolis on the African continent, a helter-skelter city of more than 17 million people.

It certainly feels that way in **downtown Cairo** (Wist al-Balad), the jumble of traffic, pedestrians, and high-rise buildings along the right (east) bank of the Nile downstream

Take a tour inside the step pyramid of Djoser in the Saqqara necropolis.

Don't forget to look up at the ornate ceilings of the Muhammad Ali mosque.

from the pyramids. Rather than a legacy of early Islamic days, the downtown area developed in the mid-19th century under the progressive Khedive Ismail, who wanted to modernize Cairo along the lines of a grand European city.

One of his creations was **Tahrir Square**, that swirl of motion beside the Nile where Egyptians have long gathered to celebrate or protest, most recently during the Arab Spring of 2011. Dominating the square's north side is the terra-cotta facade of the old **Egyptian Museum** (opened in 1902). Although many of its treasures have been moved to the new GEM, the building still houses thousands of "leftovers" from the dynastic and Greco-Roman

LAY YOUR HEAD

• **Mena House:** Located right beside the Giza Pyramids, this landmark British colonial abode opened in 1886 and has hosted many luminaries over the years, from Charlie Chaplin to Frank Sinatra; restaurants, bar, spa, swimming pools, golf course; from $134; marriott.com.

• **Villa Belle Époque:** A gorgeous 1920s mansion in the posh Maadi district is the venue for this boutique hotel with a fruit tree–filled garden near the neighborhood's fashionable boutiques and cafés; restaurant, bar, swimming pool, sundeck; from $200; villabelleepoque.com.

• **Windsor Hotel:** Born in 1890s as a Turkish bath for Egypt's royal family, the Windsor still exudes Old World—a time capsule with a roof terrace that overlooks downtown Cairo; restaurant, bar, Egypt's oldest elevator; from $45; windsorcairo.com.

• **Cairo Marriott & Omar Khayyam Casino:** Unveiled in 1869 as a palatial guesthouse for dignitaries attending the opening of Suez Canal that year, this elegant riverside hotel blends neo-Moorish architecture and French Empire decor; restaurants, bars, garden, swimming pool, spa, fitness center; from $129; marriott.com.

The modern Bibliotheca Alexandrina pays homage to the Great Library of Alexandria, though it is thoroughly modern.

eras that couldn't fit into the new museum.

Khedive Ismail also built the enormous **Abdeen Palace**, Egypt's royal court until 1952, when the monarchy was abolished. Many of its 500 rooms are now part of museums dedicated to weapons, presidential gifts, and royal silverware. Another Ismail creation, the superb **Museum of Islamic Art** offers artworks and artifacts from as far away as Morocco and Western China, a world-class collection that safeguards more than 100,000 objects.

Largest of the city's Nile islands, **Gezira** is both a posh residential area and Cairo's modern cultural heart. **Cairo Opera House** (opened in 1988) is home base for the city's symphony, ballet, opera, modern dance, and Arab music troupes, while the nearby **Museum of Egyptian Modern Art** showcases the nation's contemporary painters and sculptors. The pencil-thin (614-feet/187-m) **Cairo Tower** offers panoramic views from its viewing deck, while **El Sawy Culture Wheel** stages a wide variety of modern Egyptian music, dance, and theater performances.

Islamic Cairo lies farther inland, a warren of mosques, markets, and medieval ramparts that formed the heart of the city for 1,000 years. **Al-Azhar Mosque** and its madrassa—one of the oldest and most esteemed centers of learning in the Muslim world—was one of the first things the Fatimids built when they founded the city. **Sultan Hassan Mosque** is even larger, a 14th-century masterpiece renowned for both its immense size and innovative architecture. The two mosques bookend **Al-Azhar Park**, a rambling green space with outdoor cafés and gardens developed in the early 21st century on the site of a 500-year-old mountain of trash.

Commanding a bluff overlooking medieval Cairo is the brawny **Citadel**. The great warrior-sultan Saladin started the massive fortress in 1176; during his war against the crusaders and for the seven centuries that followed, it was Egypt's power base. The most significant of the many structures inside the bastion is the immense silver-domed **Mosque of Muhammad Ali**. In the flatlands below the bastion, the **Gayer-Anderson Museum** preserves a pair of exquisite 17th-century homes including the Beit el-Kiridiliya or "House of the Cretan Woman" (built in 1632). Restored by the museum's namesake, a retired British Army officer, the rooms are decorated with period furnishings, fashion, and artwork.

The renowned **Khan el Khalili** market starts right across the road from the al-Azhar Mosque. Along with the Grand Bazaar in Istanbul, this is the greatest emporium of the Islamic world, a labyrinth of streets, lanes, and narrow alleys where it seems like everything on planet Earth is offered for sale. Sure it's easy to get lost, but half the fun is navigating your way out of the market past hundreds of shops and stalls selling foods, household items, and Egyptian handicrafts. If you only venture once into the Khan el Khalili, make sure it's after dark and along **Al Moez Street** when the facades of the 15th-century **Al-Ashraf Mosque**, the 13th-century **Qalawun complex**, and other Mamluk-style buildings are artfully illuminated.

With an estimated two million Copts, Cairo boasts more Christians than any other Middle Eastern city. The **Coptic Quarter** shelters the city's oldest Christian shrines, including the seventh-century **Hanging Church**, built atop the remains of a Roman gatehouse, and the fourth-century **Church of St. Sergius and Bacchus**, erected on the spot where it's believed the holy family sojourned during their biblical trek to Egypt. The nearby

Family relics and possessions are on display at the Abdeen Palace Museum.

Coptic Museum displays icons, manuscripts, frescoes, and other early Christian artifacts.

As the population grows, Cairo continues to expand into the desert on both sides of the Nile to accommodate its new residents, especially its burgeoning middle class. Located around 40 minutes east of Tahrir Square (if there's not too much traffic),

New Cairo City is the flashiest of these new suburbs, a land of detached homes, golf course resorts, and air-conditioned shopping and entertainment complexes like the **Cairo Festival City Mall**. Yet among all this new is something even older than the pyramids: the **Maadi Petrified Forest Protected Area** and its 35-million-year-old stone trees. ∎

ALTER EGO

Alexandria

If Cairo is the Paris of the Nile, then surely **Alexandria** (or "Alex") is the Marseilles of the eastern Mediterranean, a hardworking port city with an impressive waterfront promenade and delicious seafood.

Egypt's leading metropolis during Greek and Roman eras—when Alexandria was the home of Cleopatra and other Ptolemy rulers—the city safeguards ancient landmarks like Pompey's Pillar, the Greco-Roman ruins of Kom el Dikka, and the creepy Kom el Shoqafa catacombs.

The fabled Library of Alexandria was destroyed during a rebellion against Roman rule. But the city carries on with its erudite heritage with **Bibliotheca Alexandrina**, a stunningly modern building with museums dedicated to science, Islamic manuscripts, Anwar Sadat, and ancient Egypt.

Alexandria's many beaches offer a chance to dip in **the Med**, while day trips along the coast lead to the World War II battlefield at **El Alamein** and the **Aboukir Bay**, where Horatio Nelson defeated Napoleon's fleet in the 1798 Battle of the Nile.

Algiers
Algeria

The onetime haunt of pirates and pashas, Algeria's captivating capital offers a glimpse of North Africa's past and present on a coast blessed with a refreshingly temperate Mediterranean climate.

Wrapped around the broad **Bay of Algiers** near the western end of the Mediterranean, Algeria's capital city has always had an intimate relationship with the sea. Settled by Phoenician mariners and a base for the notorious Barbary pirates, the old harbor is lined with the elegant whitewashed terraces and colonnades that gave the city its French colonial nickname—La Ville Blanche ("The White City").

Among the many landmarks along the waterfront are the 11th-century **Djamaa el Kebir** (Great Mosque)—the city's oldest place of Muslim worship—and the **Palais des Raïs**, a sprawling Ottoman complex converted into an arts and culture center that hosts concerts, art exhibits, and one of the city's best handicraft outlets.

Crawling up the slopes behind the terraces is the city's legendary **Casbah**, a labyrinth of cobblestone lanes and staircases, many of them too narrow for motor traffic. Among its architectural gems are the 17th-century **Ketchaoua Mosque** and the **Palace of the Dey**, where the city's Turkish overlords lived (and the second largest palace in the Ottoman Empire after the Topkapi in Istanbul). Another Ottoman palace provides an atmospheric venue for the **National Museum of Arts & Popular Traditions**.

Directly south of the old town is **Alger Centre**, a bustling downtown area that often feels more like Paris than urban Africa. Perched beneath the office buildings and posh apartments are some of the city's most popular cafés and best restaurants, an eating scene that flaunts French, Mediterranean, and North African cuisine. Turn back the clock with an evening that starts with dinner at **Tantonville Café** and an opera, concert, or dance performance at the Baroque Revival–style **National Theatre of Algeria** across the street.

La Rue Didouche Mourad, the city's main shopping street, provides

THE BIG PICTURE

Founded: fourth century B.C.

Population: 2.7 million

Size: 170 square miles (440 sq km)

Languages: Arabic, French

Currency: Algerian dinar

Time Zone: GMT +1 (CET)

Global Cost of Living Rank: 198

Also Known As: Icosium (Roman), La Ville Blanche (The White City), El Behdja (The Joyful One)

The Ketchaoua Mosque in Casbah is a UNESCO World Heritage site.

Stroll through the open markets for crafts and food on the French colonial side of the city.

an easy-to-follow route across downtown from the French colonial **Grande Poste d'Alger** and its palm-shaded plaza to the **National Museum of Antiquities** with its Roman and early Islamic artifacts. Along the way is radically modern **Sacré-Cœur d'Alger** cathedral, where relics of African saints are safeguarded beneath a soaring concrete dome that symbolizes a passageway to heaven.

Modern Algiers continues along the south side of the bay, and so does the avant-garde architecture. Soaring 302 feet (92 m) into the North African sky, the **Maqam Echahid** (Martyrs' Memorial) commemorates those who gave their lives during Algeria's long struggle for independence from France.

Just behind the towering monument is the **Riad El Feth**, a Western-style shopping mall with a food court, cinema, and performing arts center that presents plays and concerts. The huge green space on the bay-side flatlands below the monument is the lush **Jardin d'Essai du Hamma**, which does double duty as the national botanical garden and the city's finest park.

Given two centuries of French rule, it's not surprising that Algiers boasts another extravagant church. The **Basilica of Notre-Dame d'Afrique** overlooks the sea on the other side of the Casbah from downtown, an exotic-looking neo-Byzantine structure with commanding Mediterranean views from its hilltop location. ■

BACKGROUND CHECK

• The name Algiers drives from an Arabic term for "islands"—a reference to the city's initial location on islands on the Bay of Algiers.

• Spanish novelist Miguel de Cervantes was held for ransom in Algiers for five years (1575 to 1580) after being captured by Barbary pirates—an episode that inspired two of his plays and part of *Don Quixote*.

• Commodore Stephen Decatur led a flotilla of 10 U.S. warships into the Bay of Algiers in 1815 and forced the Dey into a treaty that ended the Barbary Wars.

• The Casbah was a hotbed for the anti-French insurgency during the Algerian Civil War (1954–1962). *The Battle of Algiers* (1966) is considered the definitive statement on the conflict and one of the 500 greatest movies of all time.

Marrakech
Morocco

Born almost 1,000 years ago as a modest oasis town on the edge of the Sahara, Marrakech grew into a superstar of art, architecture, and intellect, a heritage that endures in its crowded and colorful Medina Quarter.

THE BIG PICTURE

Founded: 1062

Population: 1.3 million

Size: 44 square miles (114 sq km)

Languages: Arabic, Berber

Currency: Moroccan dirham

Time Zone: GMT +1 (CET)

Global Cost of Living Rank: Unlisted

Also Known As: Red City, Daughter of the Desert

"Colored cottons hang in the air, Charming cobras in the square, Striped djellabas we can wear," wrote Graham Nash in a 1960s anthem that inspired thousands of young people to hop aboard the *Marrakech Express*. The fabled Moroccan city is no longer a stop on the hippie trail. But more than half a century later, Nash's lyrics continue to reflect the swirl of sights and sounds that unfold each day in the North African metropolis.

This is especially true in the **Djemaa El Fna**, the old town's main square, where Morocco's various ethnic groups mix with visitors from around the globe. The area is a daily street fair that features snake charmers and trained monkeys, Gnawa music troupes and outlandishly dressed water bearers, comedy acts and acrobats enveloped by handicraft stalls and open-air eateries. There's nothing else on planet Earth that better replicates how it must have felt to visit one of those great markets of medieval times. In fact, Djemaa El Fna is so incredibly unique that it inspired UNESCO to create a whole new category called Masterpieces of the Oral and Intangible Heritage of Humanity.

The helter-skelter square is the centerpiece of the city's sprawling **Medina**, its numerous palaces, gardens, mosques, and markets surrounded by medieval walls with enormous gates. With around 200,000 residents, the old town is also a crowded living quarter populated with many of the artisans who make all those wonderful things found in Medina's ancient souks.

A tree-lined pedestrian row called the Avenue Djemaa El Fna runs west from the square past a horse-drawn **caliche station** to Medina's tallest structure and architectural icon—the rectangular, ruddy-colored minaret of **Koutoubia Mosque**. Constructed in the 12th century, the shrine takes its name (Booksellers Mosque) from the fact it was once surrounded by scores of vendors selling the Koran and other texts. Nowadays you're more likely to encounter feral cats that inhabit the mosque's palm-shaded gardens.

Roads radiate from the Djemaa El Fna in nine directions, with one of them (Rue Riad Zitoun el Kedim) running south to the pleasant little **Place des Ferblantiers**, a convenient jumping-off point for explorations of the historic Mellah Jewish Quarter and the royal **Kasbah**.

Once the most lavish residence in

Stroll through the colorful and tropical Majorelle Garden in Marrakech.

The world-famous Djemaa El Fna market in the Medina Quarter is a bustling spot to find traditional crafts, food, and textiles.

North Africa, the 16th-century **El Badi Palace** now lies in ruins around a large reflecting pool. But size alone gives an indication of how it was once the "palace of the incomparable." The nearby **Saadian Tombs**—rediscovered and restored in the early 20th century—contain the graves of powerful Sultan Ahmad al-Mansur and those who ruled Marrakech after his death in 1603. The exquisite **Chamber of the Twelve Columns** in the necropolis is a masterpiece of Moorish architecture.

The area's other regal landmark, **Bahia Palace** was built by the grand viziers of Marrakech over the latter half of the 19th century. Trying to re-create the grandeur of the city's golden age, the meeting halls, private quarters, and harem are decorated with lavish painted woodwork,

intricate stucco patterns, glazed tiles, and Carrara marble.

In olden days, Marrakech nurtured a thriving Jewish community. Many of them left for Israel and elsewhere after World War II, but a handful continue to reside in the **Mellah** neighborhood. **Lazama Synagogue**, the city's oldest and

largest place of Jewish worship, was built in the 1490s, when Jews were expelled from Spain by the Inquisition. The nearby **Miaara Jewish cemetery** is the largest in Morocco.

Medina's main **market area** is just north of the Djemaa El Fna, a warren of narrow alleys and small squares with a variety of souks

---

HIDDEN TREASURES

- **Bab Debbagh:** This Marrakech version of the Fez tannery yard features scores of colorful dye vats in a compound perched in the northeast edge of Medina.

- **Musée de la Palmeraie:** Contemporary Moroccan art is the focus of this museum tucked into the palm groves on the northern edge of Marrakech.

- **Cactus Thiemann:** Succulent fans and photographers flock to this desert garden that features more than 150 varieties of cacti against a backdrop of the Atlas Mountains.

- **Orientalist Museum:** The brand-new museum in Medina features European painters who romanticized the Arab world in 19th- and early 20th-century paintings.

selling carpets, cashmere scarves, ceramics, leather bags, brass lanterns, djellabas, babouche slippers, kilim textiles, spices, essential oils, and so much more. Dickering is fully expected, so hone your bargaining skills before diving into the fray.

Farther north is another cluster of historic structures in an area that was once the city's intellectual heart. When it opened in the 16th century, the **Medersa Ben Youssef** was Morocco's largest school of theology and scientific study, the equivalent of today's universities. Nowadays it's another of Medina's architectural gems, room after room of striking woodwork, ceramic tiles, kufic script, and geometric or floral patterns rendered in stucco.

The medersa is surrounded by other vintage buildings like the 19th-century **Mnebhi Palace**, which now houses the **Museum of Marrakech** and artistic treasures from all around Morocco, and the 12th-century **Almoravid Koubba**, the city's oldest structure and only example of Almoravid architecture.

Until the 20th century, much of the area outside Medina was a patchwork of farm fields, palm groves, and desert. When the French began their occupation in 1912, the city expanded outside the walls into a new town with neighborhoods like **Gueliz** with European-style boulevards, squares, and neatly delineated blocks.

Centered around the big **Place 16 Novembre** roundabout, Gueliz remains the city's administrative center, economic hub, and modern cultural heart. Among its landmarks are the **Marrakech Museum of Art**

The Hassan II Mosque is the largest in all of Africa.

& Culture (MACMA) and the **Gare de Marrakech**, a recently refurbished Moroccan art deco–style train station where real-life versions of the *Marrakech Express* depart for Casablanca, Tangiers, and other cities. Opposite the station, the **Theatre Royal** is much younger than it first appears, a showcase for global music, dance, and drama that was actually built in the early 21st century.

Yet the neighborhood's showpiece is the marvelous **Majorelle Garden**, a botanical wonderland and outdoor art space created by French painter Jacques Majorelle over a period of 40 years after moving to Marrakech in 1923. After several decades of neglect, the garden was purchased

---

and restored by French fashion maestro Yves Saint Laurent and his business partner Pierre Bergé in the 1980s.

The garden revolves around **Villa Oasis**, a Cubist structure painted a unique hue called Majorelle blue. Paths lead through cactus, bamboo, and palm gardens splashed with water features and benches for quiet contemplation. Scattered around the compound are a bookshop, boutique, café, and a small **Berber Museum**. Just up the road is a new **Yves Saint Laurent Museum** and events center with its mementos of the legendary designer.

Another neighborhood that blossomed during French times was the **Hivernage**. Located just south of the railway station, the area once housed the colonial-era upper class but is now home to many of the city's top hotels, restaurants, and nightclubs. If the summer heat becomes too oppressive, duck into the district's **Menara Mall** for a dose of air-conditioned shopping, eating, and even ice-skating.

The urban area continues to expand into the **Palmeraie**, a five-mile (8 km) stretch of groves along the **Tensift River** that harbors around 180,000 palms. Upscale residential areas and half a dozen resort hotels have sprouted amid the palms, complemented by a water park, trendy beach club, and the 27-hole **PalmGolf Marrakech Palmeraie** country club.

Beyond the palm groves is the **High Atlas**, the often snow-topped range that divides coastal Morocco from the **Sahara**. The mountains start to rise about an hour's drive from the city center, while the celebrated desert city of **Ouarzazate** can be reached in around four hours. ∎

Indian Ocean Cities

Port Louis, Mauritius; Antananarivo, Madagascar; Victoria, Seychelles

Scattered along the western edge of the Indian Ocean, the capital cities of three island nations offer a completely different take on the African urban experience.

Old cannons line the promenade at Caudan Waterfront in Port Louis.

PORT LOUIS

Population: 150,000
Size: 18 square miles (47 sq km)

Founded by the Dutch after they came across the uninhabited island off Madagascar, and colonized by successive French and British regimes, Port Louis didn't take on its own personality until after Mauritian independence in 1968.

Le Caudan Waterfront, a redeveloped dockyard area, anchors the city's waterfront with restaurants, craft market, hotel, casino, and the **Blue Penny Museum** (which safeguards two of the world's rarest postal stamps). Farther down the shore is the **Aapravasi Ghat** World Heritage site, where thousands of indentured servants were brought from the Indian subcontinent during British colonial times.

The elegant palm-lined **Place d'Armes** flows between the waterfront and **Government House,** an 18th-century French colonial structure fronted by a stern statue of **Queen Victoria**. The recently renovated **Natural History Museum** offers a comprehensive look at the dodo, the island's giant flightless bird, last sighted in 1662 by Dutch sailors.

Port Louis's **Chinatown** is a warren of shops, shrines, and eateries that dates back to the 1820s when migrants arrived. The **Central Market** blends the sights, smells, and tastes that form multicultural Mauritius.

Visitors should be aware that Port Louis doesn't boast any of the island's famous beaches. The nearest places to soak up a little sun, sea, and sand are **Turtle Bay** on the north side of the capital, or **Albion** and **Flic en Flac** to the south.

ANTANANARIVO

Population: 2.2 million
Size: 70 square miles (181 sq km)

Set in Madagascar's central highlands, "Tana" is the centerpiece of a Malagasy culture that traces its roots to ancestors who migrated across the Indian Ocean around 1,500 years ago.

It's also one of the few cities that existed before the advent of colonial rule, founded in the early 17th century by the Merina dynasty, which reigned over Madagascar until the French invasion and abolition of the monarchy in the 1890s.

The city center revolves around **Lake Anosy** and its island monument to Malagasy troops who died during World War I while serving for France. Farther north is helter-skelter **Analakely Market**, where hundreds of vendors hawk everything imaginable, including Malagasy handicrafts.

The Merina kings and queens ruled from the **Haute-Ville**, the city's highest point, where they built the fortified **Rova Palace** (which safeguards many of the royal tombs) and the nearby **Andafiavaratra Palace**, where royal regalia and other relics are on display.

Farther down the slope, the **Andohalo** neighborhood is home to the neo-Gothic **Cathedral of the Immaculate Conception** (finished in 1890) and the excellent **Musée de la Photographie**, which documents Madagascar history and culture through films and photos.

Gare de Soarano, the city's vintage French colonial train station, has been renovated into an office and retail complex that includes handicraft stalls and the **Pok Pok Madagascar** fashion boutique.

The **Royal Hill of Ambohimanga**, the old Merina capital, occupies a commanding position on a mountaintop 15 miles (24 km) north of downtown Tana. The UNESCO World Heritage site attracts a steady stream of pilgrims to its royal tombs and other sacred sites.

VICTORIA

Population: 27,000
Size: 7.8 square miles (20 sq km)

One of the world's smallest capital cities, Victoria is the hub of a Seychelles archipelago that embraces 150 islands around 900 miles (1,448 km) off the East African coast.

French colonists settling the previously uninhabited island founded the city as L'Établissement in 1778. The city was renamed for Queen Victoria after the British captured Seychelles during the Napoleonic Wars.

Spot ring-tailed lemurs in Antananarivo.

The queen also lends her name to the city's most celebrated landmark, a miniature version of Big Ben called the **Victoria Clock Tower**. The vintage timepiece rises beside the old **Supreme Court** building, which now houses the new **National Museum of History**.

Among the other landmarks of downtown Victoria are the **Immaculate Conception Cathedral**, the bustling **Sir Selwyn Selwyn-Clarke Market**, the chromatic **Sri Navasakthi Vinayagar** Hindu temple, and

the **Kaz Zanana** art gallery inside an early 20th-century Creole home.

But Victoria's main attraction is really **Mont Fleuri Botanical Garden**, a tropical wonderland that features **coco de mer palms**, an **Aldabra giant tortoise** habitat, and tropical birds endemic to the Seychelles.

Morne Seychellois National Park rambles across the jungle-covered mountains behind Victoria. **San Soucis Road** twists and turns its way up the ridge to the jumping-off point for the **Copolia Trail** with its bird's-eye views of the city and harbor.

The road continues to the historic **Mission Lodge**, **SeyTé Tea Factory**, and the west coast of **Mahé Island** and its gorgeous boulder-strewn beaches: **Grand Anse**, **Anse Boileau**, **Anse la Mouche**, **Baie Lazare**, and **Anse Intendance**.

Sainte Anne Marine National Park renders snorkeling, scuba diving, and desert island picnics. The speedy ferries of **Cat Cocos** expedite day trips to equally exotic **Praslin** and **La Digue** islands. ■

Visit the colorful Arul Mihu Navasakthi Vinayagar Temple in the Seychelles.

Middle East &
Central Asia

The Ottoman-era Blue Mosque in Istanbul (p. 192) was constructed between 1609 and 1616.

Dubai
United Arab Emirates

No metropolis on the planet has changed more than Dubai over the past 40 years, a startling transition from an unassuming and largely unknown Arabian seaport into a sprawling, state-of-the-art global city.

Nothing quite prepares visitors for their first encounter with Dubai. It really is the miracle of the Middle East, a city that as late as the 1970s was little more than a desert backwater primarily known for pearling in the Persian Gulf.

Two events determined Dubai's future: the discovery of massive oil and gas reserves in the southern gulf and the formation of the United Arab Emirates from seven sheik-doms, an agreement that created a brand-new nation and set the stage for all of them to step into the modern age.

Dubai did so with far more vigor than the others, transforming itself what seems like overnight into one of the world's most astonishingly modern urban areas, a wonderland of architecture, the region's number one tourism destination, a safe haven in an area renowned for conflict and chaos.

The best place to start any visit to Dubai is looking down on the city from the summit of the **Burj Khalifa**, the world's tallest building. Looking like a giant rocket ship that's ready to blast off into outer space, the structure rises 2,722 feet (830 m) into the emirate sky—nearly twice as high as the Empire

THE BIG PICTURE

Founded: early 18th century

Population: 3.7 million

Size: 580 square miles (1,502 sq km)

Languages: Arabic

Currency: UAE dirham

Time Zone: GMT +4 (UAEST)

Global Cost of Living Rank: 42

Also Known As: City of Gold

State Building. Perched on the 124th floor, the **At the Top** outdoor observation deck offers a "falcon's-eye view" of the city, the Arabian Desert, and Persian Gulf. The building also hosts the world's highest restaurant: **At.Mosphere** on the 122nd floor.

Burj Khalifa anchors a sprawling arts, entertainment, and retail complex that includes the **Dubai Fountain**, which bursts into a sound-and-light show every half hour between 6 and 11 p.m. In addition to loads of shops and restaurants, the adjacent Dubai Mall harbors the **Dubai Aquarium & Underwater Zoo**, which features a saltwater crocodile habitat, a walk-through shark tank, and a special exhibit on the **UAE's Night Creatures**. For those who feel the need to perform an arabesque or camel spin in the Arabian Desert, the mall is also home to the **Dubai Ice Rink**. The state-of-the-art **Dubai Opera**—another structure that looks like something from a distant planet—offers a year-round slate of music, dance, comedy, and drama performances.

Dubai's other engineering marvel is the **Palm Jumeirah**, a colossal palm tree–shaped archipelago created with seven million tons of desert rocks and nearly 3.5 billion cubic feet (100 million m³) of sand

Floor-to-ceiling tanks show endemic species at the Atlantis Hotel in Dubai.

Catch a panoramic view of the remarkable Dubai skyline from the world's largest frame, covered in gold plating.

dredged from the Persian Gulf. The Palm blends thousands of residential units and more than two dozen hotels, including the massive **Atlantis Dubai** resort complex with a 1,500-room hotel, celebrity chef restaurants, multiple beaches and swimming pools, the **Lost Chambers Aquarium**, and **Aquaventure Waterpark**. The adjacent **Helicopter Tour Dubai** offers aerial explorations of the emirate.

Sheikh Zayid Road—the expressway that cuts across the new city from north to south—is flanked by dozens of buildings that are helping to redefine modern architecture. The new **Museum of the Future**, an eccentric oval-shaped building with a huge elliptical void (aka hole) in the middle. The **Dubai Frame** is a gigantic picture frame that doubles as an observation tower and monument to Dubai's urban evolution. Looking through the frame from

north to south, one views olden Dubai along the creek; gazing from the opposite direction, one sees the steel-and-glass towers of modern Dubai with the Burj Khalifa front and center.

The thing that makes the **Mall of the Emirates** special isn't the overall

design but the fact that it features an indoor skiing, snowboarding, and tubing run called **Ski Dubai**. Transitioning from alpine to equatorial, **The Green Planet** is an enormous bio-dome where more than 3,000 plants and animals live in a simulated tropical environment. Shaped

BACKGROUND CHECK

• Dubai was part of Arabia's "pirate coast" until 1820, when eight emirates signed truces with the Royal Navy and became a British protectorate called the Trucial States (the future UAE).

• The forerunner to British Airways (BOAC) refused to serve Dubai until the mid-1960s, because the city didn't have a paved runway; Dubai now boasts the world's largest airport terminal.

• Dubai ruler Sheikh Mohammed bin Rashid Al Maktoum is a highly

successful thoroughbred breeder whose horses have won more than 5,000 races worldwide.

• Don't speed in Dubai: The police force is equipped with the world's fastest police cars, including a Bugatti Veyron that can reach a mind-blowing 253 miles (407 km) an hour.

• Nobody really knows how Dubai got its name or even what it means; theories range from "baby locust" to "they came with lots of money."

Abu Dhabi

Ninety-three miles (150 km) down the coast, **Abu Dhabi** is emerging as a travel destination in its own right thanks to human-made projects on two islands.

Yas Island is all about action: It features the **Yas Marina** auto racing circuit where Abu Dhabi's Formula One Grand Prix is staged each year, as well as the **Ferrari World theme park** and **Warner Bros.**

World motion picture theme park.

Saadiyat Island is all about culture, with the dazzling **Louvre Abu Dhabi,** renowned for its architecture and art exhibits; the **Zayed National Museum** of local natural and human history; and the future **Guggenheim Abu Dhabi.**

But Abu Dhabi's pride and joy is **Sheikh Zayed Grand Mosque.** Visitors of any faith can explore the vast structure via guided tours or audio e-guides.

---

like an origami cube, the structure features a bat cave, sloth exhibit, Australian Walkabout, and other wildlife encounters spread across indoor areas that emulate a tropical forest from floor to canopy.

The old town flanks both sides of **Dubai Creek**, where the city was born in the 18th century as a remote fishing village and pearling port. The creek's eastern bank is dominated by the **Deira Old Souk**, a traditional Middle Eastern market featuring a **Gold Souk**, **Spice Souk**, and other specialized areas spread across a warren of narrow lanes. Lateen-rigged **dhows**, wooden sailing crafts that once ventured across the Indian Ocean, line the bank. Some are still used for cargo, but many have been converted into **sightseeing boats** that cruise the creek and Persian Gulf—an excellent way to get an overall view of the Dubai skyline, especially at sunset or early evening when the entire city glistens.

Water taxis flit across the creek to the west bank, where the tourist-oriented **Bur Dubai Souk** offers rugs, antiques, handicrafts, and Middle Eastern fashion. From the waterfront bazaar it's only a short walk to the **Dubai Museum**, which revolves around traditional everyday life and culture in the emirate prior to the discovery of oil. Galleries are located inside the 18th-century **Al Fahidi Fort** (the city's oldest building) and a modern annex.

The west bank also nurtures Dubai's oldest neighborhoods. **Al Fahidi Historical District** preserves more than 50 historic structures built by early 20th-century Persian merchants, many of them crowned by the wind towers that provided

On a cruise down the River Souq, gaze at the fronts of traditional buildings constructed in Emirati architectural style.

Explore the pristine interior of the Sheikh Zayed Grand Mosque, the largest mosque in the UAE.

natural air-conditioning in the days before mechanical cooling. Teahouses, handicraft shops, and art galleries occupy many of the bygone buildings. **Sheikh Mohammed Centre for Cultural Understanding** offers guided walking tours of the neighborhood and **Jumeirah Mosque** along the waterfront, as well as Bedouin-style cultural meals.

There's more vintage architecture down around the mouth of Dubai Creek, where **Shindagha Historic District** safeguards another slice of old Arabia. Among its landmarks are the **House of Sheikh Khalifa Bin Saeed Al Maktoum** (built in the 1890s as the primary residence of the royal family), the **Saruq Al-Hadid Archaeology Museum,** and the new **Al Shindagha Museum**.

Dubai's ever changing **beach zone** stretches around 14 miles (23 km) between Jumeirah Mosque and **Dubai Marina**, a mix of free public beaches and pay-as-you-play private beach clubs. **Kite Beach** is renowned for its often ideal kitesurfing conditions, but is also popular for water sports such as sea kayaking, paddleboarding, and beach volleyball. Farther down the shore, the sail-shaped **Burj al Arab**, one of the world's most opulent hotels, looms high above **Sunset Beach**. The marina offers big-ticket beach attractions like **Dubai Skydiving**, a huge inflatable water park called **AquaFun** that floats in the gulf, the **Ain Dubai** big wheel on Bluewaters Island, and a totally offbeat dining experience called the **Flying Cup** that

lifts an open-air table 130 feet (40 m) above the shoreline while you dine.

Beyond the emirate's urban zone, the vast **Rub' al Khal** (Empty Quarter) section of the Arabian Desert offers plenty of space for dune bashing in 4x4 vehicles or bikes, camel rides, sandboarding down giant dunes, and Bedouin barbeques.

The desert flats between Dubai and Abu Dhabi host a cluster of theme parks that includes **Legoland Dubai**, the movie-focused **Motiongate Dubai**, and **Bollywood Parks**. Yet there's also solitude: the **Dubai Desert Conservation Reserve**, a national park with 86 square miles (223 sq km) of indigenous flora and fauna including Arabian oryx and gazelle, caracals and sand cats, gerbils and jerboas. ■

Jerusalem
Israel

The adoring focus of three great faiths, Jerusalem is truly a city for the ages, a 5,000-year-old urban area where relics of the biblical, Roman, Byzantine, Arab, Ottoman, and modern eras are literally built atop one another.

THE BIG PICTURE

Founded: 3000 B.C.

Population: 900,000

Size: 90 square miles (233 sq km)

Language: Hebrew

Currency: New Israeli sheqel

Time Zone: GMT +2 and +3 (IST and IDT)

Global Cost of Living Rank: Unlisted

Also Known As: Yerushalayim (Hebrew), Al-Quds (Arabic), Zion (Biblical), The Holy City

There's nothing quite like the Old City in Jerusalem at dawn, right after a summer shower, as dark clouds are rolling away and the first rays of a new day are glistening off the Dome of the Rock.

Rain has left the air clean and crisp, and you can actually catch the aroma of the 2,000-year-old stone all around. Vendors are just starting to roll up the metal doors of their shops and artfully place their wares beside the street. Best of all, tourists and pilgrims haven't appeared from their digs outside the walls to crowd the streets. For a fleeting moment,

it's as if you've got the Old City all to yourself.

But it doesn't last long, because Jerusalem remains one of the globe's most popular destinations. There's all the history, the shopping isn't half bad, and the food is divine. Yet the thing that makes the city so incredibly popular is the fact that it's the lodestone for three great faiths: Christianity, Judaism, and Islam.

Over the centuries, the **Old City** evolved organically into **Christian, Muslim, Jewish**, and **Armenian Quarters**. Though the lines have blurred over the years—the Muslim

Quarter has many Christian shrines, for instance—the areas are still known by their age-old names. The dissection also emphasizes the fact that Jerusalem as a whole is still very much a divided city: Jews in the west and primarily Palestinian Arabs in the east, on either side of the armistice line created in 1949 after the state of Israel came into being.

Imposing stone **ramparts** envelop the old town, erected in the 16th century at the behest of Suleiman the Magnificent during the city's long period of Ottoman rule. Two stretches (totaling around two miles/3.2 km) are open to the public. Elevated walkways provide phenomenal views of both old and new Jerusalem, as well as an excellent orientation to exploring the Old City.

Far and away the most magnificent of the 11 portals that breach the walls is **Damascus Gate** on the old town's north side. Immediately inside the gateway is the **Arab Souk**, a classic Middle Eastern market filled with scores of shops and stalls selling a variety of items for tourists and residents. Animated, aromatic, and absorbing, the bazaar continues along several cobblestone streets into the heart of the Old City and a

The streets of Jerusalem are lined with shops selling souvenirs and local fare.

Icons of Jerusalem include the Western Wall and Dome of the Rock, both significant in multiple religions.

warren of other narrow lanes leading off in the cardinal directions.

Christians tend to make haste for the **Church of the Holy Sepulchre**, which overlays a hilltop called **Calvary**, where it's believed Jesus Christ was crucified, buried, and then resurrected. Although consecrated in A.D. 335, the colossal structure mostly dates from the 11th century after the crusaders captured Jerusalem. However, Jerusalem's best example of medieval European architecture is the **Church of St. Anne** in the Muslim Quarter, revered as the place where the Virgin Mary was born and where Jesus healed a sick man. The **Via Dolorosa** (Stations of the Cross)

LOCAL FLAVOR

• **Machneyuda:** Three celebrity chefs are the brain trust behind a market-to-table menu of modern Israeli dishes with most ingredients sourced from nearby Mahane Yehuda Market; *Beit Ya'akov Street 10, Nachlaot;* machneyuda.co.il.

• **Mona:** Classic Mediterranean cuisine with an Israeli twist in a building that once housed the Bezalel Academy of Arts and Design; *Shmu'el ha-Nagid Street 12, Rehavia;* monarest.co.il.

• **Blue Hall Music:** The cave-like interior of a vintage building sets the scene at a kosher restaurant with full bar and live music; *Yo'el Moshe Salomon Street 12, Downtown;* kikar-hamusica.com/bluehallmusic.

• **Chakra:** Kebabs with tahini and fire-roasted vegetables, and gnocchi with lamb and beef-butter sauce flavor a menu created by chefs Ilan Grossi, Eran Peretz, and Roger Moore, the founding fathers of modern Jerusalem cuisine; *King George Street 41, Mamilla;* chakrarest.com.

• **Moshiko Falafel:** Falafel and shawarma made to order with fresh ingredients in a tiny storefront restaurant; *Ben Yehuda Street 5, Downtown.*

Tel Aviv's harbor of Jaffa leads into the Old City, full of 19th-century homes converted into cafés and art galleries.

starts just down Lion Gate Street from St. Anne's.

The nearby **Temple Mount** is thought to be the site of King Solomon's ancient temple and Herod's Second Temple demolished by the Romans. After the seventh-century Arab conquest of the Holy Land, its Muslim leaders erected the golden **Dome of the Rock** over the traditional spot where the Prophet Muhammad ascended on his night journey and Abraham offered to sacrifice his son Isaac. Temple Mount is open to non-Muslims Monday through Thursday via the **Gate of the Moors** (Mughrabi).

All that remains of Herod's temple is the **Western Wall** (Kotel), a massive limestone retaining wall along the western edge of the Temple Mount. Basically an open-air synagogue, the prayer site is open around the clock throughout the year, but segregated into male and female sectors. Visitors can descend underground to view the **Western Wall Tunnels** along an excavated

A man prays at the Western Wall, also known as the Wailing Wall, one of Judaism's holiest prayer sites.

portion of the barrier along street level as it existed 2,000 years ago. **Jerusalem Archaeological Park** and its **Davidson Center** museum expose Roman, Byzantine, and early Islamic ruins along the Temple Mount's southern wall.

The Jewish Quarter largely revolves around **Hurva Square** and its eponymous synagogue, and a

partially reconstructed Roman-Byzantine colonnade called the **Cardo Maximus**, nowadays lined by upscale art galleries and jewelry shops. Suffused with hanging lamps, golden icons, and the aroma of incense, 12th-century **St. James Cathedral** is the centerpiece of Armenian Jerusalem. Towering above the quarter is the **Tower of David**,

---

ALTER EGO

Tel Aviv

Set along the warm and laid-back Mediterranean coast, **Tel Aviv** offers an entirely different take on the Israeli urban experience.

Sixteen beaches offer a variety of scenery and seashore lifestyles from **Metzitzim** (shallow, good swimming) and **Hilton** (LGBTQ scene, copious water sports) to **Frishman** (volleyball courts, fitness trail) and **Banana** (soft sand, bodysurfing, glitzy bar scene). The **Tayelet coastal promenade** spans nearly four miles (6.4 km) of Tel Aviv coast, lined with cafés, bars, and restaurants catering to hikers and bikers.

The oldest part of the Tel Aviv metropolis, **Jaffa** offers a seaside

version of Jerusalem's Old City, a blend of Jewish, Muslim, and Christian residents and their respective places of worship. Many of the old houses in nearby **Neve Tzedek**, a 19th-century Jewish settlement, have transitioned into stylish cafés, boutiques, and art galleries.

Away from the shore, the **White City** conservation zone boasts the world's largest collection of Bauhaus architecture, more than 4,000 total structures erected in the 1930s and '40s that have been declared a UNESCO World Heritage site.

Tel Aviv's emporiums include the traditional **Shuk HaCarmel** (Carmel Market) and **Nachalat Binyamin**, a trendy twice-weekly pop-up market with more than 200 arts and crafts stalls.

an Ottoman citadel that shelters the **Museum of the History of Jerusalem** and historic evening shows like the **Story of King David**.

Beyond the walls are other sites of historical or biblical importance. Rising in the south is Mount Zion, crowned by the handsome **Dormition Abbey**, **King David's tomb**, and the **Cenacle** (Dining Room) where it's believed the Last Supper took place. Many well-known people, including Holocaust hero **Oskar Schindler**, are buried in the Christian cemeteries on Mount Zion's lower slopes.

Exiting the old town via the **Lions' Gate** leads into the rocky **Kidron Valley** and the **Garden of Gethsemane**, where the **Church of All Nations** arose in the 1920s over the spot where it's thought that Jesus prayed before his arrest by Pontius Pilate. Hovering above the garden is the fabled **Mount of Olives** and its vast **Jewish Cemetery**, where as many as 150,000 people have been laid to rest since around 1000 B.C.

Moving around to the northern outskirts of the old town, the quite good **Rockefeller Archaeological Museum** exhibits decorated human skulls from the Neolithic era; gold, silver, and ivory treasures from the Bronze Age; and thousands of other artifacts unearthed in Palestine during the British Mandate (1919–48). The nearby **Museum on the Seam** (MOTS) takes its name from

The Church of the Holy Sepulchre is where Jesus' body is said to have been anointed before burial.

the old armistice line (or seam) that divided East and West Jerusalem. This self-described "socio-political contemporary art museum" stages often provocative exhibits in a former Israeli Army outpost that still bears the scars of the 1967 Six-Day War.

Busy **Jaffa Road** connects the Old City with **Zion Square**, nexus of modern **downtown Jerusalem**. Running west from the square, the **Ben Yehuda Street** pedestrian zone is flush with restaurants, coffee shops, dessert places, and souvenir stores. Farther up Jaffa Road, **Mahane Yehuda Market** offers an even larger menu of food and beverage outlets, as well as a busy after-dark bar scene.

The hilly **Givat Ram** neighborhood west of downtown harbors many of the nation's most important institutions, including the **Knesset** parliament building, the superlative **Israel Museum** of art and archaeology from around the globe, including a separate **Shrine of the Book** that safeguards the **Dead Sea Scrolls**. More artifacts (and cool scale models) are on tap at the private **Bible Lands Museum**, which specializes in the history and culture of the ancient Near East.

There's another cultural cluster in the upscale **Rehavia-Talbiya** area

The Montefiore Windmill was a functional flour mill, built in 1857.

south of downtown that includes the excellent **Museum of Islamic Art**, which showcases a wide variety of items created across the Muslim world, from carpets and clocks to weapons, fashion, jewelry, and game pieces. Housed inside a bold modern structure with multiple halls and exhibit spaces, the **Jerusalem Theatre** also covers a broad range: stage, screen, visual and performing arts.

Montefiore Windmill spins above historic **Mishkenot Sha'ananim**, established in the 1860s as the first Jewish residential area outside the Old City. A longtime magnet for artists and authors,

the old neighborhood hosts the **Cinémathèque** movie theater (which hosts the annual **Jerusalem Film Festival)** and an Ottoman-era caravansary restored into the eclectic **Khan Theatre**, a well-respected repertoire company that stages five new shows a year.

The **Jerusalem Light Rail** line runs right up the middle of Jaffa Road, curling around to **Mount Herzl**. Paths lead uphill through the **National Cemetery**, where many of Israel's modern leaders are buried, to the **Yad Vashem** memorial and museum for those who perished in the Holocaust. ∎

LAY YOUR HEAD

• **King David Hotel:** Winston Churchill; King George V; Prince Charles; Jordan's King Hussein; American presidents Nixon, Ford, Carter, and Clinton; and celebrities from Elizabeth Taylor to Madonna are among those who've slept at Jerusalem's most illustrious hotel; restaurants, bars, pool, spa, fitness center; from $504; danhotels.com.

• **Mamilla:** Stylish modern comfort in ancient surroundings are the hallmark of this five-star hotel just outside the Old City; rooftop restaurant and bar, pool, spa, fitness center; from $410; mamillahotel.com.

• **Hotel Alegra:** Romantic villa decorated with contemporary Middle Eastern motifs located in tranquil Ein Kerem on the western edge of

Jerusalem; restaurant, pool, terrace; from $238; hotelalegra.com.

• **Capsule Space-X:** This budget accommodation in the city center offers 14 modern but cozy capsules stacked two high that can be closed for privacy; shared bathrooms, Wi-Fi, air conditioning; from $50; capsule-space-x.thejerusalemhotels.com/en.

Istanbul
Turkey

Poised on the cusp of Asia and Europe, Istanbul has lived through 2,000 years of caesars and sultans, all the while distilling aspects of East and West into an urban brew that preserves the city's storied past and plunges boldly into the 21st century.

THE BIG PICTURE

Founded: ca 660 B.C.

Population: 13.8 million

Size: 525 square miles (1,360 sq km)

Language: Turkish

Currency: Turkish lira

Time Zone: GMT +3 (TRT)

Global Cost of Living Rank: 173

Also Known As: Byzantium, Constantinople, Nova Roma (New Rome), Kostantiniyye (Ottoman)

More than any other global city, Istanbul deserves to be called a crossroads. Consider the fact that it's the only metropolis that spans two continents. And it doesn't do so lightly: Roughly two-thirds of the population live on the European side, the remainder on the Asian shore.

From the ancient Greeks, Romans, and Persians to the Byzantines, Arabs, crusaders, Venetians, Ottomans, and modern Turks, many of the world's great civilizations have left their mark on Istanbul. Sometimes it was just fleeting, crossing from one continent to another. Others stayed for centuries and occasionally changed the city's name to something completely different (does any other major city have so many names?).

All of these factors merge in a modern metropolis that offers the globe's largest historic marketplaces, one of the greatest churches ever erected and one of the most incredible mosques, a cuisine that draws on all the great food cultures of the Middle East and Eastern Mediterranean, and a populace that rightly views their hometown as one of the urban superstars of the 21st century.

None of this would be possible without the **Bosporus**, the storied waterway that divides a city, a nation, and two continents. Stretching between the **Black Sea** and the **Sea of Marmara**, it's only 19 miles (31 km) long. Oh, but the stories it could tell—8,000 years of people battling to possess the strategic passage and cultivating civilization along its shore. Nowadays it's crossed by three great spans—including the monumental **15 July Martyrs Bridge** near the city center—as well as the **Avrasya** (Eurasia) vehicle tunnel and **Marmaray** rail tunnel.

Yet the best way to get up close and personal with the Bosporus is a means that has been around for thousands of years: ferries. In addition to regular commuter service, **Şehir Hatları** (City Boat Lines) offers scenic cruises along the waterway between the **Eminönü** waterfront in the city center and **Anadolu Kavağı** near the strait's confluence with the Black Sea. **Turyol**, another local ferry line, renders regular service to the scenic **Princes' Islands** in the Sea of Marmara.

Besides the obvious advantages

Catch a performance by traditional whirling dervishes from the Sufi Mevlevi Order.

Take a cruise in Golden Horn Bay to see the city from the water, including the Golden Horn Metro and the Suleymaniye Mosque.

like fresh air, sunshine, and hassle-free travel, ferries are also one of the best ways to view many of the city's historical landmarks and architectural treasures—in particular, the *yali* **mansions** along the Bosporus constructed by Ottoman-era grand viziers and wealthy merchants. The oldest date from the 17th and 18th centuries, while the most extravagant count among the world's most expensive homes—like the **Erbilgin Yalısı** in Yeniköy, once valued at $100 million.

Latter-day sultans also built their homes along the Bosporus: 19th-century pleasure domes like the extravagant 285-room **Dolmabahçe Palace** and a slightly more modest summer residence called the **Beylerbeyi Palace**, which boasts a waterfront bathing pavilion reserved for ladies of the harem.

Ground zero for Bosporus ferry services is **Sultanahmet** in the city center. Surrounded by water on three sides, the hilly peninsula was Istanbul's birthplace and power center until the 20th century, when the city began its rapid outward expansion. Most neighborhoods would be satisfied with a single global treasure. But Sultanahmet boasts three of the most incredible structures ever built.

Hagia Sophia started life as a Byzantine cathedral. Upon its completion in 537, it was the world's largest building of any kind. The Ottomans converted the massive structure into a mosque after they

captured the city in 1453. It became a secular museum in 1935 but was recently converted back into a mosque. Visitors can still tour the remarkable building, which reflects the pinnacle of Byzantine and Ottoman design. Hagia Sophia is especially rich in mosaics that feature both Islamic and Christian motifs.

Although it looks similar, the nearby **Sultan Ahmed Mosque** (Blue Mosque) is much younger, erected in the early 17th century at the behest of Sultan Ahmed I. A masterpiece of Ottoman Islamic design, its huge central copula is surrounded by a dozen lesser domes and six missile-like minarets, the interior garnished with thousands of gold, blue, and white glazed tiles and 200 stained-glass windows.

Fronting the Blue Mosque is tree-lined **Sultanahmet Square**, its elongated footprint determined by the shape of the Roman **Hippodrome** that once occupied the spot. The chariot races are long gone, but the square still boasts the **Egyptian obelisks** that charioteers once sped around. Arrayed around the square are restaurants, small hotels, park areas, and the **Museum of Turkish & Islamic Arts**.

The last magnum opus is **Topkapi Palace**, the primary residence of Ottoman sultans, their families, and harems for nearly 400 years. Construction kicked off just six years after the Ottoman conquest and continued for centuries. Arrayed around four courtyards, the palace features many highlights, including a 300-room **Harem** where hundreds of concubines and eunuchs once

Marvel at the ornate interiors of the Hagia Sophia, which also has artifacts on display.

lived, and the jewel-encrusted swords, daggers, and other royal trinkets in the **Imperial Treasury**. Almost lost amid all this splendor are the bravura views of the Bosporus from the palace walls.

More treasures await in the **Istanbul Archaeology Museums** beside the Topkapi, which harbor artifacts from various ages discovered throughout Turkey. Neighboring **Gülhane Park** offers the best green escape in the city center, a great place to rest and recuperate after a foray into the **Grand Bazaar**. Offering just about everything under the sun, the world's most renowned covered market shelters 4,000 shops along more than 60 lanes, streets, and alleys.

Down by the ferry docks is the aromatic **Mısır Çarşısı** (Egyptian Bazaar), which specializes in spices. The famous **Galata Bridge**—its lower decks reserved for open-air restaurants—spans a branch of the Bosporus called the **Golden Horn**.

On the other side of the bridge is **Beyoğlu,** the first area settled outside the old town and now Istanbul's focal point for dining, drinking, and dancing the night away. The area's creative juices flow through outlets like the **Istanbul Modern** museum of art, the **Pera Museum** with its classic Orientalist paintings, and the far-out **Museum of Innocence**, which revolves around fictional characters created by Turkish novelist Orhan Pamuk.

Those skyscrapers you see in the distance are **Levent** and **Maslak**, two new business, entertainment, and shopping districts. If you can possibly tear yourself away from the old Istanbul, these modern neighborhoods represent the dynamic new Turkey that has sprouted since the 1980s. The observation deck atop 54-story **Istanbul Sapphire** offers the city's highest vantage point—and the only place where you can see the entire length of the Bosporus in a single swing of your head. ■

URBAN RELICS

Troy and Ephesus

Long thought to be a legend rather than a real city, the ruins of **Troy** were discovered in 1822 by Scottish journalist Charles Maclaren. Two amateur archaeologists—British diplomat Frank Calvert and German businessman Heinrich Schliemann—began excavating the site in the 1860s, work that continues in the 21st century.

Located about a five-hour drive from Istanbul near the confluence of the Dardanelles strait and the Aegean Sea, Troy Historical National Park offers ruins uncovered over the past 200 years, a replica Trojan Horse, and an excellent new museum that displays a collection of ancient artifacts from around western Turkey.

Far more impressive ruins are found farther down the Aegean coast at **Ephesus**. Founded in the 10th century B.C. by Greek colonists from Attica and Ionia, the city flourished during the Roman Empire when it became a major trading city with perhaps a quarter of a million people.

The elegant two-story facade of the Library of Celsus highlights the modern-day Ephesus archaeological park, which also includes various temples, gateways, and fountains, an agora market area, two outdoor theaters, and the foundation of the Temple of Artemis, one of the Seven Wonders of the Ancient World.

Tbilisi
Georgia

Few places live up to the crossroads label as well as Tbilisi, a city that sits on the brink of both Europe and Asia and combines aspects of both into an urban landscape complemented by the extraordinary natural beauty of the Caucasus Mountains.

THE BIG PICTURE

Founded: A.D. 455

Population: 1.1 million

Size: 95 square miles (246 sq km)

Language: Georgian

Currency: Georgian lari

Time Zone: GMT +4 (GT)

Global Cost of Living Rank: 207

Also Known As: Tiflis, Iveria (ancient)

Founded more than 1,500 years ago along a strand of the Silk Road through the Caucasus region, Tbilisi has long been influenced by East and West, a heritage that visitors can experience with all their senses.

You can hear the city's crossroads legacy in traditional music that combines early Christian and Middle Eastern influences, you can taste and smell it in a wondrous Georgian cuisine that blends Persian and Mediterranean elements, and you can see and touch it in the architecture of **Dzveli Tbilisi** (Old Tbilisi) in the city center.

Many of the homes are arranged around **"Italian courtyards"** adorned with balconies, bridges, wooden arcades, and spiral staircases. Their capricious design was inspired by Silk Road caravanserai rather than anything Italian, although the courtyard lifestyle—families and neighbors of all generations gathering to eat, drink, talk, sing, and play—is something that Tbilisi residents have relished for centuries.

Some of the courtyards have been converted into boutique hotels and home stays in recent years, but the best way to discover them is a guided stroll with the likes of **Tbilisi Free Walking Tours**. In addition to their gratis tour, the company offers paid themed tours that revolve around food, wine, photography, and Soviet-era relics.

Or you can strike off on your own. Old Tbilisi sprawls along both sides of the **Mtkvari River.** Among its focal points are some of the Christian world's oldest churches—sixth-century **Anchiskhati Basilica** and seventh-century **Sioni Cathedral**—and the **Abanotubani** ("Bath House") district where sulfur hot springs spas like **Chreli Abano** offer private soaking rooms, massage, peeling, and even a red wine treatment.

Yet everything in the old town isn't ancient. The steel-and-glass **Bridge of Peace** leaps the river to **Rike Park** and a futuristic **Rike Concert Hall & Exhibition Center** in two giant metallic tubular structures.

Rising behind the old town's jumble of red-tiled roofs is **Sololaki hill** and the largely ruined **Narikala Fortress**, a medieval citadel constructed over centuries by Georgia's various rulers. **St. Nicholas Church** is the only part of the fort that's been

Domes cover historic sulfur-rich baths, once built for royalty, at traditional bathhouses.

Red roofs mark many of the buildings in Tbilisi's skyline.

rebuilt. Although you can hike to Narikala, the most scenic passage is a modern **aerial tramway** from Rike Park that flies high above the river and old town. The colossal Soviet-era **Kartlis Deda** (Mother of Georgia) statue, the **National Botanical Garden**, and **Zipin Tbilisi** outdoor adventure circuit are also accessed via the tramway.

That other peak looming over the city's western edge is **Mount Mtatsminda**. Declared a public park by Soviet authorities in the 1930s, Tbilisi's highest point (2,500 feet/ 762 m) features a roller coaster, Ferris wheel, wax museum, and other family-oriented attractions. The **Tbilisi Funicular** runs up and down the mountain from **Vilnius Square** in the old town.

Freedom Square and its golden **St. George Statue** mark the southern end of the "new" old town created during the 19th century by the Russian imperial authorities and expanded during Soviet days. Many of the city's government and cultural cornerstones are located along tree-lined **Rustaveli Avenue**, including the **Georgian National Museum**, **Parliament of Georgia**, **Rustaveli Theatre** (renowned for its Shakespeare productions), and the Moorish Revival–style **Opera and Ballet Theatre**. ■

HIDDEN TREASURES

• **Mtskheta:** Located about 12 miles (20 km) north of Tbilisi, Mtskheta was the capital of Georgia's ancient Kingdom of Iberia between the third century B.C. and fifth century A.D. Among its relics are Svetitskhoveli Cathedral, Jvari Monastery, and Samtavro Necropolis.

• **Tbilisi National Park:** Georgia's oldest national park sprawls across the rugged Saguramo Range north of the capital and counts lynx, deer, and bear among its wildlife inhabitants. The park offers three developed hiking trails and the scenic Tbilisi-Tianeti Road.

• **Rezo Gabriadze marionette theater:** With plays ranging from romantic comedies to war stories, this is puppetry for grown-ups rather than toddlers. Tucked down a back street in the old town, Rezo's theater is also a feast for the eyes (and camera).

Samarkand
Uzbekistan

One of the most fabled cities of the ancient world preserves its Silk Road legacy with architectural riches and hospitality that harken back to its caravan days.

THE BIG PICTURE

Founded: ca seventh–eighth century B.C.

Population: 580,000

Size: 195 square miles (505 sq km)

Languages: Uzbek

Currency: Uzbekistani so'm

Time Zone: GMT +5 (UZT)

Global Cost of Living Rank: Unlisted

Also Known As: Marakanda (ancient Greek)

"Everything I have heard about Samarkand is true," declared Alexander the Great upon visiting the city, "except it is even more beautiful that I had imagined." A sentiment that's just as valid today as it was in the fourth century B.C. when the city was already a lodestone along the Silk Road.

For more than 2,000 years, Samarkand has been the cultural crossroads of Central Asia, renowned for its scholars, merchants, and artisans, and even more for its architectural grandeur.

Restored to their former glory in modern times, the city's once ruined mosques, madrassas, and mausoleums were added to UNESCO's World Heritage list in 2001.

With its blue domes, colossal archways, and elaborate geometric motifs, the **Registan** is a masterpiece of Islamic design. Construction started in the early 15th century after Timur (Tamerlane) made Samarkand the capital of his vast Central Asian empire. By the time it was finished in the 17th century, the Registan comprised a large square

flanked by three massive colleges—**Ulugh Beg Madrasah**, **Sher-Dor Madrasah**, and **Tilya-Kori Madrasah**—where subjects like theology, philosophy, astronomy, and mathematics were taught. The square is especially attractive at sunrise and sunset, and during the evening when the elegant facades are fully illuminated.

After amassing an empire that stretched from Syria and Turkey to northern India—via bloody conquest that earned him the sobriquet "Scourge of God"—Timur died in 1404 and was buried at **Gūr-e Amīr mausoleum**. The towering facade and dome display their own exquisite workmanship, while Timur's stone tomb is the source of legends that promise a deadly curse on whoever dares to tamper with it.

The third of Samarkand's architectural triptych is the **Shah-i-Zinda**, a name that translates to "Living King"—ironic because the complex contains the tombs of numerous deceased rulers. Erected between the 11th and 19th centuries, these hillside mausoleums boast what some experts hail as the finest tile work in the Islamic world.

Though the glorious past is

Traditional textiles are displayed at the Samarkand-Bukhara Silk Carpets workshop.

Historic Registan Square was the heart of the ancient city of Samarkand during the Timurid Empire.

undoubtedly the main attraction, contemporary Samarkand offers its own charms. The modern burg (sometimes called the "Russian city") revolves around the **Markaziy Bog'i** (Central Park) with its fountains, statues, and flower beds. An old whitewashed **Russian church** in the park is now an art gallery, and scattered around the fringe are restaurants, teahouses, and a beer garden.

Although little known beyond the region, **Uzbeki cuisine** is a succulent blend of Persian and Turkish influences with rice *plov* (pilaf), tandoor-cooked meats, and *manti* (dumplings) among the signature dishes. Uzbeki cuisine is the specialty at both **Karimbek** and **Besh Chinor**, the latter located in an old Soviet-era mansion.

After bisecting the park, **Mustaqillik Street** runs north through

modern Samarkand to Soviet-era **GUM Department Store** and the **Teatr Istoricheskogo Kostyuma**, which stages a nightly "El Merosi" cultural show that features traditional Uzbeki music, dance, and fashion.

Samarkand is justly famous for its **silk carpets** and other beautiful and detailed handicrafts. The Registan complex is surrounded by dozens of rug shops and souvenir stalls, with bartering as part of the buying process. ∎

HIDDEN TREASURES

• **Khovrenko Winery:** Founded in 1868, the city's oldest winery offers a small museum and a tasting room that offers samples of their cognacs and sweet dessert wines.

• **Gumbaz Synagogue:** The lovely 19th-century shul is still an active place of worship for the city's small Jewish community.

• **Meros:** Handmade stationery is the forte of this workshop in Koni-Gil village on the outskirts of

Samarkand, which revived the ancient art with help from UNESCO.

• **Afrosiyob Bullet Train:** Take a day trip to Tashkent on Central Asia's fastest train—a Spanish-made Talgo 250 with a top speed of 155 miles (250 km) an hour.

• **Pulsar Brewery:** Founded in the 1880s, Samarkand's historic suds factory makes Czech-style beer that visitors can sip at the adjacent Bochka Pivo Bar.

Haifa

Israel

Wedged between Mount Carmel and the Mediterranean Sea, Israel's third largest city complements its busy harbor and high-tech might with a multicultural vibe that spreads across age-old neighborhoods and stylish modern suburbs.

The glimmering sea on one side and a huge, forested mountain on the other, Haifa flaunts one of the most attractive settings in the entire Middle East. But the Israeli city is more than just a pretty face.

For hundreds of years, Haifa was the world's gateway to the Holy Land, a bustling seaport that welcomed most of the country's Jewish migrants until the advent of mass jet travel in the 1960s. And it's one of the prime engines of the Israeli economy. "Haifa works, Jerusalem prays, and Tel Aviv plays," goes an old adage—which is only partially true, because Haifa knows how to let its hair down, too.

Although the majority of the populace is Jewish, Haifa also boasts a good many Christians and Muslims, as well as Druze and Bahá'í residents, an ethnic diversity expressed in neighborhoods like **Wadi Nisnas**, where Muslims and Christians live, work, and pray in close proximity at **St. John's Church** and **Al Jarina Mosque**.

In addition to its many cafés, bars, and music clubs, Wadi Nisnas also functions as the city's cultural hub, home to the **Haifa Museum of**

THE BIG PICTURE

Founded: first century A.D.

Population: 1.1 million

Size: 55 square miles (142 sq km)

Language: Hebrew, Arabic

Currency: New Israeli shekel

Time Zone: GMT +2 and +3 (IST and IDT)

Global Cost of Living Rank: Unlisted

Also Known As: Pearl of the North, Shikmona (ancient)

Art and the **Beit HaGefen Arab–Jewish Cultural Center**, which offers tours, workshops, and encounters that reflect the city's multicultural makeup.

Haifa's compact (and easily walkable) old town boasts other intriguing areas. Founded in the 1920s, **Hadar** grew into the nexus of local government, education, and commerce during the British Mandate era. **Haifa City Hall** and the **Haifa Municipal Theater** anchor the area, along with **Talpiot Market**, where fruit, vegetables, and other edibles are on display inside a vintage Bauhaus market pavilion. Hadar offers several spots to board the **Carmelit underground funicular railway**, which chugs up the mountainside to Carmel Center.

But Hadar's pride and joy is **Madatech**, the national museum of science, technology, and space. The museum features permanent exhibits on a wide range of topics from green energy and Leonardo da Vinci to puzzles, magic, aviation, and a gallery on local birds called **Between Mount Carmel and the Sea**. Two palm trees in front of the main entrance were planted by Albert Einstein and his wife during a visit in 1923. Looking

Take a dip in the Mediterranean Sea from the sandy shores of Glashanim Beach.

From the top of Mount Carmel, take in the cityscape and port of Haifa.

something like a cross between a crusader castle and a Turkish palace, the museum building was constructed during the final years of Ottoman rule over Palestine. For many years it was home to Technion, a national university of technology that's often called the "MIT of Israel." When Technion moved to a large modern campus in 1983, Madatech moved into its old digs. The university's David and Janet Polak Visitors Center offers exhibits and guided tours.

The **German Colony** neighborhood—founded by 19th-century Templar Protestants who were largely expelled from Palestine during World War II because of Nazi sympathies—is another magnet for great food and drink. Local history comes alive at the **Haifa City Museum**, located in old stone buildings that once served as a Templar school and community center.

Beyond the harbor is **Bat Galim**

("Daughter of the Waves"), a seaside suburb with sandy strands and a shoreline promenade, as well as the **National Maritime Museum** and the **Clandestine Immigration &**

Naval Museum that tells the fascinating story of how a ragtag armada smuggled thousands of Jews into Palestine during the British Mandate. On the hillside behind the

URBAN RELIC

Caesarea National Park

Located about halfway between Haifa and Tel Aviv, **Caesarea National Park** shelters the impressive ruins of a seaside city founded by Herod the Great in the first century A.D. that grew into the Roman and Byzantine capital of Judea province. The archaeological park also boasts early Arab, crusader, and Ottoman landmarks.

Among its many vestiges of the past are a Roman theater, hippodrome, double aqueduct, bathhouse, harbor warehouses, and the Reef Palace, as well as crusader walls, a Byzantine church, and a Bosnian mosque.

Largely used to entertain visiting

sailors, the second-century outdoor theater was built by order of Emperor Vespasian. Some of the original seats and marble decorations are still visible. The 10,000-seat hippodrome was used for chariot races and gladiatorial combat.

In addition to a visitors center, the national park offers guided daytime and candlelight tours, restaurants and shops, and three multimedia attractions: the Caesarea Experience, Time Tower, and the interactive Stars of Caesarea, which affords visitors an opportunity to virtually meet historical figures from the city's past. Much of the excavation was carried out in the 1950s and '60s but is ongoing as more remains are discovered.

museums is **Elijah's Cave**, believed to be the place where the Old Testament prophet once took shelter.

Haifa Aerial Cable Car carries passengers on a five-minute journey from the beach area to **Stella Maris Monastery** on the side of Mount Carmel. Erected in the 1830s, the church is renowned for its mural-covered ceiling, intricate stained glass, and its own version of Elijah's Cave in a grotto beneath the nave. Across the street, **Stella Maris Observation Deck** affords awesome views along the Haifa coast.

From the monastery, it's just a five-minute drive along Tchernikhovski Street to the **Bahá'í Gardens**, which climb Mount Carmel as a series of 19 terraces with ever increasing panoramic views of the city and Mediterranean coast. Halfway up is the domed **Shrine of the Báb**, where the 18th-century Persian prophet who inspired the Bahá'í faith was laid to rest.

Above the gardens is the upscale **Carmel Center** (Merkaz) neighborhood where many of the city's best restaurants and hotels are located. It's also celebrated for bakeries like **Gal's** and **Shemo** that help stoke Haifa's reputation as the nation's pastry capital. The uppermost station of the Carmelit funicular disgorges passengers beside **Gan Ha'Em** park and **Haifa Zoo**, which offers a botanical garden and archaeological museum.

Down the road is **Mané-Katz Museum** in the villa where the Ukrainian-French artist lived and painted until his death in 1962. And then something completely unexpected—the **Tikotin Museum of**

Terraced gardens and the golden shrine of Báb mark the hillside of the port city of Haifa.

Japanese Art—which revolves around one of the foremost collections of Japanese traditional creativity outside of Japan. Tikotin offers samurai swords, woodblock prints, netsuke carvings, painted screens, and even a Zen garden.

Moriah Street and Abba Khoushy Avenue (Route 672) climb higher up the mountainside to the lofty campus of **Haifa University** and the **Hecht Museum** of art and archaeology, which displays an incredible range of items from the fifth century B.C. **Ma'agan Mikhael shipwreck** to paintings by Van Gogh, Monet, and Modigliani. Thirty-story **Eshkol Tower**, the university's highest building, is topped by an observation deck with the highest views across Haifa.

Bordering the campus is **Mount Carmel National Park**, which sprawls across the summit of the legendary peak and down its southern slopes. Israel's largest national park renders plenty of opportunity for hiking, biking, and camping in a wilderness that safeguards Israeli

common oak, Jerusalem pine woodland, and 670 other plant species, as well as indigenous wildlife and more than 250 archaeological sites.

The park's **Hai-Bar Carmel Nature Reserve** breeds captive rare and endangered species like the Persian fallow deer, griffon vultures, and wild sheep for possible reintroduction into the wild. Nearby **Nesher Park** offers short hikes to a pair of pedestrian-only swinging bridges across a deep gully. The park also features four dedicated mountain biking routes and a 16-mile (26 km) all-terrain-vehicle route called the **Mount Carmel Trail**.

An island of urbanization in the middle of the national park holds two Druze villages: **Isfiya** and **Daliyat al-Carmel**. Both settlements offer outlets for traditional Druze foods and handicrafts, while the Carmel-Druze Heritage Center in Isfiya showcases the culture of an Arabic-speaking minority group with a faith that draws on diverse Mediterranean and Middle Eastern philosophies. ■

LAY YOUR HEAD

• **The Schumacher Hotel:** This 40-room boutique hotel in a historic Templar building overlooks the Baha'i Gardens and urban Haifa; adults only, spa, wine at check in; from $201; theschumacher.co.il.

• **Carmel Forest Spa Resort:** Israel's largest health spa unfolds as an adult-only oasis in the middle of Mount Carmel National Park; restaurant, bars, pool, spa, fitness center, yoga; from $410; isrotel.com.

• **Dan Carmel Haifa:** Located about halfway up the city side of Mount Carmel, this small high-rise hotel offers beautiful gardens, bay views, and quick access to the Bahá'í

Gardens and Gan Ha'Em park; restaurants, bars, pool, gym, spa; from $240; danhotels.com.

• **Templers Hotel:** Boutique digs in the German Colony in an 1870s building with minimalist but comfy decor about halfway between Bahá'í Gardens and the Haifa City Museum; restaurant, bar, massages, some kitchenettes; from $253; goldencrown.co.il.

• **Bat Galim Boutique Hotel:** Relaxed beach vibe and walking distance to the shore in the seaside Bat Galim district; continental breakfast, garden; from $108; batgalim-boutique-hotel.co.il.

Amman

Jordan

One of the fastest-growing cities in the Middle East, Jordan's capital buzzes with new highways and high-rise buildings, yet retains much of its Arabian mystique.

Belying its modern steel-and-glass facade, Amman is actually one of the world's oldest urban areas, with a history that stretches back more than 9,000 years.

Artifacts from the Stone Age—and all the other epochs of local history—are on display inside the excellent **Jordan Museum**, part of a new municipal complex in the Ras al-Ain neighborhood that also includes Amman's City Hall. Among the museum's many treasures are pages from the Dead Sea Scrolls and Neolithic statues from Ayn Ghazal.

Amman's most impressive antiquities are just up Quraysh Street from the national museum, a cluster of well-preserved Roman ruins that includes a large **amphitheater**, the smaller **Odeon** theater, and the **Nymphaeum** fountain arrayed around the **Hashemite Plaza**.

Looming high above the amphitheater is the sprawling **Amman Citadel** (Jabal al Qala'a), a stronghold from Assyrian and Babylonian days right up through World War I, when the British fought the Ottoman Turks for control of the hilltop bastion. Among the many structures of the Citadel are the Roman **Temple of Hercules**, the sixth-century **Byzantine Church**, and the **Umayyad Palace** constructed by

THE BIG PICTURE

Founded: 7250 B.C.

Population: 4 million

Size: 650 square miles (1,680 sq km)

Language: Arabic

Currency: Jordanian dinar

Time Zone: GMT +2 or +3 (EET or EEST)

Global Cost of Living Rank: 94

Also Known As: Rabbath Ammon (biblical), Philadelphia (classical)

the Islamic caliphate that ruled Amman in the eighth century.

Old Arabia endures in **Al Balad**, the city's crowded downtown district, which huddles around the confluence of King Hussein, King Faisal, and Sha'aban 9 Streets in another of the valleys below the Citadel. A bustling market area since Hellenistic times (when Amman was called Philadelphia), Al Balad is a feast for all the senses—an aromatic, appetizing, and melodious blend of Arab clothing boutiques, spice stores, jewelry shops, coffeehouses, shawarma stalls, and legendary dessert places like **Habibah Sweets**.

Tucked among all the shops and stalls is **Duke's Diwan**, a private museum at 12 King Faisal Street that preserves one of Amman's oldest homes. Built in 1924, the rooms are furnished and decorated much as they would have been nearly 100 years ago.

Nearby **Rainbow Street** reflects the modern side of Amman, a hip strip that includes rooftop restaurants and gastro pubs, burger joints and ice cream parlors, as well as the historic **Rainbow Art House Theatre** (opened 1957) with its ongoing slate of documentaries and feature films

See classic cars, including a 1952 Rover P4-75, at the Royal Automobile Museum.

Roman ruins take center stage in the ancient citadel park near the city center of Amman.

(including occasional screenings of *Lawrence of Arabia*).

For those into motorized movement, Amman offers two substantial vehicle collections. The **Royal Automobile Museum** features a wide variety of wheels used by Jordan's monarchs, from a 1916 Cadillac and vintage Cord convertible to the Harley-Davidson motorcycles that the late King Hussein piloted through the desert. From a military perspective, the **Royal Tank Museum** exhibits more than 120 armored vehicles from World War I through modern times.

Several of Jordan's other iconic sights are a quick half-day trip from Amman. An hour's drive north of the central city, the Greco-Roman ruins of **Jerash** are among the largest and best preserved in the Mediterranean world. A similar distance south of the city is **Mount Nebo**, where Moses died and was buried after leading the Israelites to the Land of Canaan.

Just west of the city, a UNESCO World Heritage site at **Bethany** enshrines a tranquil spot beside the Jordan River, where it's believed that John the Baptist baptized Jesus Christ. Bethany is easily combined with a dip in the **Dead Sea** at one of the waterfront resorts in **Sweimeh**. ∎

URBAN RELIC

Petra

Chiseled into a maze of desert canyons between Amman and the Red Sea, **Petra** was founded by the Nabataeans in the fourth century B.C. as a caravan stop on busy trade routes between Arabia, Egypt, Asia Minor, and Mesopotamia.

The "rose-red city" reached a dazzling peak during Roman times as a hub of their military might, political control, and commercial stretch in the Middle East. By the end of the Byzantine era, Petra was largely abandoned and forgotten by the outside world until its rediscovery by European travelers in the 19th century.

Named one of the New Seven Wonders of the World in 2007, Petra has morphed in modern times from a ghost town into one of the world's most iconic archaeological sites. Although the exquisite Treasury (Al Khazneh) is the city's most photographed landmark, scores of other structures were carved into the canyon walls, including Nabataean tombs that Romans turned into lavish underground villas.

Nicosia
Cyprus

The only global city that functions as the capital of two entirely different territories, Nicosia is the largest city of both the Republic of Cyprus and the Turkish Republic of Northern Cyprus.

The infamous **Green Line**—a United Nations buffer zone that divides Nicosia—was established in 1963, after intercommunal violence between the island's Greek and Turkish ethnic groups racked Cyprus. A decade later, the no-man's-land became a permanent fixture when Turkey invaded Cyprus, capturing around 38 percent of the land area, including the northern part of Nicosia. The border reopened in 2003, allowing visitors to explore both sides of the divided city.

South of the Green Line, the Greek Cypriot city revolves around a tightly packed **Old Town** surrounded by stout **Venetian Walls** that were erected during the Renaissance when the Italian city-state controlled the island. Renowned for its Greek Orthodox icons, the 15th-century **Panagia Chrysaliniotissa** is the city's oldest church, while 17th-century **St. John's Cathedral** is known for its fine frescoes.

Three of Nicosia's best museums cluster around the cathedral, including the **Byzantine Museum** in the

THE BIG PICTURE

Founded: seventh century B.C.

Population: 300,000

Size: 47 square miles (122 sq km)

Language: Greek, Turkish

Currency: Euro and Turkish lira

Time Zone: GMT +2 and +3 (EET and EEST)

Global Cost of Living Rank: Unlisted

Also Known As: Lefkosa, Ledra (ancient), North Nicosia (Turkish section)

Archbishop's Palace, the **Cyprus Folk Art Museum**, and the **National Struggle Museum** that examines the island's long fight for independence. The nearby **House Hadjigeorgakis Kornesios** preserves the lavish mansion of a late 18th-century Ottoman despot.

The tallest structure in the Old Town, the **Shacolas Tower** offers lofty views of Nicosia and a short video on the city's history. A few blocks to the east, **Hamam Omerye** offers steam rooms, massages, facials, clay wraps, and other treatments in a 16th-century Turkish bath built during the island's Ottoman occupation.

Flush with souvenir shops and tavernas, the **Laiki Geitonia** market area and surrounding **Laïki Yitonia** neighborhood are the Old Town's only touristy corner. But the narrow lanes and vintage buildings are also infused with a bygone feel. Past **Tripoli Bastion** (and just outside the walls) is **Cyprus Museum,** where the **Aphrodite of Soloi** and relics from the **Royal Tombs of Salamis** highlight a superb archaeological collection.

Modern shops line the **Ledra**

Mosaics and frescoes line the walls of the Panagia Chrysaliniotissa Church.

Stop for a rest at the café in the courtyard of the Büyük Han (Great Inn) caravansary.

Street pedestrian zone, which ends at one of only two places where people can cross the Green Line. After almost a half century of separation, North Nicosia seems far more Middle Eastern than Mediterranean.

One of the first things you encounter on the other half of the Old Town is the **Büyük Han** (Great Inn), a 16th-century Ottoman caravanserai that's been nicely restored and converted into a cluster of cafés and handicraft shops with a courtyard where live music, folk dancing, and other events are staged.

If you're wondering why **Selimiye Mosque** looks far more Gothic than Ottoman, it's because the hulking 13th-century structure was the Cathedral of Saint Sophia before its conversion into the city's foremost place of Muslim worship. The

adjacent **Bedesten** is another convert: an ancient Byzantine church that's now a venue for **whirling dervish** performances. Opposite the Bedesten is the main entrance to the **Bandabuliya** covered market, a mini version of Istanbul's Grand Bazaar chockablock with produce stalls and souvenir stands.

With Ottoman-era landmarks like the **Dervish Pasha Mansion** (built in 1807), the **Arab Ahmet Quarter** is one of the more atmospheric neighborhoods. ∎

HIDDEN TREASURES

• **Aes Ambelis:** The closest vintner to Nicosia—and its Maratheftiko, Shiraz, and Chardonnay wines—lies 17 miles (27 km) from the city center near the village of Kalo Chorio.

• **Machairas Monastery:** Established by 12th-century hermits, this lofty retreat and its marvelous icons hovers at around 3,000 feet (914 m) in the Troodos Mountains about an hour's drive from Nicosia.

• **Athalassa National Forest Park:** Perched on the city's southern edge, this leafy green space offers native plants and animals, as well as lakeshore viewpoints, along 13 miles (21 km) of trails.

• **Cyprus Classic Motorcycle Museum:** A 1914 New Hudson Deluxe and military bikes from World War II are just a few of the two-wheeled jewels in a collection devoted to heavy metal thunder.

Middle East Cities

Doha, Qatar; Riyadh, Saudi Arabia; Muscat, Oman

The main street of Souk Waqif features stunning architecture and a thumb statue.

Dubai might be the most famous, but many Arabian urban areas have morphed from desert backwaters to bustling 21st-century cities over the past 40 years.

DOHA

Population: 1.8 million
Size: 222 square miles (575 sq km)

Few people outside the Middle East had heard of Qatar until the tiny Arabian emirate was chosen to host the 2022 FIFA World Cup. Following the announcement, Doha scrambled to fashion itself into a global metropolis worthy of the games.

One of the milestone projects is the $36 billion **Doha Metro**, a subway system that went online in 2019.

The three lines connect the airport, the old town around **Souq Waqif**, the scenic **Corniche** waterfront, and the high-rise **West Bay** area.

Cultural icons are also on the menu. The **Museum of Islamic Art** boasts a massive collection of art and artifacts spanning 1,400 years and three continents. Inside a building inspired by local desert rose crystals, the **Qatar National Museum** unveils everything from Bedouin culture to the natural history of the Persian Gulf region.

Located on reclaimed land in the gulf, **Pearl Island** and the adjacent **Katara Cultural Village** offer beaches, waterfront restaurants, and a year-round slate of entertainment including concerts, shows, and exhibitions.

Guided tours of **Al Shaqab equestrian center** afford an insider's look at the breeding and pampering of Arabian horses at an ultramodern complex that includes horse treadmills and swimming pools. Public art is another passion, some of it surprising in a conservative Muslim country. **"The Miraculous Journey,"** a controversial ensemble of 14 bronze sculptures at Sidra Medical & Research Center by Damien Hirst, represent the phases of a human fetus. **Salwa Road** is adorned with 52 murals by French-Tunisian street artist eL Seed. Richard Serra's **"East-West/West-East"** installation in **Brouq nature reserve** northwest of Doha has no address, just a GPS location.

RIYADH

Population: 6.9 million
Size: 646 square miles (1,673 sq km)

Saudi Arabia's desert capital was off-limits to ordinary travelers until September 2019, when the kingdom suddenly announced it was opening its doors to tourism. Though the unrelenting **Arabian Desert** and World Heritage sites like the ancient Nabataean city of **Madâin Sâlih** and **Jubbah rock carvings** are the main attractions, Riyadh is also worth exploring.

One of the world's wealthiest cities, Riyadh has come a long way from being a way station on a desert trading route. Its skyline features ultramodern towers like the 99-story **Kingdom Centre** (featuring a catenary arch **Sky Bridge**) and the avant-garde **Burj Al Faisaliah**, crowned by a huge golden ball of the **Global Experience** viewing platform.

King Abdul Aziz Historical Center harbors the **National Museum** and its comprehensive look at Saudi Arabia from prehistoric through modern times, as well as the restored **Murabba Historical Palace** (built in the 1930s) and **King Abdul Aziz Grand Mosque**. Mud-brick **Masmak Fort** featured prominently in the 1902 Battle of Riyadh, which solidified Al Saud control over the region.

Locals gather in the cool of the evening in **King Abdullah Park** to stroll, chat, and view the nightly dancing fountain show. **Old Dir'aiyah**, another World Heritage site on the outskirts of the city, was the original home of the Saudi royal family.

Megaprojects currently in the works will add to the city's tourism appeal. The $22.5 billion, six-line **Riyadh Metro** is nearly finished,

Peer into the Sultan's Palace through cast-iron and gilded gates.

while the **Qiddiya** sports and entertainment zone will include a Formula One racetrack, 20,000-seat stadium, golf course, and several theme parks. Also in the planning stage, **King Salman Park** is slated to be the world's largest city park (5.1 square miles/13.2 sq km).

MUSCAT

Population: 1.2 million
Size: 148 square miles (383 sq km)

Sandwiched between the **Gulf of Oman** and the **Jibāl al-Ḥajar** (Rocky Mountains), Muscat is one of the few Arabian capitals to maintain a semblance of its pre-petroleum days.

Lithographs from the 19th century depict an **Old Muscat** waterfront not altogether different from today, with **Al Mirani Fort** and **Al Jalali Fort** bookending the bay. The latest reincarnation of **Al Alam Palace** dates from the 1970s, but its Arabian architecture is very much in keeping with the bygone mood.

Farther west along the waterfront are **Al-Riyam Park** with its giant

ornamental incense burner and popular funfair, and the two-mile (3 km) **Mutrah Corniche**, a great place for watching people, spinner dolphins, or dhows at anchor in the bay. The old **Muscat Fish Market** is perched at the western end of the promenade, and behind the shore is **Mutrah Souk**, a labyrinth of alleys ripe for haggling on Omani souvenirs.

Steps lead upward to the 16th-century **Mutrah Fort**, erected by the Portuguese to defend against attacks by the Ottoman army. Higher up the slope are the **Old Watch Tower** and a trailhead for the **C38 Hike**, which affords amazing views of Muscat.

Muscat's pride and joy is **Sultan Qaboos Grand Mosque**, a gift from its namesake to the people of Oman on the 30th year of his reign.

Bait al Baranda Museum, which housed the 1930s American Mission and 1970s British Council buildings, traces the natural and human history of both the city and nation, while the **Royal Opera House** in Al Qurum district is the epicenter of musical arts and culture. ■

The Kingdom Centre has an observatory bridge at its top, looking over Saudi Arabia from 992 feet (302 m).

Eastern & Southern Asia

The majestic Bentendo Hall of Daigo-ji Temple is a UNESCO World Heritage site in Kyoto, Japan (p. 212).

Kyoto
Japan

The political, commercial, artistic, and religious capital of Japan for more than 500 years, Kyoto preserves the nation's rich traditions through its buildings, gardens, and the many cultural experiences accessible to visitors.

THE BIG PICTURE

Founded: 794

Population: 1.5 million

Size: 319 square miles (826 sq km)

Language: Japanese

Currency: Japanese yen

Time Zone: GMT +9 (JST)

Global Cost of Living Rank: Unlisted

Also Known As: Kyō (ancient), Heian-kyō (medieval), City of History

Many of the customs that we think of as quintessentially Japanese—from geisha and samurai to sumo, woodblock prints, and those ubiquitous beckoning cats in shops and restaurants—were either born or developed in Kyoto during medieval times.

Though Tokyo may have snatched its political and economic power, Kyoto has remained the major force in traditional Japanese culture—a city where it often feels like you're ambling through the past rather than the streamlined, avant-garde Japan of the present.

Nowhere is this more apparent than the **Gion** district on the east side of the Kamo River, a warren of ancient streets flanked by traditional wooden *machiya* townhouses with restaurants, art galleries, and old-fashioned *ochaya* teahouses, where geisha *(geiko)* continue to entertain patrons as part of a tradition that stretches back more than 1,000 years.

Many of Gion's lanes are pedestrian only, including historic **Hanamikoji-dori** and super-narrow **Kiyomoto-cho** just south of the petite **Tatsumi Bridge**. The district's cultural cornerstones include **Minamiza Kabuki theater** (founded in 1610); the traditional music, dance, and tea ceremony presented at **Gion Corner**; and the chromatic **Yasaka Shrine**, founded in 656 and home base for the annual **Gion Matsuri** festival.

Many of Kyoto's architectural treasures perch on the heavily wooded foothills beyond the Gion district, Buddhist and Shinto shrines that form the core of the city's UNESCO World Heritage collection. With around 2,000 total temples and shrines, it would literally take months to explore all of them properly.

But a few literally rise above the others. Like **Kiyomizu-dera Temple** with its panoramic views across the Kyoto Basin, the cities of Osaka and Kobe in the hazy distance. **Kōdai-ji Zen Temple** is renowned for both its stone garden and many national treasures, including exquisite gold-flecked lacquerwork and Hideyoshi's *jinbaori,* a 16th-century shogun's coat sewn with gold and silver thread.

The largest wooden gateway in Japan towers over the entrance to **Chion-in Temple**, which also boasts the nation's single largest

Pay homage at the Great Buddha in the Todai-ji temple.

Called the "Golden Pavilion," the Kinkaku-ji Buddhist temple is one of the most iconic buildings in Japan.

bell. **Eikando Zenrinji** temple complex is lauded as one of the best places in Kyoto to capture fall colors with your eyes and camera. It's also a jumping-off point for walks along the leafy **Tetsugaku-no-michi** (Philosopher's Path) to **Ginkaku-ji Temple**, which started as the estate of a powerful 15th-century shogun.

Farther south along the ridgeline, **Fushimi Inari Taisha** is famed for the hundreds of bright red *torii* (gates) that frame the hillside walkways behind the main shrine. **Tōfuku-ji** is the place to go for bona-fide Zen meditation sessions and blazing autumn colors via the many maple trees. On the other hand, **Sanjusangen-do** shrine is best in spring when the cherry blossoms explode—although any season is

great for contemplating the **1,000 statues of Kannon** (goddess of mercy) inside a temple that ranks as Japan's longest wooden structure.

There's also a northern temple precinct that includes what's

possibly the most photographed building in Japan—**Kinkaku-ji**, the famed Golden Pavilion—its glimmering facade reflected in the surrounding water. Farther west are two of Kyoto's great botanical

---

URBAN RELIC

Nara

Located around 25 miles (40 km) south of Kyoto, **Nara** was founded in 710 and is widely considered the cradle of classic Japanese civilization.

Foremost among the city's ancient structures are the Seven Great Temples. Several of them are Japanese national treasures, including the Great Buddha Hall of Tōdai-ji, which houses the world's largest bronze statue—the 550-ton (500 metric tons) Daibutsu Buddha.

Nara National Museum safeguards many of the religious artworks originally created for the city's temples, including the famous Hell Scroll, a vividly illustrated depiction of the afterworld that awaits sinners. It is also home to numerous Nara deer, the beloved symbols of Nara itself.

The compact city is easy to explore on foot. Shops around the main train station rent regular and e-bikes, and several outfitters offer guided cycle tours of Nara.

A larger than life bamboo forest lines the walkways in Sagano.

treats: the late-blooming Omuro Sakura cherry trees of **Ninna-ji Temple** and the **Arashiyama Course**, a pathway through the magical **Sagano Bamboo Grove**, past tucked-away temples and stones engraved with haiku poems, to the 400-year-old **Togetsu-kyo Bridge**.

Back in the central city, the Kamo-gawa marks a watery divide between spiritual Kyoto in the east and the secular city that evolved along the western side of the river where shoguns and emperors held sway.

Surrounded by double walls and two moats, 17th-century **Nijō Castle** was home to the legendary Tokugawa shoguns and endured as a royal bastion until 1939. The interior features numerous structures, including **Ninomaru Palace** with its gold-leaf decor and *ugui-subari* ("nightingale") floors with boards that chirp like birds to alert the shogun to assassins or other intruders.

Though Kyoto shoguns were the de facto rulers of Japan during much of the Middle Ages, the emperor remained the ceremonial and spiritual head of state. Sequestered inside the nearby **Imperial Palace** (Kyoto Gosho), their life was lavish but basically powerless. Extending across an area equivalent to almost 100 city blocks, the royal compound is a virtual city within a city that embraces several palaces and shrines surrounded by the **Kyoto Gyoen National Garden** and natural treasures like the **Demizu-no-Ogawa** mini river and **Gyoen Bairin** plum forest. Kyoto Gosho sheltered the royal family for nearly 500 years until 1869, when the Japanese capital relocated to Tokyo.

Katsura Imperial Villa on the city's western edge and **Shugakuin Imperial Villa** on the side of Mount Hiei provide another intimate glimpse of Kyoto's imperial legacy. Built during the 17th century, these sumptuous country estates provided the royal family with places to "get away from it all" in silence and seclusion.

Away from the royal compounds,

LOCAL FLAVOR

• **Kikunoi:** Often described as edible art, the 14-course *kaiseki* feast served in this time-honored Kyoto restaurant traces its roots to the imperial court cuisine of medieval times; *459 Shimokawaracho;* kikunoi.jp/kikunoi web.en/top.

• **Usagi no Nedoko:** Rock hounds flock to a restaurant decorated with artistically arranged crystals, shells, and skeletons with a menu that features gemstone-inspired desserts like the Gardenquartz Tiramisu and Amethyst Panna Cotta; *37 Nishinokyo Minamiharamachi;* usaginonedoko .net/en/kyoto/cafe.

• **Honke Owariya:** Owned and operated by the same family for 16 generations, this soba (buckwheat noodle) restaurant near the Imperial Palace first opened its doors in Kyoto in 1465; *322 Niomontsukinukecho;* honke-owariya.co.jp/en.

• **Menbaka:** "Fire ramen" is the forte of this fun eatery that takes flambé to a whole new level; they also serve *gyoza* (dumplings) and Japanese-style fried chicken; *757-2 Minamiiseyacho;* fireramen.com/menbaka.

• **Nishiki Market:** Pick and choose from thousands of seafood, vegetable, and baked treats at this five-block-long covered street market that opened in 1310; *609 Nishidaimon-jicho;* kyoto-nishiki.or.jp.

the west side's architectural treasure is the five-story **Tō-ji Pagoda**—Japan's tallest wooden structure—on the grounds of the eponymous temple. On the 21st of each month, the grounds provide a venue for the **Kōbō-san flea market**, named for the temple's ninth-century abbot.

Located near the flavorsome **Nishiki Market** in downtown Kyoto, the **Samurai and Ninja Museum** offers original katana, wakizashi, and tanto swords and reproduction armor suits, as well as a sword-fighting demonstration and workshops on samurai and ninja skills.

Kyoto offers dozens of other ways to experience traditional Japanese culture: **cooking**, **calligraphy**, and *ikebana* (flower-arranging) classes; *kendo* (stick-fighting) workshops; traditional **tea ceremonies**, and close encounters of the geisha kind. There are scores of old-fashioned (but modernized) *ryokans,* inns offering rooms with *tatami* floors, futons, sliding rice-paper doors, and *ofuro*-style bathtubs.

And, of course, there's rice wine. Located in Fushimi district on the city's south side, **Gekkeikan Ōkura Sake Museum** explains the sake-making process and history of a distillery founded in 1637 that's now the planet's leading producer of the popular libation. The museum exits through a shop where visitors can taste and purchase sake.

For an even higher take on the cityscape than any of the temples afford, ride an elevator to the observation deck atop the 430-foot (131 m) **Kyoto Tower**, or hop a cable car to hiking trails around the summits of **Mount Kurama** (where Reiki healing was born) and **Mount Hiei** (where French Impressionist paintings inspired a garden). ∎

HIDDEN TREASURES

• **Otagi Nenbutsu-ji Temple:** Tucked into a wooded valley west of the city, this eighth-century temple safeguards more than 1,200 moss-covered *rakan* statues, every one of them unique (and somewhat creepy).

• **Kyoto International Manga Museum:** Japanese comic books (manga), original comic art, and graphic novels are the forte of this unique museum that also features creative demonstrations, live *kamishibai* performances, and reading areas to peruse the collection's 300,000-plus comics.

• **Saiho-ji Moss Temple:** Zen reaches a green extreme at a 1,300-year-old shrine celebrated for its lush moss garden arranged around Ougonchi Pond, shaped like the Chinese character for "heart"—designated as a Special Place of Scenic Beauty by the Japanese government.

• **Yokai (Monster) Street:** Officially called Ichijo-dori, this offbeat street in Kamigyo Ward derives its popular name from the dozens of human-size supernatural beings *(yokai)* from Japanese folklore that merchants display in front of their shops and restaurants.

Dressed in traditional kimonos, women walk toward the Kiyomizu-dera Temple.

Tokyo
Japan

As a confluence of the endearingly old and shockingly new, the globe's most populous urban area can change on a dime, from neon-splashed canyons and skyscraper forests to tranquil temples, parks, and village-like neighborhoods that seem light-years removed from the biggest of all cities.

THE BIG PICTURE

Founded: 1603

Population: 38 million

Size: 3,178 square miles (8,231 sq km)

Language: Japanese

Currency: Japanese yen

Time Zone: GMT +9 (JST)

Global Cost of Living Rank: 4

Also Known As: Edo (medieval), the Big Mikan

Devout Buddhists and avid baseball fans, suit-wearing salarymen and leather-clad *bosozoku* bikers, the trendy cosplay crowd and weathered old tuna fishermen, kimonoed ladies and Gothic Lolitas—Tokyo often seems more like a collection of tribes than a homogeneous city.

That cultural stew is exactly what makes the Japanese capital such a fascinating place to discover, explore, and people-watch. Sometimes by design and often by accident, Tokyo's 38 million residents have crafted one of the planet's most intriguing cityscapes, a metropolis that has totally rebuilt itself twice over the past century in the aftermath of a huge 1923 earthquake and the destruction of World War II.

Tokyo is full of surprises. You truly never know what you're going to find around the next corner. It might be something straight from the future—like the robotic androids at the **Miraikan National Museum of Emerging Science & Innovation** or the towering **Tokyo Skytree**, Japan's tallest building at 2,080 feet (634 m). Or it could be something incredibly ancient—like the seventh-century **Sensōji Temple**, oldest of the city's 3,000-plus religious shrines, or the solid stone **Fujimi-Yagura Watchtower**, the only remaining keep from otherwise ruined Edo Castle, where emperors ruled during Japan's medieval days.

No matter which Tokyo (or which tribe) you're questing, the likely starting point is **Tokyo Station** in the **Marunouchi** central business district. The station is ground zero for trains and buses from the city's two airports as well as Tokyo's spiderweb of subway and commuter rail lines. One of the most useful is the **Yamanote (Green) Line**, which loops around central Tokyo with stations in many of the most popular neighborhoods, including Akihabara, Ueno, Ikebukuro, Shinjuku, Harajuku, and Shibuya before returning to Tokyo Station.

The station is also within easy walking distance of several iconic

The famous pedestrian scramble is a crossing outside Shibuya Station.

Cherry blossoms bloom in Chidorigafuchi Park along the moat of the Imperial Palace during springtime in Tokyo.

attractions. The emperor and his family still reside inside the sprawling **Imperial Palace** and its beautiful parklike surrounds. The Imperial Household Agency offers free guided tours of the private inner compound, but the **East Gardens**, **Nijubashi Bridge**, and **Edo Castle** ruins are always open to the public.

The famed **Ginza**—the city's neon-cloaked shopping district—is also a short walk from the station. On weekends, vehicle traffic is banned along **Chuo Dori,** which becomes one of the world's largest pedestrian streets. Newcomers include a **Ginza SIX** shopping

HIDDEN TREASURES

- **Gōtokuji Temple:** Edo-era Buddhist temple in Setagaya City thought to be the birthplace of the *maneki-neko* (beckoning cat). Devotees continue to add to the thousands of lucky statues already on display.

- **2D Café:** The decor at this Shinjuku restaurant flattens the 3D world into a comic book–like monochrome world. Tokyo's other anime eateries include the Pretty Guardian Sailor Moon musical restaurant, Pokémon Café, and Square Enix Café.

- **Sugamo:** Known locally as Obachan's (Grandma's) Harajuku, this very local neighborhood north

of the city center is a great place to glimpse ordinary Japanese life. Maruji, one of its trademark stores, oddly sells nothing but bright red underwear.

- **Shiro-Hige's Cream Puff Factory:** The only place in the world that sells fully licensed Totoro-shaped cream puffs based on the iconic animated character from Hayao Miyazaki's 1988 film *My Neighbor Totoro*. Go early; they sell out fast.

- **Chatei Hatou:** Like any traditional *kissaten* (coffee shop), this old favorite in Shibuya serves tea, coffee, breakfast, and desserts.

complex with more than 240 trendy stores and **Ginza Place** with its Nissan and Sony showrooms. **Ginza Wako**, a vintage Japanese-style department store, kicked off the area's upscale shopping cred in 1932.

Kabukiza Theatre, the city's legendary Kabuki playhouse, is also located in Ginza. The nearby **Shinbashi Enbujō** theater stages Kabuki, melodramatic *shinpa,* and other traditional Japanese genres. If you're out late in Marunouchi or Ginza, follow your nose to **Yakitori Alley** and its gaggle of grilled-meat stalls beneath the old railway arches near **Yurakucho Station**.

Tasty places are also abundant in nearby **Tsukiji**. The famous fish market has relocated to modern quarters in **Toyosu** on an artificial island in **Tokyo Bay**. That's where the famous early morning tuna auctions take place nowadays. But many of the old market's satellite services remain: amazing little seafood eateries and stalls selling top-quality knifes and other Japanese kitchenware in the streets behind the river.

In addition to a huge bay, Tokyo also grew up along an inland waterway. The **Sumida River** was once indispensable to local culture and commerce. It featured in haiku poetry and Kabuki plays, and was a favorite subject of Edo-era paintings and woodblock prints. The best way to venture upriver is hopping aboard one of the glass-topped **Tōkyō-to Kankō Kisen** sightseeing boats that ply the Sumida between **Hamarikyu Gardens** near the old fish market and Asakusa district, around five miles (8 km) upstream.

Asakusa might be the most

Traditional torii gates line the Shinto Hie Shrine in the center of Tokyo.

captivating area in a city that boasts many fascinating neighborhoods. The area first became a place of pilgrimage in the seventh century after a statue of Kannon (the Japanese goddess of mercy) was discovered in the Sumida River. During medieval times, it also developed into a thriving entertainment zone. And it remains so today. The sprawling **Asakusa** temple complex harbors the aforementioned Sensōji Temple and five-story pagoda. But Asakusa is also home to the crowded but colorful **Nakamise Shopping Street** and **Hanayashiki**, the city's oldest amusement park (founded in 1853).

In its entertainment guise, Asakusa is also a traditional *hanamachi* (geisha district). Although the elegantly clad ladies are more common along the cobblestone streets of

the **Kagurazaka** area near Iidabashi Station. Here you might just see a geisha or two at twilight as they commute to work at a high-end dinner to engage in erudite conversation, play instruments like the *shamisen,* recite poetry, dance, or even play drinking games. The **Nihonbashi Information Center** offers a number of cultural activities, including an **"Omotenashi experience"** that includes a kimono or *yukata* rental, tea ceremony, a geisha performance, and a chance to play party games with a geisha.

It's also notoriously difficult to see a **sumo** wrestling match in Tokyo because they are so often sold out. However, a number of sumo training houses are located in the **Ryogoku** area on the east bank of the Sumida River. Some of them open their morning training sessions to the public. Sumo matches take place at the **Ryōgoku Kokugikan** arena, where the **Sumo Museum** displays paraphernalia and photos of grand champions.

Beside the sumo arena, the **Edo-Tokyo Museum** details the city's history from medieval through modern times via scale models of events and neighborhoods, including a reconstructed Kabuki theater. Its sister museum in the western suburbs, the **Edo-Tokyo Open Air Museum**, preserves dozens of historic buildings from around Japan.

Another north-central neighborhood that begs exploration is **Akihabara** ("Electric City"), a shopping hub famed for electronics retailers like **Tsukumo Robot Kingdom** and the giant **Yodobashi Camera** store. The area is also known for its cosplay cafés (where staff dress as maids or butlers) and places like **Radio Kaikan**, **Animate**, and **Kotobukiya** that specialize in manga comic

Shop along the Nakamise Street, which leads to the Sensōji temple and blooms with sakura trees in spring.

books and anime swag. The bright red **Kanda Myōjin**, which enshrines two of the seven gods of fortune, is popular with those seeking wealth and techies hoping to ward off damage to their electronics.

From Akihabara, the Yamanote Line chugs due north to **Ueno Park**, where visitors can rent boats for a float across the ponds, gawk at the giant pandas of the **Ueno Zoo**, browse the **Tokyo National Museum** (which exhibits thousands of Japanese cultural treasures from kimonos to samurai swords), or catch an opera, symphony, or ballet at the modern **Tokyo Bunka Kaikan** concert hall.

Japanese companies like Nintendo, Sega, and Sony were among the pioneers of video games, an industry now valued at some $152 billion worldwide. Tokyo's gaming hotbeds are the arcades of **Ikebukuro**, an area that also offers anime and manga stores as well as the cosplayers of nearby **Naka-Ikebukuro Park**. Those who want to watch or test their gaming skills should visit **Super Potato**, **Hirose Entertainment Yard**, **Tokyo Joypolis**, or **Mikado** with its retro games.

From Ikebukuro, it's a straight shot (four stations) down the Yamanote Line to **Shinjuku**, one of Tokyo's skyscraper districts and another popular shopping area. It's also one of Tokyo's more raucous nightlife districts, especially the tiny bars and eateries of **Omoide Yokocho** ("Memory Alley") and **Kabukicho**. Among Shinjuku's sky-high landmarks are the 50-story **Mode Gakuen Cocoon Tower** (2008's global "Skyscraper of the Year") and the twin towers of the **Metropolitan Government Office** (Tokyo's city hall), each of them topped by free public observation decks.

If there was ever a place that feels like a nonstop fashion runway, it's nearby **Harajuku**. **Takeshita Street** is the main drag and the best place for people-watching, but many of the side streets feature their own quirky delights—like Purikura photo booths and shops that specialize in *kawaii* (Japan's culture of cuteness) and cosplay trends. **Cat Street** is lined with higher-end boutiques.

Harajuku is also the place to try crazy snacks like the rainbow cheese sandwich at **Le Shiner** and animal-shaped ice cream cones at **Eiswelt Gelato**. The mirrored entryway at **Tokyu Plaza Omotesando Harajuku** is on the bucket list of every social media influencer. West of busy Harajuku Station are the **Meiji Shrine**, dedicated to the first emperor and empress of modern Japan, and big leafy **Yoyogi Park**. A former military parade ground and postwar U.S. Army barracks, the popular green space offers walking paths, street performers, and one of the city's best *sakura* (**cherry blossom**) shows in spring.

The next stop down the Yamanote Line is **Shibuya**, another popular retail and entertaining area. Among its many shopping options are **Shibuya Hikarie** and **Shibuya Stream** skyscrapers, and **Shibuya 109** department store with its cylindrical tower. The new **PARCO** megamall includes Japan's first official **Nintendo** store and **Pokémon Center**. If you see an inordinate number of medieval warriors wandering the streets of Shibuya, it's because **Samurai Armor** photographic studio offers rentable costumes.

The legendary **Shibuya Crossing** intersection flaunts copious neon signs, giant electronic billboards, and a reputation as the globe's busiest pedestrian crossing—as many as 2,500 people at a time. Gaze down on the human crush from the new **Shibuya Scramble Square**, where the 47 floors of shops and restaurants are topped by the **Shibuya Sky** observatory and its vertigo-inducing outdoor **Sky Edge**. The roof area also offers hammocks for cloud-watching and an observation compass that identifies major landmarks.

Shibuya's **Center-Gai Street** offers a range of restaurants, bars, and nightclubs. If sake soothes your soul, cruise the *izakaya* (Japanese pubs) of nearby **Nonbei Yokocho** (Drunkards' Alley). The neighborhood also hosts **Shibuya Blue Cave**, a popular winter illumination attended by more than two million people annually.

Rising just east of Harajuku and Shibuya are the **Roppongi Hills**, an upscale business and residential area with a penchant for the finer things in life. The district's ever changing **National Art Center** and the nearby **Mori Art Museum** (on the 52nd and 53rd floors of the Roppongi Hills Mori Tower) offer the nation's best collections of modern art, while the **Suntory Museum of Art** endeavors to fuse Japanese tradition and modernity in a relaxing homey environment. ∎

Find popular restaurants and a thriving nightlife scene in the Shimbashi neighborhood.

Yogyakarta
Indonesia

More than any other Indonesian city, Yogyakarta retains vestiges of the days when powerful dynasties and sultans ruled the island of Java, erecting monuments to both themselves and the great faiths that washed across the land.

THE BIG PICTURE

Founded: 16th century

Population: 2 million

Size: 90 square miles (233 sq km)

Language: Bahasa Indonesia

Currency: Indonesian rupiah

Time Zone: GMT +7 (IWT)

Global Cost of Living Rank: Unlisted

Also Known As: Yogya, Jogjakarta, City of Scholars, Gudeg City

Of the hundreds of monarchies large and small that once ruled over the vast Indonesian archipelago, the only one that remains is the Sultanate of Yogyakarta on the south side of Java island. The city owes its special status to its ardent support for the 1940s war of independence that expelled the Dutch.

Modern-day "Yogya" is also a university town with tens of thousands of students, as well as a thriving handicraft center with skilled artisans who create batik fabric, silverwork, shadow puppets, intricate wooden masks, and other precious items. Yet its most fascinating aspect is a chance to glimpse a royal lifestyle that was once prevalent through Southeast Asia.

The city revolves around the sprawling **Kraton**, a walled palace compound where the royal family and thousands of retainers—guards, artisans, musicians, dancers, etc.—continue many of the traditions established in the 18th century when it was first constructed.

Much of the Kraton is open to the public, including the ornate **Bangsal Kencana** (Golden Pavilion) and the adjacent **Gedhong Praba-yeksa** with its royal heirlooms, as well as the partially ruined **Taman Sari** water garden and the **Museum Kereta Keraton** with its flamboyant royal carriages. The palace also hosts daily performances of Javanese dance, gamelan music, and *wayang kulit* puppetry by the esteemed royal troupes.

Starting from Kraton's main gate, a crazy, colorful, crowded street called **Jalan Malioboro** runs north through the city center, flanked by scores of eateries and handicraft shops. Along the way are the **Sonobudoyo Museum** of Javanese culture and archaeology—which presents a two-hour puppet show most evenings—and 18th-century **Fort Vredeburg**, an old Dutch bastion that now houses a museum devoted to Indonesian independence.

Indonesia's largest and most significant ancient monuments are found in the lush countryside around Yogyakarta.

Northwest of the city, the impeccably restored **Borobudur** temple is the largest stone structure in the Southern Hemisphere *and* the world's biggest Buddhist shrine of any kind. The massive shrine was erected during the eighth and ninth centuries by the Shailendra dynasty. It mirrors the shape of a mandala, a

Old horse-drawn carriages of Kraton are displayed at the Bangsal Pagelaran.

Stone structures of the Borobudur temple are worth visiting, particularly during sunsets.

gradual ascent up stone stairs and ramps past 500 statues, 72 smaller stupas, and more than 1,400 masterful stone reliefs.

Around the same time as Borobudur was rising, a rival dynasty called the Sanjayas were building **Prambanan**, a sprawling Hindu religious complex east of Yogyakarta. Only a few of the hundreds of shrines that once marked the site have been restored, most notably the **Shivagrha Trimurti** temple with a richly decorated *prasada* tower. A daytime visit pales in comparison to an after-dark presentation of the *Ramayana* epic on an open-air stage with the temples as a backdrop (May to October).

Rising high above the temples is **Mount Merapi**, the nation's most active volcano. Owing to the significant danger, it's now forbidden to hike to the summit. However, the **Bukit Klangon** observation tower and Merapi Volcano Museum farther down the mountain are still open.

The big blue **Indian Ocean** lies just 18 miles (29 km) south of the city center. **Parangtritis** with its black-sand strand and beachside cafés is the best place to hit the shore. There are no big Bali-style resorts but plenty of *losmen* and other small hotels for a stay at the shore. ∎

ALTER EGO

Jakarta

Indonesia's massive, sprawling, helter-skelter capital, **Jakarta**, is a hard and often frustrating city to navigate. It's best to limit explorations to two main areas.

The stretch of **Jalan Sudirman** between **Merdeka Square** and the **Patung Pemuda Membangun monument** harbors many of the finest hotels, restaurants, and shopping centers, as well as the **National Museum**.

Old Jakarta revolves around **Fatahillah Square** and a trove of Dutch colonial buildings that shelter the **Jakarta History Museum**, **Wayang Museum of Indonesian puppetry**, and atmospheric **Café Batavia** (opened in 1805).

Traditional Bugis schooners still dock at nearby **Sunda Kelapa**, where the **Jakarta Bahari Museum** chronicles Indonesia's long seafaring heritage.

Beijing
China

With one foot planted firmly in a past that features seven World Heritage sites and another stepping boldly into the technological future, China's capital city offers a fascinating lesson in how new and old can coexist in perfect harmony.

How can a place be one of the world's oldest cities and one of its newest all at the same time? Ask the residents of Beijing, who live that dichotomy on a daily basis.

Founded around 3,000 years ago, it's one of the world's oldest continuously occupied cities. It's one of the few urban areas that can boast *seven* UNESCO World Heritage sites. But in the same breath, Beijing has embraced the future to a much larger degree than other global cities by diving headlong into artificial intelligence, facial recognition, and other new technologies.

This yin and yang is exactly what makes China's capital such a fascinating place to discover in this day and age, especially for those looking to avoid the "temple fatigue" that can arise in cities with so many historical treasures.

The well-known **Forbidden City** in the heart of Beijing offers a natural starting point for exploring the city. For nearly 500 years (1420–1912), it was home to China's supreme rulers and their royal courts. The planet's largest collection of preserved wooden buildings, the compound harbors 980 structures

THE BIG PICTURE

Founded: 1045 B.C.

Population: 19.4 million

Size: 1,600 square miles (4,144 sq km)

Language: Mandarin

Currency: Chinese renminbi (yuan)

Time Zone: GMT +8 (CST)

Global Cost of Living Rank: 9

Also Known As: Peking

inside a wide moat and 26-foot-high (8 m) walls.

Most of the Forbidden City's major attractions—and various galleries of the **Palace Museum**—are located along a central axis that stretches between the **Meridian Gate** in the south and the **Gate of Divine Might** in the north. Among the most spectacular structures are the **Hall of Supreme Harmony**, where large or important events unfolded, and the **Palace of Heavenly Purity**, built as a residence for Ming rulers but later converted into a throne room and audience hall. The adjacent **Hall of Mental Cultivation** was the main royal residence during the Qing era, while an imperial retirement "villa" called the **Palace of Tranquil Longevity** now safeguards the famous **Nine-Dragon Screen** and a **Treasure Gallery** with "extant accoutrements for palace life" made from jade, gold, pearls, and other precious materials.

One way to avoid the crowds that spill into the Forbidden City (especially on Chinese holidays) is exploring recently opened areas like the **Palace of Compassion and Tranquility** (Cining Gong), home to dowager empresses in days gone by and now the Palace Museum's official **Sculpture Gallery**.

China's CCTV building is a 51-story skyscraper and an iconic Beijing landmark.

A lion statue guards the Gate of Heavenly Peace at Tiananmen Square in the city center of Beijing.

Outside the Forbidden City, the much photographed **Gate of Heavenly Peace** (with its giant portrait of Mao Zedong) overlooks vast **Tiananmen Square**, which dates from the 17th century but nowadays is more associated with the communist era. The **Great Hall of the People**, **National Museum of China**, and the other severe, largely unadorned Soviet-style buildings that flank the square are in stark contrast to the flamboyant imperial structure inside the Forbidden City. **Chairman Mao Zedong's Mausoleum**—where his embalmed corpse is on display—dominates the southern end of the square.

But just around the corner is the gleaming **National Centre for the Performing Arts** (aka "The Egg"),

an ultramodern cultural space that doubles as a powerful statement of the new China that's emerged since

the 1980s. Shaped like an egg half submerged in water and covered in titanium cladding, the center offers

GATHERINGS

• **Chinese New Year:** In Beijing, like everywhere else in China, the Lunar New Year festivities in January or February are a joyous time, celebrated with temple fairs, lion dances, and spectacular fireworks displays.

• **Beijing Kite Festival:** Yuanboyuan Park is the venue for this April event that features hundreds of kites from around China, some of them shaped like tigers, pandas, and dragons.

• **Great Wall Marathon:** Considered one of the world's most grueling footraces, the May marathon, half

marathon, and fun run are staged on the Huangyaguan section of the Great Wall northeast of Beijing; great-wall-marathon.com.

• **798 Arts Festival:** Beijing's role as a center for cutting-edge contemporary art comes into focus during this September event in the hip 798 Art District; 798district.com/en.

• **Mid-Autumn Festival:** Feast on moon cakes, Chinese opera, and temple fairs during this September harvest celebration beneath the full moon.

The Summer Palace was an imperial garden in the Qing dynasty.

a year-round slate of music, dance, opera, and other shows by artists from around the world.

Other modern masterpieces tower above Beijing's new central business district directly east of the Forbidden City. Rising 18 stories in four towers, **Galaxy SOHO** shopping mall features high-rise bars and restaurants with great city views. Even more radical is the 51-story **China Central Television**

(CCTV) headquarters, an arch-shaped structure that locals have dubbed "the big pants" because of its oddball shape.

Several of the city's best historical and art collections are also housed in contemporary buildings, including the excellent **Beijing Capital Museum** that spins urban stories via thousands of cultural artifacts and the small but astonishing **Poly Art Museum** that showcases

Chinese relics recovered from overseas museums and private collections.

Prior to the modern age, Beijing's most prominent "high-rise" was the exquisite **Temple of Heaven**. Another of the city's World Heritage sites, it arose in the early 15th century as a shrine where the emperors staged elaborate ceremonies to ensure good harvests. The 660-acre (267 ha) temple park is popular for tai chi, martial arts, folk dancing, and other traditional pastimes.

An even more spectacular garden surrounds the **Summer Palace** on the city's northwest side. Developed over several centuries as a rural retreat for the royal family, the sprawling compound boasts numerous palaces, pavilions, and temples scattered amid natural features like **Kunming Lake** and **Longevity Hill**. Among its landmark structures are the **Long Corridor** beside the lake and the **Tower of Buddhist Incense** crowning the hilltop.

Though ancient and modern Beijing generally try to keep their distance from one another, the one area where old and new overlap is the city's *hutong* (alleys), where commoners lived during imperial times. Rich or poor, they lived in traditional *siheyuan*-style houses with central courtyards arrayed one after another along the alleys. Since the turn of the 21st century, many of these homes have morphed into cafés, bars, souvenir shops, and art galleries in hip hutong neighborhoods around the Forbidden City. **Shijia Hutong Museum** renders a great overview of the cultural significance before diving into the alleyways and their sundry delights. Many of the best-known hutongs are located just north of the Forbidden City, including **Nanluogu Xiang** (Centipede Alley), **Mao'er**

Hutong with its stylish mansions, and **Guozijian Street**, home to the impressive **Temple of Confucius** and **Imperial Academy**.

Creative solutions are also reviving old industrial buildings. The trendy **798 Art District** nestles in a Bauhaus-style factory complex designed by East Germans during the Great Leap Forward. Many of the city's leading galleries—**UCCA Center for Contemporary Art**, the **Beijing-Tokyo Art Projects** (BTAP), and the **Faurschou Foundation** have found a home in the cavernous space still marked with Maoist propaganda.

Several leading landmarks, old and new, are found around the periphery of the central city. Constructed for the 2008 Summer Games, **Beijing Olympic Village** features iconic structures like **Bird's Nest Stadium** and the adjacent **Water Cube** aquatics center. Farther north, the **Ming tombs** are another of the city's World Heritage sites. The mausoleums of 13 emperors are reached via a the four-mile (7 km) **Spirit Way** lined with statues of luminaries and legendary animals.

There's more World Heritage on the south side, where the **Grand Canal Forest Park** in Tongzhou includes a walkway along the northernmost stretch of the **Grand Canal**, one of the world wonders of hydrological engineering. Way out west, **Zhoukoudian site museum** preserves the cave where a 750,000-year-old human ancestor called Peking Man (*Homo erectus pekinensis*) was discovered in 1929.

Despite its utter failure in keeping foreigners at bay—it was a mere bump in the road to Kublai Khan's Mongol army—the **Great Wall of China** endures as a global icon. Famously one of the few human-made objects that can be seen from outer space, the barricade snakes its way across the countryside just north of Beijing. The section between **Mutianyu** and **Badaling** is closest to the city center, therefore the most visited. Far fewer people hike more remote segments like **Huanghuacheng** and **Simatai** that are still largely in ruins and far more evocative of medieval times. ∎

The Great Wall of China stretches continuously for more than 13,000 miles (20,921 km) across historical northern borders.

Shanghai
China

Shanghai is definitely the déjà vu city, a place that first sprang to prominence as a 19th-century treaty port, hit fever pitch as an entrepôt between the World Wars, and is now in the midst of its latest incarnation as one of the world's great trading places.

THE BIG PICTURE

Founded: ca 4000 B.C.

Population: 22.1 million

Size: 1,550 square miles (4,014 sq km)

Language: Mandarin Chinese

Currency: Chinese renminbi (yuan)

Time Zone: GMT +8 (CST)

Global Cost of Living Rank: 6

Also Known As: Módū (Magic City), Paris of the East, Queen of the Orient, Nevernight City

There's no better indication that China's largest city is one of the globe's most forward-thinking urban areas than stepping off an airplane at modern Pudong International Airport (one of the top 10 busiest in both cargo and passengers) and onto the **Shanghai Transrapid**. The world's first and only commercial magnetic levitation (maglev) train is also the speediest, zipping along at 268 miles (431 km) an hour—faster than even the fastest Formula One or NASCAR racer.

It's part of the amazing transformation that has overtaken Shanghai since the late 1980s, when China opened its economy to the outside world. With the possible exception of certain Middle Eastern emirates, no global city has changed as much since then. And it's not the first time it's experienced this phenomenon: The first growth spurt occurred in the middle of the 19th century after European powers forced China's Qing emperor into declaring Shanghai a treaty port for foreign trade.

Even more than the speedy airport train, the gleaming **Lujiazui** business district (aka Pudong) on the east bank of the **Huangpu River** reflects Shanghai's modern transformation. A generation ago, the area hosted riverside factories, warehouses, and the occasional rice paddy, sampans huddled along its old wooden docks. But all that was washed away by a rising tide of skyscrapers that includes the **Oriental Pearl Tower** with its revolving restaurant and **Space Capsule** observation deck, the 101-floor **Shanghai World Financial Center**, and tallest of all, the twisting, 121-story **Shanghai Tower**.

Much closer to the ground, Lujiazui renders modern attractions like the **Shanghai Ocean Aquarium** and the **Aurora Museum** of Chinese antiquities, which specializes in ancient jade, ceramics, and Buddhist sculptures. A long riverside walkway wraps around **Pudong Point**, with assorted waterfront eateries along the way like the German-style beer garden called **Bao LaiNa**.

On the other side of the river is the legendary **Bund**, a progeny of Shanghai's treaty port and for

The Shanghai Transrapid maglev train races up to 268 miles (431 km) an hour.

Take in views of the Pudong skyline and Huangpu River from the Bund waterfront area in the historical district of central Shanghai.

many years the epitome of Asian exotic. You can cross the Huangpu via underwater metro and vehicle tunnels, or ferries that flit back and forth between **Dongchang dock** on the Pudong side and **Jinling East wharf** on the Bund side. But undoubtedly the most unusual way is the **Bund Sightseeing Tunnel**, a driverless train that takes passengers through a psychedelic light show worthy of the 1960s.

The name Bund derives from an old Persian term for embankment, and the Bund is just that: a riverside road and walkway rather than a neighborhood per se. Back in the day it was lined with foreign banks and trading houses and luxury hotels. But there was also an underbelly of casinos, nightclubs, opium dens, and assorted other vices. As American author John

Gunther observed in 1939, the Bund was notable for two things: "money and the fear of losing it." In other words, the original sin city.

Neglected during the Cultural Revolution as a relic of dreaded capitalism and China's acquiescence to Western powers, the Bund found

LAY YOUR HEAD

• **Wanda Reign on the Bund:** High-tech amenities and comfort with a decidedly Chinese flair in the heart of Shanghai; restaurants, bar, pool, salon, spa, gym, karaoke; from $379; wandareignonthebund.com.

• **Fairmont Peace Hotel:** Unveiled in 1929, this legendary hotel overlooks the Huangpu River along the Bund. Recent renovations have restored this art deco masterpiece to its original opulence while adding contemporary twists; restaurants, bars, spa, museum; from $317; fairmont.com/peace-hotel-shanghai.

• **New World Shanghai Hotel:** Spacious rooms and suites overlooking Zhongshan Park in the Changning district, popular with foreign residents and international companies; restaurants, bar, lounge, outdoor pool, gym, spa; from $79; shanghai.newworldhotels.com/en.

• **citizenM Shanghai Hongqiao:** Hip modern hotel with a cheerful-but-functional vibe in the south-city Minhang district; restaurant, living room, bar, from $66; citizenm.com.

new life in the 1990s as Shanghai's leading tourist attraction. The opulent **Peace Hotel**, an art deco masterpiece opened in 1929, recalls the bygone days better than anything. But there are plenty of other 1920s relics along the same stretch: the neoclassical **Hong Kong & Shanghai Bank**, the **Customs House** with its trademark clock tower, and the wonderfully restored **Bund18** with its upscale French restaurants and the rooftop **Bar Rouge** that brings Shanghai's rowdy nightlife era into the 21st century.

Starting from right beside the Peace Hotel, **Nanjing Road East**, a vehicle-free thoroughfare flanked by neon-studded restaurants and department stores, heads away from the river. The pedestrian zone peters out at **People's Park**, a green space that started life in 1862 as a thoroughbred horse track for the city's foreign residents.

People's Park is also home to several of the city's foremost cultural institutions. With its distinctive sloping roof, the futuristic **Shanghai Grand Theatre** hosts music, drama, and dance from around the world. The **Museum of Contemporary Art Shanghai** (MOCA) occupies a stunning glass structure originally built as a giant greenhouse. With perhaps the best collection in all of China, the **Shanghai Museum**—its design inspired by ancient bronze cauldrons—showcases treasures from every era of Chinese culture.

During treaty port days, the Bund, Nanjing Road, and People's Park were part of the Shanghai International Settlement, jointly administered by the British and Americans. The powers that be in Paris carved out a separate enclave called the **French Concession**. Still flush with elegant colonial-era buildings, the area has reinvented itself as a hub for gourmet dining, upscale shopping, craft breweries, and retro cocktail bars.

Take a walk on the Jade Water Corridor bridge in the Yuyuan Garden, part of Shanghai's Old City district.

Ironically, this onetime bastion of foreign imperialism provides a domicile for the **Propaganda Poster Art Centre** and its vintage Maoist-era proletarian art. The **Shanghai Dramatic Arts Centre**, which stages more than 150 domestic and international productions each year, also calls the French Concession home.

During treaty port days, Chinese residents were confined to the **Old City** south of the Bund. Once surrounded by a circular wall and moat, the area traces its roots to a village founded around 6,000 years ago and now harbors many of the city's oldest features. The 15th-century **City God Temple** (Chenghuang Miao) honors Shanghai's elevation to municipal status.

A century later, a wealthy Ming dynasty family created the adjacent **Yuyuan Garden** with its koi ponds, artistic rockeries, and **Huxinting Teahouse** (built in 1855). Anyone hunting for souvenirs, keepsakes, and curios should browse the **Yuyuan Bazaar**, which fans out along Yuyuan Old Street, Ninghui Road, and Middle Fangbang Road near the temple and gardens.

Like big cities everywhere, much of Shanghai's modern urban sprawl is of little interest to visitors. But here and there in the far reaches of the city are sights and attractions that beg a long subway or taxi ride.

Disused factories provide a provocative setting for the avant-garde **M50 Arts District** and its multiple creative spaces: the **M97** photo gallery, the eclectic contemporary creations of **ShangART**, and the innovative Liu Dao Art Collective's **island6**.

Shanghai Disneyland Park (SDP) is a must-see with kids in

A woman practices traditional methods of drying green tea leaves in the Longjing village of Hangzhou.

tow, if for no other reason than to compare the Chinese version to theme parks back home. Unveiled in 2016, the park's **Enchanted Storybook Castle** is the largest and tallest of its kind in the entire Disney realm. Based far more on Disney films than its predecessor, SDP features seven themed areas with an eighth on the way. ∎

ALTER EGO

Hangzhou

Described by Marco Polo as the "finest and most splendid city in the world," **Hangzhou** lies just 112 miles (180 km) inland from Shanghai at the confluence of the Grand Canal and the Qiantang River.

One of China's seven ancient capitals—with a history dating back to the Qin dynasty (221–206 B.C.)—Hangzhou is renowned for its copious water features, many of them part of the **West Lake Cultural Landscape World Heritage site**.

Divided into five sections by three causeways, the city's legendary West Lake district comprises numerous temples and pagodas, bridges and artificial islands, gardens and pathways, many of the landmarks illuminated after dark.

Among the must-see sights are **Leifeng Pagoda**, **Three Pools Mirroring the Moon**, **Yue Fei Temple** lakeshore, as well as outlying **Lingyin Temple** and **Xixi Wetland Park**.

Hangzhou is also the birthplace of Longjing "Dragon Well" tea, ranked among the top 10 of China's 70 tea varieties. The aromatic brew is celebrated at the **China National Tea Museum** and the annual **Longjing village spring festival** (April–May) amid the tea plantations near the lake. The **China National Silk Museum** honors another local tradition.

Hangzhou's latest forte is e-commerce. As the hometown and headquarters of Jack Ma—founder of the massive Alibaba global trading company—the city has morphed into China's Silicon Valley.

Hong Kong
China

Still one of the world's great crossroads, Hong Kong's supersize skyline and bustling harbor offer an eye-catching contrast to its ancient temples, boisterous street markets, and outlying wilderness peaks and parks.

THE BIG PICTURE

Founded: 1842

Population: 7.4 million

Size: 110 square miles (285 sq km)

Language: Cantonese, English

Currency: Hong Kong dollar

Time Zone: GMT +8 (HKT)

Global Cost of Living Rank: 2

Also Known As: Fragrant Harbour, Pearl of the Orient, Asia's World City

Hong Kong is a hybrid: both a colonial creation and a quintessential Chinese city. Or more precisely, Cantonese, because the local language, food, and other customs are steeped in the age-old ways of China's subtropical south coast.

The city was famously founded in 1842 by the British, who transformed a rocky, sparsely inhabited island into their gateway to the lucrative China trade. Hong Kong didn't grow into a megacity until after World War II, when the population exploded with refugees fleeing the communist mainland. In 1997, after protracted negotiations, the British handed the territory back to China.

Once called Victoria after the British monarch who reigned during the colony's birth, **Central District** reflects Hong Kong's past, present, and future. **Flagstaff House Tea Museum** in an 1846 colonial mansion, the **Duddell Street Steps** with its Victoria-era gas lamps, and the **Old Supreme Court Building** (finished in 1912) exist in a time warp that seems so out of place with skyscrapers that now dominate the one-time heart of British Hong Kong.

Legendary architects played a huge role in this reach for the sky,

including I. M. Pei who designed the sleek blue **Bank of China Tower** and Norman Foster who created the modular **HSBC Main Building** with its massive atrium and feng shui features. The **International Finance Centre,** a 1,361-foot (415 m) behemoth that hovers above the ferry piers, was contrived by Argentina's César Pelli.

Central District is the jumping-on point for three iconic Hong Kong transportation modes: the green-and-white **Star Ferry** (founded in 1888) that flits back and forth to mainland Kowloon; the cute two-story **trams** (launched in 1904) that *clang-clang-clang* along Queensway and Des Voeux Road; and a **Peak Tram** funicular railway (also started in 1888) that slowly chugs to the summit of 1,811-foot (552 m) **Victoria Peak** for a view that remains Hong Kong's best despite all those fancy new skyscrapers.

Much newer is the **Central-Mid-Levels Escalator**—the world's longest outdoor covered escalator system—which rises from the flatlands to **Hollywood Road** with all its antique shops, a new creative cauldron called the **Tai**

Catch a race day at the Happy Valley Racecourse.

The Victoria Peak Viewpoint on Hong Kong Island is a great spot for pictures with the city's skyline.

Kwun Centre for Heritage & Arts on the grounds of a restored colonial-era police station and prison, and a hip **SoHo** (south of Hollywood) entertainment district flush with restaurants, bars, and the **TakeOut Comedy Club**.

A breezy promenade rambles 2.4 miles (3.9 km) along the central and western waterfront between the **Golden Bauhinia** statue outside the futuristic convention center and **Sun Yat Sen Memorial Park**. Along the way are the **Hong Kong Maritime Museum**, the **Hong Kong Observation Wheel** (a giant Ferris wheel with air-conditioned cabins), and ferry piers with service to the **Outlying Islands**.

By the time the waterfront walkway reaches the **Macau Ferry Terminal**, it's deep into **Sheung Wan**, one of the city's oldest and most evocative neighborhoods. Redbrick,

Edwardian-style **Western Market** has transitioned in recent years from a pungent meat, fish, and produce market into a cluster of cafés and fabric stalls. Uphill from the market is colorful **Cat Street** (Upper Lascar Row) with its chockablock antique and curio stalls. Dedicated to the Taoist gods of literature and martial

arts, nearby **Man Mo Temple** is one of Hong Kong's oldest and most atmospheric shrines (opened 1847).

The waterfront east of Central District is spangled with other crowded, colorful neighborhoods. Nowadays, once notorious **Wan Chai** is better known for excellent eating places than rowdy sailor bars.

BACKGROUND CHECK

• Along with nearby Macau—which Portugal relinquished in 1999—Hong Kong is one of two special administrative regions of China.

• With a population density of 67,600 people per square mile (26,100 per sq km), Hong Kong ranks as one of the world's most crowded cities, along with Dhaka, Bangladesh, and Mumbai, India.

• Hong Kong easily outpaces other high-rise cities like New York and Dubai in the category of most skyscrapers (more than 350 buildings of more than 492 feet/150 m).

• In addition to a large mainland area, Hong Kong boasts 263 islands.

• As of 2020, Hong Kong ranked sixth on the list of countries or territories with the most billionaires—more than Britain, France, or Japan.

Ten Thousand Buddhas Monastery is built on a bamboo-forested hillside.

Beyond the junks and yachts that crowd into its famous typhoon shelter, **Causeway Bay** is a vibrant shopping and nightlife area that spills over into **Happy Valley Racecourse** for thoroughbred ponies and 40,000-seat **Hong Kong Stadium** where the fabled **Hong Kong Sevens** rugby tournament plays out.

Roads lead over the mountains and around both ends of **Hong Kong Island** to the south side, renowned for the floating seafood restaurants of **Aberdeen Harbor**, the roller coasters and water rides of **Ocean Park**, and touristy but tempting **Stanley Market**, where it sometimes seems like everything under the sun is for sale.

The Star Ferry may be the most romantic way to cross **Victoria Harbour**, but you can also ride the **MTR subway** or cruise through an underwater tunnel in one of the city's ubiquitous red-and-white taxis. **Kowloon** looms on the other side, a peninsula that spills down from the jagged "Nine-Dragon" peaks that give the area its name.

Ferry passengers spill onto waterfront walkways leading to the nearby **Hong Kong Cultural Centre** for performing arts, the **Hong Kong Space Museum**, and a giant elongated shopping mall called **Harbour City** with hundreds of shops and eating places.

Just north of Harbour City, a large tract of reclaimed land is being transformed into the **West Kowloon Cultural District**, a decade-long project that already includes the **Xiqu Centre** for traditional Chinese music and opera; the **M+ museum** of modern design, architecture, and moving images; and the **Freespace** complex for contemporary music, dance, and other live events. Over the next few years, the district will also debut a new **Palace Museum** featuring relics from the Forbidden City in Beijing.

Best place to snatch a bird's-eye view of the avant-garde ensemble is the **Sky100** observation deck on the 100th floor of the **International Commerce Centre**. The territory's tallest building (1,587.9 feet/ 484 m) also features the **ICC Light & Music Show**, which is displayed twice each evening on the building's soaring facade.

HIDDEN TREASURES

- **MacLehose Trail:** Hikers flock to a 62-mile (100 km) route through the rural mountains and valleys of the New Territories between the Sai Kung Peninsula and Tuen Mun—with incredible views of the city along the way.

- **Ten Thousand Buddhas Monastery:** A steep walkway flanked by life-size golden Buddhas leads from Sha Tin station to a temple compound with an impressive pagoda and far more than 10,000 statues.

- **UNESCO Global Geopark:** Established in 2015, this wilderness area on the Sai Kung Peninsula highlights Hong Kong's volcanic and sedimentary rock origins with a new visitors center and trailside exhibits.

- **Hong Kong Wetland Park:** Feathered friends and human bird-watchers flock to a nature preserve in the New Territories that features a freshwater marsh, tidal mudflats, butterfly garden, and mangrove boardwalk.

- **Ping Shan Heritage Trail:** Hong Kong's last remaining ancient pagoda is one of more than a dozen vintage buildings along a one-mile (1.6 km) historical route through three villages in the New Territories.

Chaotic **Nathan Road** cuts across the heart of Kowloon, a neon-lit path to legendary shopping spots like **Temple Street Night Market**, **Yau Ma Tei Jade Hawker Bazaar** beneath the overpass on Kansu Street, the fashion-centric **Ladies' Market** along Tung Choi Street, and the electronics shops that cluster in the high-rise buildings of **Mongkok**. Two of the city's best museums—the **Hong Kong Museum of History** and **Hong Kong Science Museum**—are located just east of Nathan Road in the **Tsim Sha Tsui East** neighborhood.

Back on the Hong Kong side, a multitude of ferries connect the central city with outlying islands like **Cheung Chau**, **Peng Chau**, and **Lamma** with their family farms, tiny temples, and over-the-water restaurants. However, the largest and most intriguing of the Outlying Islands is **Lantau**, which can also be reached by road via the immense **Kap Shui Mun** and **Tsing Ma** suspension bridges.

Nearly twice as large as Hong Kong Island, Lantau hosts some of the city's newest and oldest attractions. The island's eastern end is home to **Hong Kong Disneyland** theme park and **Hong Kong International Airport**, which replaced old inner-city Kai Tak Airport in 1998.

The rest of Lantau is dominated by a mountainous spine with hiking trails that ramble up **Sunset Peak**, **Lantau Peak**, and other summits with views across the **Pearl River estuary** in the west and nearly all of Hong Kong to the east. More lofty views await at **Po Lin Monastery**, a hilltop sanctuary crowned by the **Tian Tan Buddha**, one of the world's largest bronze statues. The monastery can also be reached on the **Ngong Ping 360 Cable Car**, which begins its ascent from a station near the airport.

Poised near the western end of Lantau, **Tai O** village affords a rare glimpse of old-time Hong Kong. Sampan boats crowd a small harbor surrounded by *pang uk* (stilt houses) perched above the water. All four of the village temples date from before the 19th century. ∎

The Temple Street Night Market is the longest-running night market in Hong Kong.

Manila
Philippines

One of the world's 10 most populated urban areas, the Philippines capital is making a bold leap into the 21st century without sacrificing its historical heart, bygone charm, or the traditions of millions who have come to live here from around the archipelago.

THE BIG PICTURE

Founded: 1571

Population: 25 million

Size: 700 square miles (1,813 sq km)

Language: Tagalog, English

Currency: Philippine peso

Time Zone: GMT +8 (PST)

Global Cost of Living Rank: 78

Also Known As: Pearl of the Orient, Queen City of the Pacific

Will the real Manila please stand up? Is it the walled city of **Intramuros** with its Spanish colonial treasures? The high-rise hotels and notorious nightclubs of the **Malate** district? Or futuristic **Bay City** with its malls, casinos, and entertainment outlets?

Much of what's truly old Manila falls within the walls of Intramuros, founded in 1571 by Spanish conquistador Miguel López de Legazpi. **Fort Santiago** still guards the mouth of the **Pasay River** as it did hundreds of years ago when the Manila Galleon treasure ships docked there. Philippines hero **José Rizal** is recalled with a museum and memorial inside the fort.

Hulking **Manila Cathedral** (consecrated in 1581) rises nearby, destroyed and rebuilt several times in the wake of earthquakes and warfare. But the old town's architectural star is **San Agustin Church**, a baroque masterpiece completed in 1607 and now the oldest continuously occupied church in the Philippines. The **Museo de San Agustin** displays religious treasures of the colonial era, while the **Cloisters** (filled with tropical plants) provides a quiet reprieve from Manila's helter-skelter streets.

Just across General Luna Street from the church, a replica 19th-century townhouse called **Casa Manila** is decorated with original furnishings and artwork. Farther down General Luna is **Silahis Art & Artifacts**, a museum-like shop packed with textiles, ceramics, basketry, woodwork, and other Filipino handicrafts.

The historic **Intramuros Golf Course**—laid out after the Spanish-American War when the U.S. occupied the Philippines—wraps around the old town in what was once the moat outside the city walls. In addition to its green spaces, neighboring **Rizal Park** harbors the **National Museum Complex** and branches dedicated to anthropology, natural history, outer space, and fine art. The complex safeguards more than three dozen **National Cultural Treasures** ranging from fossilized remains of Tabon Man (who lived between 16,500 and 47,000 years ago) to an astrolabe from the Spanish treasure ship *San Diego*.

Busy **Roxas Boulevard** and the **Baywalk** pedestrian promenade connect the old town with Bay City

Fort Santiago is a citadel built by Spanish conquistador Miguel López de Legazpi.

The San Agustin Church is located in the walled city of Intramuros in Manila.

and its many modern splendors, rising on land reclaimed from Manila Bay. The sprawling **Cultural Center of the Philippines** offers a year-round slate of music, dance, drama, film, and fine art exhibitions. The district boasts several Las Vegas–style **casino resorts**, an amusement park called **Star City**, and the "dancing" **Fountain at Okada**, which springs to life at the top of each hour. Bay City's single largest feature is the massive **SM Mall of Asia**, with more than 1,300 shops and 400 restaurants.

Manila's pivotal role in World War II is reflected in the **American Military Cemetery & Memorial**, where more than 17,000 U.S. and Allied troops are buried. Ferries make the 75-minute crossing to heavily fortified **Corregidor Island** at the entrance to Manila Bay, where American and Filipino troops held out for months against the invading Japanese. Guided tram and walking tours are offered on the tadpole-shaped island, and there's a resort hotel for those who want to split the trip into two days. ■

LOCAL FALVOR

• **The Aristocrat:** Chicken and pork barbecue are the specialties at this oldie but goodie, a Manila institution since before World War II; *423 San Andres St. and Roxas Blvd., Malate.*

• **MilkyWay Café:** The food really is out of this world at this local favorite. The MilkyWay Building boasts Thai, Filipino, Spanish, and Japanese restaurants all under one roof. The MilkyWay Café has a 50-year history and is celebrated for its chicken asparagus sandwich, crispy *hito* (deep-fried catfish), and *halo-halo* dessert; *900 Antonio Arnaiz Ave., Makati;* cafe.milkywayrestaurant.com.

• **Purple Yam:** Modern takes on traditional Filipino dishes like *escabeche* (filleted fish), *binakol* (chicken soup), and *lumpia* (spring rolls) rule the menu at this atmospheric eatery in a restored 1949 home; *603 Julio Nakpil St. at Bocobo, Malate;* purpleyamph.com.

• **Abe:** Kapampangan cuisine from central Luzon Island—including offbeat dishes like *betute* (stuffed frogs) and *camaru* (adobo-style crickets)—is the forte of this popular chain with six outlets around metro Manila; *SM Mall of Asia, Bay City.*

Hanoi & Ho Chi Minh City

Vietnam

Although they both came of age during the French colonial era, Vietnam's largest cities have evolved into strikingly different urban areas, Hanoi the somewhat sober national capital and Ho Chi Minh the budding business and entertainment hub.

HANOI

Old timers still call it "36 Streets," a name conferred on Hanoi's **Old Quarter** in the early 20th century, when each lane focused on a specific trade or craft. Although the past endures in names like **Hàng Bạc** (Silver Street), **Hàng Cá** (Fish Street), and **Hàng Gai** (Hemp Street), the artisanal alignment of olden days has evolved into a modern-day mixed bag of eating, drinking, and shopping establishments in what is easily Hanoi's most vibrant neighborhood.

Hanoi's lively **Weekend Night Market**—a jumble of food, clothing, electronics, and handicrafts—sprawls the length of Hàng Đào and Hàng Đường streets. **Lotus Water Puppet Theatre** offers a unique form of Vietnamese entertainment, while the **Hanoi Opera House** hosts ballet, symphony, opera, and the occasional rock concert in a French colonial theater built in 1911.

Visit the 18th-century Ngoc Son Temple (or Temple of the Jade Mountain) in Hanoi.

THE BIG PICTURE

Founded: Hanoi, third century B.C.; HCMC, 1698

Population: Hanoi, 8.3 million; HCMC, 11 million

Size: Hanoi, 335 square miles (868 sq km); HCMC, 635 square miles (1,645 sq km)

Language: Vietnamese

Currency: Vietnamese dong

Time Zone: GMT +7 (ICT)

Global Cost of Living Rank: Hanoi, 139; HCMC, 143

Also Known As: Hanoi—City for Peace; HCMC—Saigon, Prey Nokor (ancient Khmer)

By daylight, stroll across the red wooden **Welcoming Morning Sunlight Bridge** to the **Temple of the Jade Mountain** on an island in **Hoàn Kiếm Lake** or duck into **Bach Ma Temple** and the **Temple of Literature**, both around 1,000 years old. For a secular time trip, **Heritage House** is a restored merchant's residence in the Old Quarter.

Vietnam's tumultuous 20th-century history comes to light at sights like **Hoả Lò Prison Museum** with its famous **Flag Tower**, the weapon-filled **Military History Museum**, the rebuilt **Long Bien Bridge** across the Red River, and **Ho Chi Minh Mausoleum**.

HO CHI MINH CITY

Once called Saigon, Vietnam's largest city was renamed for the revolutionary leader at the end of the Vietnam War in 1975. After languishing for decades as the nation rebuilt, Ho Chi Minh City (HCMC) began to blossom again in the early 21st century. Cars and scooters have replaced the abundant bikes and *cyclos* of old, while

Ho Chi Minh City, once known as Saigon, is the most populous city in Vietnam.

high-rise construction continues to transform the skyline.

The past endures along tree-shaded **Đồng Khởi Street**, which traverses the French colonial old town (District One) between the historic **Hotel Majestic** overlooking the **Saigon River** and redbrick **Notre-Dame Cathedral** (consecrated in 1880) with its distinctive Romanesque bell towers. Scattered along Đồng Khởi are the elegant **Municipal Theatre** (1900), the elaborate **Central Post Office** (1891), and **Union Square**, a new cluster of luxury boutiques disguised behind a faux–French colonial facade.

A block over, a Paris-like pedestrian promenade along **Nguyễn Huệ Street** connects the river with the **People's Committee Building** (opened in 1908 as Saigon City Hall)—its flamboyant French colonial architecture a startling contrast

to the nearby skyscrapers. Snatch a bird's-eye view of the ever evolving city from the **Saigon SkyDeck** on the 49th floor of **Bitexco Financial Tower**.

On the outskirts of the central city, two vastly different neighborhoods also reflect the yin and yang of 21st-century HCMC. **Cholon**, the

city's ethnic Chinese area, is filled with atmospheric shrines like **Ba Thien Hau Temple** and **Phuoc An Hoi Quan Pagoda**, as well as the bustling **Binh Tay Market**. On the other hand, a colossal new residential, retail, and entertainment district called **The Crescent** is more like something out of modern Singapore. ■

URBAN RELIC

Huế

When Vietnam was united under a single dynasty in 1802, Emperor Gia Long mandated the construction of a new capital city along the north bank of the Perfume River.

Protected within six miles (10 km) of walls and moats, **Huế** would eventually include the Imperial Citadel and, within that, the Tu Cam Thanh (Forbidden Purple City) where

the Nguyen dynasty royal family resided until 1945, when they were deposed.

Largely destroyed during the 1968 Tet Offensive of the Vietnam War, restoration work is ongoing and likely to continue for decades. The Meridian Gate, Thế Miếu temple, and Điện Thái Hòa throne hall are among the structures that have already risen from the ashes of this UNESCO World Heritage site.

Bangkok
Thailand

The epitome of an exotic Asian city, Bangkok continues to dazzle visitors with is glittering golden pagodas, saffron-clad monks, fiery Thai cuisine, shop-until-you-drop markets, and a brand of homegrown hospitality that few urbanites around the globe can match.

THE BIG PICTURE

Founded: 15th century A.D.

Population: 16 million

Size: 1,175 square miles (3,040 sq km)

Language: Thai

Currency: Thai baht

Time Zone: GMT +7 (ICT)

Global Cost of Living Rank: 46

Also Known As: Krung Thep, the Big Mango, City of Angels

Like the other "City of the Angels" on the far side of the Pacific, Bangkok doesn't really have a single nexus, a place that can really be called the city center. Among its 50 districts, there's the nebulous central business district around **Lumphini Park**, nightlife-heavy **Ratchada**, shopaholic **Siam**, the skyscraper forest along **Silom Road**, and bustling **Sukhumvit** with its retinue of hotels and restaurants.

Though any of those places might lay claim to the title of Bangkok's heart and soul, none come close to the significance of the **Royal City** (or Phra Nakhon to use its Thai name) along the big bend in the **Chao Phraya River**. It may not have skyscrapers, five-star hotels, or glitzy malls, but that's where Bangkok's reign as Thailand's capital was born nearly 250 years ago.

Undoubtedly the city's most recognizable landmark, the **Grand Palace** has been the official residence of Thailand's monarch since 1782, when King Phutthayotfa Chulalok (Rama I) moved the capital of his kingdom to the east bank of the Chao Phraya. The tradition continues under the current monarch, King Maha Vajiralongkorn (Rama X), although for many years the royal family has actually lived in the less grandiose **Dusit Palace**.

Reflecting the pinnacle of traditional Thai architecture, the Grand Palace compound shelters more than 100 structures. Only a handful are open to the public, including **Chakri Mahaprasad** (Grand Palace Hall) with its weapons museum, **Queen Sirikit Museum of Textiles**, **Dusit Maha Prasat** audience hall, and famed **Wat Phra Kaew** (Temple of the Emerald Buddha), which safeguards a 14th-century jade Buddha whose robes are changed seasonally by the king in keeping with the Buddhist calendar.

Directly south of the palace is **Wat Pho**, the legendary Temple of the Reclining Buddha. Renowned for its giant, golden effigy at 150 feet (46 m) long, the temple was also Thailand's first public university and now serves as a leading school for traditional Thai medicine and massage (including short courses and massage treatments for

Paddle Swan Boats in Lumphini Park, surrounded by the Bangkok skyline.

The Temple of the Emerald Buddha outside the Grand Palace is just one of nearly 400 temples in Bangkok.

visitors). Having pampered your muscles, let your nose run amok at nearby **Pak Khlong Talat flower market**, the city's main source for *phuang malai* (floral garlands given as religious offerings).

North of the Grand Palace, the oblong **Sanam Luang** royal parade ground is the best place to snap panoramic pictures of the exotic palace skyline. Perched at the northern end of the green are Thailand's **National Gallery** of art, the **National Theatre**, and the **National Museum**, which includes the wonderful **Red House** residence of early 19th-century Princess Sri Sudarak.

The Phra Nakhon area is also home to **Wat Saket**—the fabulous

Golden Mount with its wide-ranging city views—and **Rajadamnern Stadium**, the nation's oldest and most renowned arena for Muay Thai (Thai boxing). The fascinating **Amulet Market** near the National Museum attracts both tourists and Buddhist pilgrims to its vast selection of statues, medallions, coins, and other trinkets meant to engender good luck or ward off the bad kind.

From several docks near the Grand Palace and Wat Pho, it's possible to hop on and off the **Chao Phraya Express** water taxis or hire one of those loud and speedy *rua hang yao* (long-tail boats) that run up and down the **Chao Phraya** to dozens of other Bangkok landmarks and neighborhoods.

Directly across the "River of Kings" from the Royal City are the **National Museum of Royal Barges** with its incredible gold leaf–covered fleet and **Wat Arun** (Temple of the Dawn) crowned by an impressive Khmer-style *prang* (tower) that towers 262 feet (80 m) above the waterfront. The west bank also hosts the offbeat **Bangkok Forensic Museum** (aka Museum of Death) at Siriraj Hospital, a treasure trove of pathological, forensic, parasitological, and anatomically abnormal specimens. Not for the faint of heart.

Farther downstream is crowded, noisy, and thoroughly enchanting **Chinatown**, one of the few places in Bangkok (other than the temples and royal relics) that seems to have changed very little with the march of time.

Looming above the area's countless shops and stalls is **Wat Traimit** (Temple of the Golden Buddha).

Wat Arun Ratchawararam is a Buddhist temple built in 1656 B.C.

Local vendors sell food and fresh fruits from boats at the Taling Chan Floating Market.

Although created in the 13th century, the six-ton (5.4 metric tons) statue was covered in plaster. Nobody realized it was solid gold until 1955, when it was accidentally dropped and the casing cracked. On the southern edge of Chinatown, the **Bangkokian Museum** illuminates upper-middle-class life in early 20th-century Thailand via three beautifully preserved teak houses.

Another couple of stops down the river is a landing for **Silom Road** and its forest of modern skyscrapers. Tallest of all is the **King Power Mahanakhon**, a 1,031-foot (314 m)

URBAN RELIC

Angkor

A one-hour flight from Bangkok, the ruins of **Angkor** in northwestern Cambodia rank with the Great Pyramids and Parthenon among the all-time great achievements of civilization.

Capital of the Khmer Empire between the ninth and 15th centuries A.D., the city sprawled across more than 150 square miles (388 sq km) and was the globe's largest urban area prior to the industrial revolution.

Attacked and looted by Thai invaders in the 15th century, Angkor fell into ruins and was lost in the jungle for hundreds of years until rediscovered by French archaeologist Henri Mouhot in 1860.

The national archaeological park and UNESCO World Heritage site that preserves the city today includes 70 individual sites accessed from nearby Siem Reap.

Most famous of these is the incomparable Angkor Wat, built as the funerary temple of 12th-century King Suryavarman II. Among its other celebrated structures are Ta Prohm, the "Tomb Raider Temple" strangled by roots of giant rainforest trees; the pink-hued Banteay Srei with its marvelous decoration; and the giant stone heads of Angkor Thom.

Vintage toy trucks and race cars are on display at the Chatuchak Weekend Market.

and lively traditional Thai puppet shows.

Away from the river, the **Siam** area along **Rama I Road** is one of the city's more intriguing neighborhoods. It's best known for the bountiful shopping around **Siam Square**, with many of the malls and retail lanes connected by pedestrian bridges that hover above the bumper-to-bumper traffic.

But there's plenty of culture, too. **Bangkok Art and Culture Centre** (BACC) brandishes contemporary art in an appropriately modern building that highlights some of Thailand's most celebrated artists, including Lampu Kansanoh, Denpong Wongsaroj, and Bundit Padungvichian.

The **Jim Thompson House Museum** honors the marvelous silk produced in Thailand and the American expatriate who resurrected the then stagnant industry after World War II before mysteriously disappearing in Malaysia. Visitors can purchase silk clothing and decorative items in the museum store, or escape the city's heat, hustle, and bustle in the lush gardens that surrounded the teakwood house.

Bangkok shopping reaches a fever pitch at the sprawling **Chatuchak**

behemoth distinguished by its unusual cuboid crown and a spiral that wraps around the middle of the building. The **Skywalk** summit features several vertigo-inducing vantage points, including a **Glass Tray** deck with a transparent floor.

This part of the river has seen a building boom in recent years. Historic structures like the 19th-century **Mandarin Oriental Hotel** are now overshadowed by massive new creations like the twin towers of **Iconsiam**, which also features an outdoor sound-and-light fountain along the riverfront and an indoor floating market called **Sook Siam**. Another new waterfront developed, **Asiatique The Riverfront** offers a similar mix of modern shops and restaurants, as well as a giant Ferris wheel, Las Vegas–style cabaret dinner show, Muay Thai demonstrations,

LAY YOUR HEAD

• **Mandarin Oriental:** This hallmark for luxury and service in the Thai capital recently underwent a huge renovation; restaurants, bars, spa, fitness center, pool, yoga, tennis courts, Muay Thai, teakwood shuttle boats; from $640; mandarinoriental .com/bangkok.

• **The Salil:** Located just off Sukhumvit Road in the trendy Thonglor area, the Salil merges modern Thai style and 19th-century European elegance; restaurant,

tearoom, shop, rooftop pool, fitness center; from $310; thesalilhotel.com.

• **Okura Prestige:** Five-star luxury in the heart of Bangkok with skyline views and triple-glazed low-E coated windows to ensure a quiet haven; restaurants, bar, pool, spa, fitness center; from $192; okurabangkok .com.

• **Anantara Riverside:** Set amid 11 acres (4.4 ha) of greenery on the west bank of the Chao Phraya,

this urban oasis features low-rise accommodations and plush tropical gardens; restaurants, bars, spa, pool, fitness center, tennis courts, kids' club, alms giving; from $123; anantara.com/en/riverside-bangkok.

• **Phranakorn-Nornlen:** Eco-friendly boutique hotel with 31 rooms down a quiet alleyway in the Phranakorn district; organic restaurant, workshops; from $51; phranakorn -nornlen.com.

Market. Reputedly the world's largest outdoor market—and easily the biggest in Thailand—the weekend-only market covers an incredible 35 acres (14 ha). Its 15,000-plus stalls are divided into 27 sections offering just about any and everything from clothes to electronics to food, furniture, plants, handicrafts, and antiques.

Chang Chui Plane Night Market in **Thonburi** is equally eclectic but far more hip, a contemporary take on the traditional Thai market that features live music, craft beer, and epicurean delights, as well as art galleries, handcraft stalls, and vintage clothing. Why the strange name? Because the market revolves around an old Lockheed L-1011 TriStar jetliner that hovers above the perfectly named **Runway Bar**.

Several other worthwhile sights are scattered around the periphery of Bangkok. **Samut Prakan Ancient City**, a folksy theme park near the Gulf of Thailand, features more than 100 scaled-down models of the nation's major temples and monuments spread across a landscaped garden shaped like the map of Thailand.

Papaya Studio is a massive warehouse showroom in the **Wang Thonglang** district that features what seems like an infinite number of antiques, handicrafts, decorative arts, and curiosities sourced from all over Thailand. Whether you're buying or just browsing, it's a jaw-dropping spectacle.

Although **Damnoen Saduak Floating Market** is the most well known, it's far too touristy these days. A more authentic aquatic emporium is **Amphawa Floating Market**, around 55 miles (89 km) southwest of Bangkok, or **Taling Chan Floating Market** in Thonburi.

Other than riverboats, one of the best ways to get around Bangkok is the **BTS Skytrain**, a clean, efficient, and refreshingly air-conditioned mass-transit system with 48 stations across six lines. Just for fun, visitors should also take at least one jaunt in a *tuk-tuk* three-wheeler or **motorbike taxi**. ∎

A woman follows traditional methods to craft Thai silk at the Jim Thompson House Museum.

Luang Prabang

Laos

Deemed the best preserved traditional city in Southeast Asia, Luang Prabang is a throwback to the days when it was the royal capital and Buddhist epicenter for the ancient kingdom of Laos.

THE BIG PICTURE

Founded: seventh century A.D.

Population: 55,000

Size: 3.1 square miles (8.2 sq km)

Language: Lao (Laotian)

Currency: Lao kip

Time Zone: GMT +7 (ICT)

Global Cost of Living Rank: Unlisted

Also Known As: Muang Sua (7th–11th centuries); Xieng Dong Xieng Thong (11th–14th centuries)

Sprawling on a thumb-shaped peninsula between the Mekong and Nam Khan Rivers, Luang Prabang survived the conflicts that ravaged Southeast Asia during the 20th century to such an extent that the entire city was declared a UNESCO World Heritage site in 1995.

The epitome of an exotic tropical urban area, Luang Prabang renders golden stupas and riverside palaces, saffron-clad monks and French colonial villas, the mulchy aroma of the Mekong River blending with the sweet smell of incense drifting from dozens of temples.

Mount Phou Si, with its glimmering golden stupa, provides a panoramic view of the Old Town and Mekong River Valley. Arrayed around the base are smaller shrines and the popular **L'Etranger bookstore and teahouse**, which doubles as a local cultural center with art exhibits and movie nights.

Wedged between Mount Phou Si and the Mekong, the **Royal Palace** was home to the Laotian imperial family between 1904 and 1975, when the communist Pathet Lao overthrew the monarchy. In addition to a main building with a **throne room** and the king's massive **Erawan bed**, the palace complex includes a garage with vintage royal vehicles (like a Ford Edsel sedan) and a temple housing the city's namesake, the golden **Phra Bang Buddha**. Three times each week, the **Royal Ballet Theatre** on the palace grounds stages the dazzling *Phra Lak Phra Lam*—the Lao version of the Ramayana epic.

The cadres may have ousted the royals, but they couldn't quash the city's spiritual gravitas. Faith remains Luang Prabang's heart and soul, and monks its most conspicuous inhabitants, most of them bright-eyed youngsters lured from the countryside by the prospect of a free education, a solid roof over their heads, and three meals a day.

With its sweeping roofline and elaborate mosaics, **Wat Xieng Thong** is the most prominent of the city's 33 temples and monasteries. Once a regal place of worship, the "Golden City" was a venue for royal investitures and funerals. The **Royal Carriage House** on the grounds of Wat Xieng Thong temple contains an

Laotian Buddhist monks walk along the street for the morning alms procession.

The Nam Khan River flows through the city of Luang Prabang, with the Phou Si hills in the background.

ornate wooden hearse decorated with Naga snakes.

Among the other must-see shrines are the lavishly decorated **Wat Mai Suwannaphumaham** and the 16th-century **Wat Wisunarat** with its giant golden Buddha and religious art museum. Be sure to rise at dawn and watch the **alms procession** by hundreds of monks through the Old Town.

Markets are another integral part of the Luang Prabang cityscape. The aromatic **Morning Market** is a warren of narrow lanes and jam-packed shops where many locals buy their foodstuffs. Sprawling along nearby Sisavangvong Road, the city's celebrated **Night Market** blends street food and handicraft hawkers. Souvenirs are also available at **Ock Pop Tok Living Crafts Centre** and the **Traditional Arts and Ethnology Centre (TAEC)** museum shop, which features artisan items

from the nation's 20 ethnic groups.

The countryside around Luang Prabang offers treasures of its own: **Kuang Si Falls** and the nearby **Bear Rescue Center**, as well as **Living Land** community farm where visitors can learn how to plant, cultivate, and harvest rice.

Numerous outfitters offer boat trips on the **Mekong River**, including excursions to the **Pak Ou Caves**, a network of limestone caverns decorated with more than 4,000 Buddhas. **Luang Say** offers overnight riverboat cruises between Luang Prabang and the Golden Triangle. ■

LOCAL FLAVOR

• **L'Elephant Restaurant:** Tucked inside a restored art deco building, the menu features French and local cuisine, including a *menu dégustation* with eight Lao dishes; *Kounxoua Road, Ban Vat Nong;* elephant-restau.com.

• **Paste at the Apsara:** Artistically presented nouvelle Lao cuisine in a chic modern setting; *Apsara Hotel, Kingkitsarath Road;* pastelaos.com.

• **3 Nagas:** A restored 1898 French villa provides a romantic venue for an eatery that serves deep-fried Mekong riverweed with spicy buffalo jam, steamed fish in lemongrass, flambéed mango in Lao whiskey with coconut sorbet, and other local delights; *18/02 Khem Khong;* 3-nagas.com.

• **Le Banneton:** Freshly baked baguettes, croissants, crepes, and croque-monsieur flavor the menu at this Parisian-style bakery and bistro; *03/46 Sakkhaline Road;* facebook.com/lebannetonluangprabang.

Yangon

Myanmar

As it transitions into a modern metropolis, Myanmar's capital clings to its ancient ways and means, traditions that endure in the city's celebrated Buddhist shrines.

THE BIG PICTURE

Founded: 11th century A.D.

Population: 5.3 million

Size: 230 square miles (596 sq km)

Languages: Burmese

Currency: Myanmar kyat

Time Zone: GMT +6:30 (MST)

Global Cost of Living Rank: 104

Also Known As: Rangoon (British colonial)

Whether you know the city by its former moniker (Rangoon) or the revised name declared by the military junta in the 1980s (Yangon), the Burmese metropolis remains one of the planet's most exotic destinations.

Yangon is undergoing a fast-paced cultural and economic evolution that has spawned mobile phones, steel-and-glass skyscrapers, and bumper-to-bumper traffic since the turn of the 21st century.

Yet there are plenty of vestiges of old Rangoon: Incense wafting through a golden temple, the smell of spices in an open-air market, or the mulchy scent of the great rivers that frame the city—aromas alone are enough to conjure visions of bygone Asia.

But the real time warp is **Shwedagon Paya**. "A beautiful winking wonder that blazed in the sun," is how British author Rudyard Kipling described it in the 1890s. And if you don't count the escalators that carry pilgrims and visitors to its hilltop location, Shwedagon has barely changed since then, a 326-foot-tall (99 m) pagoda decorated with an estimated US$3 billion worth of gold and gemstones.

The plazas and pavilions surrounding the stupa feature fortune-tellers, flower sellers, and meditation sessions, while the saffron-clad monks or pink-clad nuns who hang out around the grounds are always game for a photo or practicing their English. And Shwedagon is especially evocative after dark during candle- and incense-lighting ceremonies.

Hundreds of other Buddhist shrines are scattered around the city. Among the most noteworthy are **Chaukhtatgyi Paya** with its 215-foot (65 m) diamond-encrusted reclining Buddha, the giant golden-winged seated Buddha of **Ngahtatgyi Paya**, and the 2,600-year-old **Sule Paya** that rises in the middle of a traffic circle in downtown Yangon.

Yangon's city center is still dominated by classic British-colonial structures like the **Strand Hotel** (opened in 1901), the sprawling red-brick **Secretariat** (built 1890–1905), and a stupendous 1930s **City Hall** that combines art deco and Burmese motifs. Largest of the downtown markets, **Bogyoke Aung San** is home to more than 2,000 stalls, including fruit and vegetables, clothing and household goods, and the city's best selection of Burmese art, antiques, jewelry, and handicrafts.

Located right beside the market, **Holy Trinity Anglican Cathedral** (built in the 1980s) renders a quiet respite from the helter-skelter street scene and a moving tribute to Allied

Purchase a traditional wooden puppet at the Bogyoke Aung San Market.

The Shwedagon Paya is the most sacred Buddhist pagoda in Myanmar. Its spire stands 367 feet (112 m) tall.

troops who lost their lives liberating Burma from Japanese occupation during World War II. A 10-minute walk to the south, **Musmeah Yeshua Synagogue** (opened in 1896) offers further testament to the city's diverse religious roots.

Looking more like a pagoda than a transport hub, the **Central Railway Station** (opened in 1877) is both an architectural wonder and the place to catch the **circular train**—both a commuter line and a great way to explore Greater Yangon. The three-hour journey makes whistle stops at 39 stations.

The muddy Yangon and Bago Rivers remain hardworking waterways rather than leisure or recreational attractions. That's because the city is blessed with amazing lakes. Shwedagon Pagoda looms above **Kandawgyi Lake**, surrounded by parkland and waterfront restaurants like the royal barge–shaped **Karaweik Palace** that stages a nightly Burmese cultural show.

In the northern suburbs, much larger **Inya Lake** offers lakeside trails and visitor boating and paddleboarding at the **Yangon Sailing Club** (established in 1924 as the Rangoon Yacht Club). ■

URBAN RELIC

Bagan

Bagan (Pagan) flourished between the ninth and 13th centuries A.D. as the capital of an empire that united, for the first time, all of what is now Myanmar (Burma).

Other than Angkor, no other Southeast Asian site boasts as many ancient structures—more than 2,200 pagodas, temples, monasteries, and lesser shrines scattered along the eastern bank of the Irrawaddy River around 400 miles (644 km) north of Yangon.

More than 3,000 historic temples stretch across the Bagan plain. Among its enduring treasures are the colossal golden Buddhas that hunker inside Ananda Temple, the massive redbrick Dhammayan, and the soaring Shwesandaw Pagoda, a popular spot to watch sunrise or sunset over the river and ruins.

Bagan can be explored on van tours, oxcart, or horse carriage, by foot or bike, and via hot-air balloon.

Singapore
Singapore

Singapore has morphed from humble beginnings into a global financial center, shopping oasis, and tourism hub while fostering its ethnic diversity and a "city in a garden" philosophy that has transformed the tropical isle into Asia's greenest city.

THE BIG PICTURE

Founded: 1819

Population: 5.7 million

Size: 200 square miles (518 sq km)

Language: English, Mandarin, Malay, Tamil

Currency: Singapore dollar

Time Zone: GMT +8 (SST)

Global Cost of Living Rank: 7

Also Known As: Lion City, Temasek (ancient), Singapura (ancient)

When Singapore achieved its independence in the 1960s, founding prime minister Lee Kuan Yew envisioned both a business powerhouse and a green metropolis where abundant trees and other flora would soften the harshness of the asphalt jungle. The tiny city-state at the tip of the Malay Peninsula has achieved both of those goals—although hardly anyone outside of Singapore seems to know about the latter.

In addition to the millions of trees that now blanket the island, the government and private enterprise have added recreational, ecological, and tourist attractions into the green mix. Singapore's national parks board manages more than 350 parks and gardens, figures that would make any urban area proud but that are especially significant in a small island state.

One of Singapore's most significant achievements is how much rainforest and mangroves it has managed to preserve in parks like **Bukit Timah Nature Reserve** with its colossal tropical trees, **Labrador Nature Reserve** along the south shore, and **Sungei Buloh Wetland Reserve**, the only place where saltwater crocodiles are still seen. New on Singapore's green scene is the 22-mile (36 km) **Coast-to-Coast Trail**, which stretches across the middle of the island between **Coney Island Park** in the Johor Strait and the recently revamped **Jurong Lake Gardens**.

But the biggest patch of jungle is protected within the confines of the **Central Catchment Nature Reserve**, which derives its name from four reservoirs that "catch" drinking water for the island. In addition to hiking and biking trails, the vast catchment area offers paddle sports, bandstand concerts, and a **TreeTop Walk** swinging bridge through the rainforest canopy. It also nurtures indigenous wildlife—macaques and banded leaf monkeys, spitting cobras and vipers, endangered mammals like the colugo and slow loris, and hundreds of bird and insect species.

See species like the sun parakeet at Jurong Bird Park, the largest of its kind in South Asia.

Cross the Helix Bridge at night to see the city in lights, including the white lotus-shaped ArtScience Museum.

The island's forested heartland also provides a home for the well-respected **Singapore Zoo** and its efforts to save tropical and equatorial species through captive breeding programs. Sprawling across 64 acres (26 ha) beside **Seletar Reservoir**, the zoo safeguards more than 2,400 animals across 300-plus species in habitats that mimic their natural environment. And some of them are quite rare: Malay tigers, Vietnamese langurs, cotton-top tamarins, and Komodo dragons.

Over the years, the zoo has hatched two nearby siblings. Visitors explore the after-dark **Night Safari** on trams or footpaths through habitats with 900 nocturnal animals from around 100 species (over 40 percent of them threatened). **River Safari**, the latest offspring, features a Disney-like boat cruise and walking trails through habitat zones like **Wild Amazonia**, the **Yangtze River**, and **Giant Panda Forest**. Singapore's zoological authority also operates the popular **Jurong Bird Park**, where giant walk-through aviaries showcase around 3,500 birds from 400 species (20 percent of them threatened in the wild).

What didn't come naturally, Singapore simply created—like the parklands and attractions located on landfill around **Marina Bay**. A mind-blowing botanical theme park called **Gardens by the Bay** features the world's tallest indoor waterfall (100 feet/30 m)—which plunges over an artificial cliff inside a massive indoor cloud forest—and canopy walks through the gigantic **Supertree Grove** inspired by the movie *Avatar*.

On the other side of **Helix Bridge**

(its spiral shape inspired by DNA), the **Marina Promenade** offers the **Singapore Flyer** observation wheel and the waterfront **Marina Bay** circuit, which hosts the annual Formula One **Singapore Grand Prix**. Guarding the strategic point where the **Singapore River** empties into Marina Bay is the famed **Merlion** statue, a strange half-lion/half-fish creature that is Singapore's official mascot. Its name and concept derives from combining the French word for sea *(mer)* and the animal that features in the city's ancient name (Singapura or "Lion City"). All of this is on reclaimed land that 50 years ago was part of the **Singapore Strait** that helps connect the Indian Ocean and the South China Sea.

Marina Bay is nearly engulfed by Singapore's famous skyline. Although bank buildings and condo blocks are taller, the structure that catches everyone's eye (and camera) is the stunning **Marina Bay Sands**. The 57-story casino-hotel complex is topped by the world's largest overhanging cantilever deck, a massive surfboard-shaped platform that includes a **SkyPark** with bars, restaurants, observation areas, and a vertigo-inducing infinity pool.

Beyond the shiny skyscrapers are the historic districts laid out by Sir Stamford Raffles when he devised the town plan after the island became a British colony in 1819. Figuring that his newborn trading post would become "a place of considerable magnitude and importance," Raffles formulated a city divided into three distinct ethnic

The Supertree Grove is a unique vertical garden of solar-powered artificial trees that "sing." You can walk the Skyway that connects the trees.

LOCAL FLAVOR

- **328 Katong Laksa:** This humble-looking shop front serves up the city's most celebrated *laksa* (spicy Peranakan noodle soup). What it lacks in ambience, the little eatery more than makes up for with incredible taste; *51 East Coast Road;* 328katonglaksa.sg.

- **Warong Nasi Pariaman:** Modest shophouse restaurant famed for its *nasi padang* (Indonesian fried rice) since 1948; *738 North Bridge Road.*

- **Candlenut:** Old family recipes feature in many of the dishes at the first Peranakan-style restaurant anywhere in the world to earn a Michelin star.

The menu includes 100 combinations of one-bite dishes and a 20-course tasting menu; *Block 17A Dempsey Road;* comodempsey.sg/restaurant /candlenut.

- **Old Airport Road Food Centre:** It may not be an appetizing name—derived from the fact that it sits on a runway at old Kallang Airport—but this assembly of 150 hawker stalls is the island's most popular food court; *51 Old Airport Road.*

- **Thevar:** Modern Indian eatery with unique twists on classic dishes like tandoori lamb, crab curry rice, and chicken roti; *9 Keong Saik Road;* thevar.sg.

---

enclaves and a government area. All of them are still alive and thriving, their historic buildings meticulously restored and many adapted to modern uses as restaurants, bars, galleries, and boutique hotels.

In a place as multicultural as Singapore, it comes as no surprise that Chinatown also harbors the city's oldest Hindu and Muslim shrines: **Sri Mariamman Temple** (opened in 1827) and **Masjid Jamae** (built in 1826), as well as the atmospheric **Thian Hock Keng,** the city's oldest Taoist temple (established in 1839). On the other hand, the **Buddha Tooth Relic Temple**—with its tranquil rooftop garden, relic-filled museum, and mandala-shaped interior—is a relatively new creation, finished in 2007.

Many of the neighborhood's restored shophouses (retail on the ground floor, residential space above) offer arts and crafts, clothing, jewelry, traditional Asian remedies, and some of the world's best Chinese food. One of the eateries—**Liao Fan Hawker Chan Chicken Rice & Noodle Stall**—has earned a Michelin star. **Lau Pa Sat market**, **Maxwell Food Centre**, **Neil Road**, and rapidly gentrifying **Ann Siang Hill** are among other places to tempt your taste buds in Chinatown. And Smith Street has so many restaurants that it's now called Chinatown Food Street.

On the northern side of the Singapore River, **Kampong Glam** is Singapore's oldest Muslim neighborhood. Originally allocated to the Malay, Arab, and Bugis communities, the area was one of the seats of the island's Malay royalty. The old **Istana** (royal palace) is now occupied by the **Malay Heritage Centre** museum, while nearby **Sultan Mosque** and its massive golden dome look much the same today as they did in the 1920s when they were constructed. Notice that many of the streets in Kampong Glam are named after cities in the Middle East, like Kandahar, Muscat, Baghdad, and Basra.

Although Raffles was designated along the Singapore River for Indian residents, migrant workers from the

Explore the city's nightlife at cafés and restaurants along the riverfront.

subcontinent eventually moved farther out to a rural area where they could raise cattle and other livestock. It's now called **Little India**, a long stretch of **Serangoon Road** flanked by banana-leaf curry and tandoor restaurants, and shops selling spices, saris, and all manner of goods imported from India. **Sri Veeramakaliamman**, one of Singapore's oldest Hindu temples, is renowned for its colorful statues that embellish the interior and central tower.

Located near the confluence of Marina Bay and the Singapore River, the **Civic and Cultural District** harbors a cluster of colonial-era buildings that now house many of the city's leading cultural institutions. The **National Gallery Singapore**

displays its significant collection of Southeast Asian art in two national monuments, the former **City Hall** and **Supreme Court Building**. **Empress Place**, an elegant Victorian-era government office building, is now the primary campus of the **Asian Civilisations Museum** and its impressive array of artifacts from around the continent. The **National Museum of Singapore**, which rambles through a gorgeous neoclassical building constructed in the 1840s, charts the nation's past, present, and future through digital and experimental exhibitions.

St. Andrew's Anglican Cathedral (consecrated in 1862) was built by Indian convicted laborers. The Madras chunam plaster covering its

stark white walls is a mix of egg white, eggshell, lime, sugar, coconut husk, and water. Even older are the **Armenian Apostolic Church of St. Gregory the Illuminator** (opened in 1836) and the Roman Catholic **Cathedral of the Good Shepherd** (finished in 1847). The **Convent of the Holy Infant Jesus** (founded in 1852) has been converted into a popular hangout called **Chijmes** that includes a craft brewery, whiskey bar, and winery tapas bar.

Rising behind the historic core is **Fort Canning Hill**, where much of Singapore's history unfolded. The 14th-century kings of Singapura had their palaces here, where Raffles built his bungalow (so he could survey his civic creation), and where the British Army based its high command for East Asia. Nowadays, the park is divided into nine historically themed gardens.

On the far side of Fort Canning and the National Museum is **Orchard Road**, the city's most popular shopping street long before any of the fancy new malls came along and still a vibrant retail corridor. Tucked among the luxury department stores are tiny snatches of the past like **Emerald Hill** and its lovingly restored Peranakan homes, many dating from the late 1800s. Peranakans—people of mixed Chinese and Malay/Indonesian heritage—were also instrumental in developing the lovely **Katong/Joo Chiat** area on Singapore's east coast.

Many other neighborhoods beg exploration, from old British military areas like **Dempsey Hill** and art deco–filled **Tiong Bahru** to modern reincarnations like **Clarke Quay**, **Robertson Quay**, and **Boat Quay** in restored shophouses and godowns (warehouses) along the Singapore River.

Singapore also boasts some interesting outlying islands. The country's last remaining kampong (traditional Malay village) is found on **Pulau Ubin**, a rural island that features a hiking and biking route through old coconut and rubber plantations. The tiny **Southern Islands** in the Singapore Strait—reached by ferry from Marina Bay—boast pristine beaches, picnic areas, Malay shrines, and an ancient Chinese temple.

Reached via causeway and cable car, **Sentosa Island** has evolved way past its days as a British military base and Japanese POW camp during World War II into a masterfully planned tourist destination. Among its many wonders and

The Asian Civilisations Museum explores the artistic heritage of Singapore.

attractions are beaches, golf courses, casinos, and 14 hotels, as well as **Universal Studios Singapore**, **Adventure Cove Waterpark**, **Singapore Butterfly Park & Insect Kingdom**, the **Sentosa Nature Discovery** museum, and **S.E.A. Aquarium**. ■

LAY YOUR HEAD

• **Raffles:** The original grande dame of Singapore hotels opened its doors in 1887 but has been substantially upgraded and renovated since its colonial glory days; restaurants, bars, spa, outdoor pool, gardens, shopping arcade; from $509; raffles.com/singapore.

• **Fullerton:** Opened in 1928 as the colony's General Post Office, this neoclassical landmark and national monument now boasts 400 elegantly understated guest rooms; restaurants, bars, spa, gym, outdoor pool, heritage gallery, luxury shops, koi pond; from $221; fullertonhotels .com.

• **The Warehouse:** Asian contemporary styling distinguishes this 37-room boutique hotel in a converted 19th-century godown at lively Robertson Quay on the banks of the Singapore River; restaurant, bar, outdoor pool; from $231; thewarehousehotel.com.

• **Vagabond Club:** Located in the heart of the Central Heritage District, the decor is quirky, the guest rooms are Chinese chic, and the Artist in Residence program means there's always something new to look at; restaurant, whiskey bar, gym; from $184; hotelvagabond singapore.com.

• **Lloyd's Inn:** Sleek and minimalist Singapore boutique with an industrial chic vibe; outdoor pantry, rooftop terrace, outdoor wading pool; from $112; lloydsinn.com/singapore.

Ulaanbaatar
Mongolia

After decades crouching behind the Iron Curtain, Mongolia's once sleepy capital has leaped into the limelight as a business boomtown, cultural epicenter, and outdoor adventure hub.

THE BIG PICTURE

Founded: 1778

Population: 1.4 million

Size: 1,816 square miles (4,703 sq km)

Languages: Mongolian

Currency: Mongolian tughrik

Time Zone: GMT +8 (ULAT)

Global Cost of Living Rank: Unlisted

Also Known As: Urga (pre-1911), Niĭslel Khüree (1911–1924)

In the years since glasnost and Mongolia's flight from the Soviet sphere, Ulaanbaatar has changed more than any city on the planet. The once vacant streets now buzz with vehicles. Nowadays the stodgy Stalinist blocks are overshadowed by modern steel-and-glass structures like **Galleria Ulaanbaatar** and the sail-shaped **Blue Sky Tower**. And the city's once reserved urbanites have become some of the world's most exuberant hosts.

One thing that hasn't changed is the pivotal role of **Sükhbaatar Square** in local politics, culture, and tourism. The giant concrete plaza doesn't boast a single tree. But there's an equestrian statue of revolutionary leader Damdin Sükhbaatar and a colonnade dedicated to national heroes Genghis Khan, Kublai Khan, and Ögedei Khan.

Arrayed around the square are the stock exchange, central post office, city hall, and immense **Government Palace**, venue for a colorful daily **changing of the guard** ceremony (11 a.m.). Live performance is also the forte of the **Mongolian State Academic Theatre of Opera and Ballet** in the pink building on the plaza's eastern side.

The warren of streets around Sükhbaatar Square hosts other cultural icons like the **National Museum of Mongolia**, the **Choijin Lama Monastery** with its treasure of religious art and Buddhist relics, and even the **Fat Cat Jazz Club**. The nearby **Central Museum of Mongolian Dinosaurs** offers one of the world's great bone collections, including the skeletons of a *Velociraptor* and *Tarbosaurus* (an Asian version of *T. rex*) uncovered at digs in the Gobi.

Given its roots as a tented monastery town, "UB" boasts other Buddhist enclaves, including **Gandan Monastery**, originally constructed in 1809 and now home to more than 500 monks and a giant golden image of **Migjid Janraisig**. The city's other 19th-century architectural gem is the **Winter Palace of the Bogd Khan**. Home to the last king of Mongolia, the sprawling complex showcases the old royal throne, his flamboyant ceremonial tent, and even the bling that adorned his pet elephant.

For a panoramic view of Ulaanbaatar and surrounding countryside, hike or drive to the hilltop **Zaisan Memorial** on the city's south side. Dedicated to Mongolian and Russian soldiers who fought together in

Rest under a pagoda in the Gorkhi-Terelj National Park on a sunny day.

The Mongolian Parliament marks Ulaanbaatar's city center.

World War II and other conflicts, the monument's huge circular mural depicts the defeat of the Nazis and other wartime events.

The Mongolian wilderness starts just beyond Zaisan. Despite its ominous-sounding name, **Bogd Khan Uul Biosphere Reserve** is actually a great place to hike or horseback ride, with trails leading along rocky ridges and through coniferous forest. Thought to be the world's oldest nature reserve, it was declared in 1778 and originally patrolled by Buddhist monks. The park also harbors **Sky Resort**, Mongolian's most popular winter sports area (downhill and cross-country skiing and snowboarding).

Around an hour east of UB, an immense stainless-steel equestrian statue of **Genghis Khan** (131 feet/40 m) crowns a bluff above the **Tuul River Valley**. An elevator rises to the horse's tail and a walkway through its body to a viewing platform in the head. The statue rises near the main entrance of **Gorkhi-Terelj National Park**, a vast expanse of forest, steppe, and mountains that can be explored by foot, horse, or camelback.

The city's biggest annual bash is **Naadam**, three days of traditional Mongolian archery, wrestling, and horse racing at the **National Stadium** in early July. ■

Seoul
South Korea

Like the blue-and-red *taeguk* symbol on the Republic of Korea flag, there's a yin and yang to Seoul, a modern city with an ancient heart as well as a massive urban area with distinct neighborhoods that make the Korean capital feel more like a cluster of towns and villages than the globe's fifth largest metropolis.

THE BIG PICTURE

Founded: 1394

Population: 24.3 million

Size: 1,060 square miles (2,745 sq km)

Language: Korean

Currency: South Korean won

Time Zone: GMT +9 (KST)

Global Cost of Living Rank: 11

Also Known As: Hanseong (ancient), Soul of Asia

Having ducked the limelight for so long, Seoul is primed to burst onto the world scene. And not a moment too soon considering the South Korean metropolis has been around for more than 600 years.

The city's newfound confidence is derived from several factors. One is sheer size—one of the planet's largest metro areas (25 million people) is home to more than 20 percent of all South Koreans. Another is economic muscle—dozens of multinationals, a thriving middle class, and increasingly well-known brand names like Hyundai, LG, and Samsung.

Yet another driving force is Seoul's recent emergence as a cultural phenomenon thanks to the Oscar-winning film *Parasite,* the K-pop music invasion, and a little ditty called "Gangnam Style" that at one point was the most "liked" video in YouTube history.

One of 25 *gu* (districts) that make up the city, **Gangnam** personifies the city's emergence onto the world scene. An area of high-rise real estate and affluent lifestyles along the south bank of the **Han River**, the neighborhood allegedly accounts for 10 percent of South Korea's property value and perhaps 90 percent of the nation's trendiness.

Gangnam is renowned for upscale shopping at **Starfield COEX Mall** (which also boasts the country's largest aquarium), **Galleria Department Store**, and the super-posh **Cheongdam-dong** area. The culinary scene features foods from around the globe, and the area's nightlife is legendary even outside of Korea for places like the techno-infused **Club Octagon** dance hall or laid-back **Once in a Blue Moon** jazz joint.

But there's plenty to see by daylight, too. **Simone Handbag Museum** offers more than 300 samples from around the globe, some of them more than 300 years old, while the **Horim Museum Sinsa** showcases an exquisite collection of Korean ceramics, calligraphy, and paintings like **"Ksitigarbha and the Ten Kings of Hell."** Gangnam's architecture also runs a broad gamut from the outrageous pop-art **Tangent** facade on the HDC Building

A "Gangnam Style" statue rests in front of the COEX Mall in the Gangnam District.

The Gyeongbokgung Palace, built in 1395, was the main royal residence of the Joseon dynasty.

to the eighth-century **Bongeunsa Temple**, an oasis of tranquility with a two-day **Templestay** program during which visitors can dress, eat, sleep, meditate, and learn calligraphy with Buddhist monks.

Gangnam even has a sports scene. The LG Twins of Korea's professional KBO League play their home games at **Jamsil Baseball Stadium**. Meanwhile, martial arts aficionados can browse the museum or watch demonstrations by Korean masters at **Kukkiwon**, tae kwon do's world headquarters.

Two of the city's best museums are located across the river from Gangnam. Housed in a massive new home dedicated in 2005, the **National Museum of Korea** on the southern edge of **Yongsan "Dragon Hill" Park** safeguards national treasures from prehistoric through modern times. Housed inside an equally stunning venue in Itaewon, the **Leeum, Samsung Museum of Art**

showcases traditional Korean art and contemporary global creations.

Itaewon hasn't always been so cultured. As the longtime location of the **U.S. Army Yongsan Garrison**, the neighborhood was renowned for its rowdy GI bars and kitschy stores selling designer knockoffs. But over the past 20 years, that once seedy area has

morphed into a multicultural hub that draws locals and visitors alike to its overseas eateries and special events like the annual **Itaewon Global Village Festival**. On a more somber note, the **War Memorial of Korea** museum details Korea's military history since World War II, including the ongoing conflict between North and South.

URBAN RELIC

Gyeongju

Located along the eastern edge of the peninsula, **Gyeongju** served as capital of an ancient Silla Kingdom (57 B.C. to A.D. 935) that grew from a small tribal city-state into the modern Korean nation.

Often called a museum without walls, Gyeongju is steeped in historic and religious significance. The city's UNESCO World Heritage site comprises Bulguksa Temple and

Seokguram Grotto as well as a Gyeongju Historic Area with multiple sites and 31 Korean national treasures.

There's also a modern side to Gyeongju, especially the area around Bomun Lake with its many hotels, restaurants, and attractions like Gyeongju Bird Park inside a giant glass aviary, Gyeongju Teddy Bear Museum, and the new K-Pop Museum. Come spring, the 9,000 cherry trees that surround the man-made lake will burst into bloom.

Stop for a bite and skyline views in the N Seoul Tower.

Rising right behind Itaewon is **Namsan Mountain**, a rugged park that seems almost primeval compared to the surrounding concrete jungle. Hiking paths meander through maples, pines, and acacias that provide habitat for squirrels, pheasants, and other urban wildlife. Crowning the summit, **N Seoul Tower** renders views across much of the urban area.

Namsan cable car provides a scenic three-minute descent into the heart of Seoul, a valley-like area that's been the nation's political, commercial, and cultural capital since the late 14th century when the Joseon dynasty rose to power. No more than a 15-minute walk from the lower cable car station, the crowded **Namdaemun Market** area and the neon-spangled **Myeong-dong** district provide plenty of options for eating and shopping. Farther along, the ultramodern **Seoul City Hall**—with a glass facade that resembles a giant breaking wave—marks the start of the

Palace District where so much of Korean history has played out.

Bastion of royal power for more than 500 years, the 14th-century **Gyeongbokgung** is the largest of the five royal palaces and venue for an impressive changing of the guard ceremony. The massive compound harbors thousands of rooms spread across 330 structures arrayed along a central axis, including the majestic **Gwanghwamun Gate**, the **Geun-jeongjeon** throne room, and the lovely **Gyeonghoeru Pavilion** on an island in the middle of an artificial lake.

Considered the most authentic of the five, **Changdoekgung** was the home of Myeongseong (Korea's "last empress") until her death in 1966. Members of the royal family lingered until the late 1980s, living in seclusion in a portion of the palace originally built for the royal concubines. The only palace that offers an evening illumination, **Deoksugung** also features a daily changing of the guard. Armed with royal banners and menacing weapons, several dozen guardsmen march through the palace's main gate to the blare of horns and conch shells and a clash of symbols and drums.

Neighborhoods around the palace quarter provide their own trips down memory lane. Originally a residential

LOCAL FLAVOR

• **Jungsik:** Two-starred Michelin chef/owner Jungsik Lim parlays his Korean roots and culinary training in New York and Spain into reinterpretations of classic Korean dishes, including multicourse tasting menus with wine pairings; *11 Seolleung-ro 158-gil, Gangnam;* www.jungsik.kr.

• **Balwoo Gongyang:** The pioneer of Seoul's temple cuisine scene adheres to strict Buddhist principles, dishes with no meat or seafood and none of the garlic or onion spices used in Korean cooking. A corkage fee is added for guests who bring their own alcohol; *56 Ujeongguk-ro, Jongno-gu;* balwoo.or.kr.

• **Gaon:** An Asian minimalist aesthetic suffuses a Michelin three-star restaurant that specializes in traditional Korean dishes served on KwangJuYo dinnerware; *317 Dosan-daero, Sinsa-dong, Gangnam.*

• **Imun Seolnongtang:** The very first Seoul restaurant to register for a business license (in 1904) is also a popular destination for locals who dine on delicious *seolleongtang* (milky beef bone soup). *38-13 Ujeongguk-ro, Gyeonji-dong, Jongno-gu.*

The Myeong-dong commercial area is filled with restaurants, shops, and a lively nightlife scene.

area for courtiers, government officials, and minor members of the royal family, **Bukchon** preserves more than 900 *hanok,* traditional houses known for their distinctive tile roofs and ingenious *ondol* (heating systems). A number of hanok have been transformed into guesthouses, art galleries, and antique shops.

Clinging to the lower slopes of Namsan Mountain, **Namsangol Hanok Village** offers traditional homes relocated from elsewhere in Seoul, as well as cultural performances at **Seoul Namsan Gugakdang** theater and workshops in calligraphy, *hanji* papermaking, and other traditional Korean crafts.

Insadong also started life as a village serving the royal households

before evolving into a wealthy residential area, postwar hangout for avant-garde artists and writers, and, most recently, a haven for art galleries, handicraft stores, street food stalls, and those seeking enlightenment at **Jogyesa Temple**. Although most visitors come for the three great golden Buddhas, the temple also features an overnight program that involves chanting, meditation, a formal monastic meal, and 108 prostrations. Former monks founded nearby **Sanchon** restaurant, where the menu includes traditional vegetarian dishes like fiery *kimchi jjigae* stew, *jeon* (pancakes made with lotus root), and deep-fried *yugwa* dessert with honey and mugwort.

Running east-west across the city

center, **Cheonggyecheon** stream was once covered by an elevated expressway. But an audacious experiment in urban renewal demolished the highway and rehabilitated the watercourse into a linear park with trees, waterfalls, pedestrian bridges, and outdoor performance spaces.

Another transformational project is **Dongdaemun Design Plaza** (DDP), the city's new fashion and design hub. Clad in thousands of aluminum panels, the world's largest 3D amorphous structure contains fashion boutiques, the **Design Museum**, and **Dongdaemun History & Culture Park**, which exhibits part of Seoul's ancient city wall and other relics discovered during DDP construction. ■

Kathmandu
Nepal

A crossroads of the Himalaya for more than a thousand years, Nepal's boisterous capital is both a gateway to outdoor adventure and a fascinating insight into one of Asia's most unique cultures, a city reaching for the future but not at the expense of its ancient ways and means.

THE BIG PICTURE

Founded: A.D. 723

Population: 3 million

Size: 77 square miles (199 sq km)

Language: Nepali

Currency: Nepalese rupee

Time Zone: GMT +5:45 (NST)

Global Cost of Living Rank: Unlisted

Also Known As: Yeṃ Deśa (Newar), Yambu (Sino-Tibetan)

Kathmandu "drinks up the sky" crooned Cat Stevens back in the day when the city's Freak Street was awash in young Western backpackers and the hash cake–serving Pleasure Palace was the place to hang (sometimes for days on end).

How times have changed. The Nepalese capital is still beset by young travelers, but nowadays they utilize the city as a portal for hiking, biking, climbing, and white-water rafting the Himalaya rather than a gateway to psychedelic Shangri-la.

Freak Street is a shadow of its former self, long ago surpassed by the helter-skelter **Thamel** district as ground zero for backpacker hostels, bargain eats, and buying everything from pashmina shawls to trekking equipment. Tucked on a quiet square on the outer edge of the district is **Kathesimbu Stupa**—the city's primary place of Tibetan Buddhist worship—and its adjoining monastery.

Farther south is the **Old City**, a clamorous, colorful melting pot of Buddhist and Hindu architecture and Newari artisan workshops. Bustling streets like **Chandraman Singh Marg**, **Makhan Tole**, and **Asan Tole** (where Cat Stevens allegedly wrote his ode to the city while in a local teahouse) flow down to **Durbar Square**, once the hub of Nepal's monarchy and still one of its most important religious sites.

More a jumble of open spaces than a single square, Durbar is surrounded by timeless treasures like the **Basantapur Tower**, **Taleju Temple**, **Hanuman Dhoka Palace**, and **Kumari Bahal**. The last is home to the young Kumari Devi ("Living Goddess"), who peers down from an ornately carved wooden balcony each morning. A number of the structures around Durbar Square were badly damaged by the 2015 earthquake, and reconstruction of the UNESCO World Heritage site is ongoing.

During the 19th century, the monarch moved from Durbar to modern digs in the **Narayanhiti Palace**. The original structure was replaced by the current palace in the 1960s, the royal family by a republic in 2006 when the palace was transformed into a museum. A self-guided

Monkeys rest at the Swayambhunath Stupa, also known as Monkey Temple.

The Boudhanath Stupa is a UNESCO World Heritage site. It is a symbol of quiet and serenity for the people of Kathmandu.

tour includes the **throne room, reception hall**, crown jewels, and the bullet-riddled spot where the royal massacre of 2001 took place.

Bookending the city on the east and west are a pair of incredible religious shrines. Erected in the fifth century A.D., **Boudhanath Stupa** is renowned for its sublime design, a whitewashed dome topped by a golden tower decorated with the "all-seeing" eyes of Buddha. The shrine is especially moving at night when thousands of lights illuminate the grounds. Out west, the 3,000-year-old **Swayambhunath Stupa** bears a similar design but offers an even more impressive view of the valley.

Kathmandu's urban sprawl has gobbled up other historic towns like **Bhaktapur** and **Patan**, once detached from the capital by pastures and rice paddies. South of the Bagmati River, **Patan** (also called Lalitpur) was founded in the third century B.C. as the valley's first urban center. With thousands of temples and other shrines—and palace squares every bit as dazzling as the one in Kathmandu—Patan and Bhaktapur have also been declared UNESCO World Heritage sites. ∎

HIDDEN TREASURES

• For those who want to claim they "climbed" in the Himalaya without actually setting foot in the mountains, **Astrek** provides 20 different routes up Nepal's highest artificial climbing wall.

• Despite its hippy-dippy name, the recently restored **Garden of Dreams** (Swapna Bagaicha) in Thamel district was created in the 1920s as the private park of a Nepalese field marshal.

• That same field marshal funded the nearby **Kaiser Library,** in the Keshar Mahal palace complex, an astonishing collection of old books, antiques, taxidermied animals, and vintage globes.

• Rather than clove oil, locals often seek relief from dental pain by nailing coins into the stumpy **Toothache Tree** (Vaisha Dev) near Bangemudha Square.

Thimphu & Paro

Bhutan

Located about 30 miles (48 km) from each other in the foothills of the Himalaya, Thimphu and Paro are the urban pillars of the Kingdom of Bhutan, one of them a modern capital, the other an ancient place of pilgrimage.

THIMPHU

What does Thimphu have in common with Ankara, Brasilia, and Canberra? It's a 20th-century capital rather than an ancient creation—no more than a collection of villages until the 1950s, when Bhutan's monarchy decided to relocate its seat of power from Punakha to a more centralized (and easier to reach) venue. In fact, Thimphu wasn't declared the capital until 1962.

Of course, looks can be deceiving. Rather than mid-century modern, Thimphu is spangled with updated versions of traditional Bhutanese architecture, a factor that gives much of the city the look (and sometimes feel) of a sprawling Buddhist monastery.

Among the standout structures are the **Gyelyong Tshokhang** (parliament building), **Tashichhoe Dzong** (a combined royal palace, government office, and monastery), the **National Library**, and the dark red **Hotel**

THE BIG PICTURE

Founded: Thimphu, 1952; Paro, ca 17th century

Population: Thimphu, ca 100,000; Paro, ca 12,000

Size: Thimphu, 2 square miles (5 sq km); Paro, 499 square miles (1,293 sq km)

Language: Dzongkha (Bhutanese)

Currency: Bhutanese ngultrum

Time Zone: GMT +6 (BTT)

Global Cost of Living Rank: Unlisted

Also Known As: N/A

Druk on Clock Tower Square. Two of the city's most sacred shrines are also recent creations. The whitewashed **National Memorial Chorten** was erected in the 1970s while the golden **Buddha Dordenma**—a colossal 177-foot (54 m)-high statue that crowns a mountaintop on the south side of town—was completed in 2015.

But not all is new. Scattered around town are Himalayan Buddhist compounds and shrines that sometimes predate the capital. The oldest surviving fortress-monastery is the 17th-century **Simtokha Dzong** (also called the "Castle of the Profound Meaning of Secret Mantras"), which now houses Bhutan's national language institute. Older still is **Dechen Phodrang** ("Castle of Great Bliss"), a monastic school that traces it roots to the 12th century. The hilltop **Changangkha Lhakhang** sanctuary (built in the 13th century) offers a sublime atmosphere and panoramic views across the city.

Traditional Bhutanese arts and crafts are nurtured at several institutions that open their doors to visitors, including the **Royal Textile**

An artist makes a wood carving at the Thimphu Art School training center.

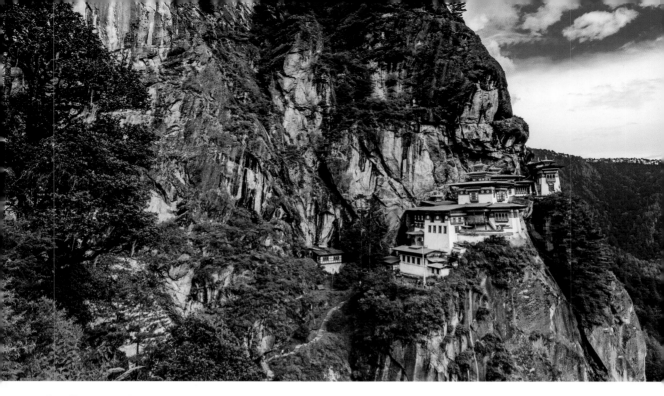

Paro Taktsang, or Tiger's Nest, is a sacred Vajrayana Himalayan Buddhist site, perched on a cliff side of the Paro Valley.

Academy, the **Choki Traditional Art School** for underprivileged students, and the esteemed **National Institute for Zorig Chusum**, where 13 traditional disciples are taught *thangka* painting, wood carving, and embroidery. Another good place to browse and potentially buy Bhutanese crafts is the **National Handicrafts Emporium** near the post office.

PARO

Visitors have two primary reasons to pass through Paro, the capital's sister city, 31 road miles (50 km) to the west. First, they have no choice. **Paro Airport** is Bhutan's only international gateway. Flanked by 18,000-foot (5,486 m) peaks, it also boasts one of the aviation world's most spectacular takeoffs and landings.

The second and far more important reason is **Paro Taktsang** monastery. The celebrated "Tiger's Nest"

monastery hugs the side of a 3,000-foot (914 m) cliff that overlooks the north end of the Paro Valley. Work on the lofty shrine commenced in 1692, and today it's both a sacred place of pilgrimage and endearing national symbol. Most visitors reach Tiger's Nest via a round-trip hike that takes four to five hours, but it's also possible to travel on horseback (90 minutes).

More Himalayan Buddhist landmarks await in town, including 17th-century **Rinpung Dzong** ("Castle on a Heap of Jewels") and 15th-century **Tamchog Lhakhang**, which is reached by crossing the Paro River on an ancient (pedestrian-only) suspension bridge. An ancient stone watchtower called the Paro Ta Dzong provides a suitably historic venue for the **National Museum of Bhutan** and its collection of traditional fashion, weapons, art and handicrafts.

Compact and easy to navigate, the city center feels more like a village, especially when the **Paro Weekend Market** is in full swing. Be sure to dress yourself in traditional Bhutanese garb if you're going to use the nearby **Rinpung Archery Ground**. And duck into restaurants along the two main streets to sample local favorites like *hentshey datse* (spinach and cheese) or pork and chicken *momos* (dumplings). ∎

HIDDEN TREASURE

• Philatelists flock to the **Bhutan National Post Office** in Thimphu to purchase limited-edition and collector stamps. Bring a photo of yourself and the bureau will print a dozen personalized stamps bearing your image that are officially authorized for use on postcards and letters.

Mumbai

India

From its birth as a backwater trading post, Mumbai has grown to embody vibrant modern India, an amalgam of business, Bollywood, and "slumdog" hustle-bustle that personifies many of the things that make South Asia such a fascinating (and at times frustrating) place to visit.

THE BIG PICTURE

Founded: 1534

Population: 24 million

Size: 340 square miles (881 sq km)

Language: Marathi, Hindi, English

Currency: Indian rupee

Time Zone: GMT +5:30 (IST)

Global Cost of Living Rank: 78

Also Known As: Bombay, City of Dreams, Hollywood of India

As part of the dowry he received upon wedding a Portuguese princess in 1662, England's King Charles II took possession of the Seven Islands of Bombay in far-off India. The scantly populated isles weren't much use to the Brits until the early 1800s, when they connected the bits and bobs with a causeway to form a deep harbor.

Within a few years, it had grown into India's gateway to the outside world.

It remains that way today, a thriving finance and factory hub, transportation center and nexus of the nation's entertainment industry. Mumbai boasts more billionaires than any other Indian city but also some of its most fetid slums. The famously chaotic traffic offers an extreme contrast to the city's urban national park and the wild leopards that stalk the streets at night.

Poised on a narrow peninsula between the harbor and Indian Ocean, **South Mumbai** serves as the city's historic core and the nerve center of business, retail, art, and other aspects of modern living. Completed in 1924, the colossal **Gateway of India** is a popular gathering place (and selfie spot) for locals and visitors alike. Even more impressive, the adjacent **Taj Mahal Palace Hotel** opened its doors in 1903 and is now fully recovered from the 2008 terrorist attack that spawned world headlines and several movies.

From docks beside the arch, ferries flit across the harbor to the UNESCO World Heritage site on **Elephanta Island**, a cluster of cave temples constructed between A.D. 450 and 750. South of the Gateway are the colorful **Colaba Causeway** shopping area (handicrafts, jewelry, fashion, accessories) and **Sassoon Dock** with its pungent morning fish market.

About a five-minute walk from the waterfront, the big **Wellington Fountain** roundabout is flanked by

Ancient Hindu statues stand tall inside the cave temples of Elephanta Island.

The Taj Mahal Palace Hotel stands next to the Gateway of India, both dazzling sites to see from a river cruise.

Bombay's premier museums, including the **National Gallery of Modern Art** and the imposing **Chhatrapati Shivaji Maharaj Vastu Sangrahalaya** (CSMVS), where more than 50,000 exhibits offer a comprehensive take on India's human and natural history.

From the roundabout, Madame Cama Road shoots across the peninsula to **Marine Drive**. Dubbed the "Queen's Necklace" after its sparkling lights, the waterfront road is lined with posh hotels, apartment blocks, and Mumbai's **cricket stadiums**. The drive curls around to **Chowpatty Beach**, which springs to life each night with food stalls and street performers. Rising behind is **Malabar Hill** with its tranquil **Hanging Gardens** and panoramic city views.

The upscale neighborhood north of Malabar Hill is home to the 27-story **Antilia** tower—the world's most expensive private residence (ca US$2 billion)—and the **National Museum of Indian Cinema**. Several local agencies offer guided **Bollywood tours**, including visits to the studios in **Film City** on the city's north side.

Film City lies just outside the boundaries of **Sanjay Gandhi National Park**, one of the world's largest urban wildlife reserves. In addition to leopards, cobras, flying foxes, and several species of deer and monkeys, the park safeguards the **Kanheri Caves**, a complex of underground Buddhist monasteries and dwellings dating from as early as the first century B.C. **Bombay Natural History Society** offers guided hikes of the reserve from its **Conservation Education Centre** in Film City. ∎

ARTBEAT

- **Best Movies:** *Deewaar* (1975), *Ardh Satya* (1983), *Salaam Bombay!* (1988), *Slumdog Millionaire* (2008), *The Lunchbox* (2013).

- **Best Books:** *Breathless in Bombay* by Murzban F. Shroff, *Midnight's Children* by Salman Rushdie, *Shantaram* by Gregory David Roberts, *Narcopolis* by Jeet Thayil, *Maximum City* by Suketu Mehta.

- **Best Art:** The surrealism of Jaideep Mehrotra, figurative works of Bhupen Khakhar, and resin sculptures and mixed-media portraits of Jitish Kallat.

- **Best Music:** *The Best of Bollywood* by multiple artists, *Nath Maza Mi Nathancha* by Shrinivas Vinayak Khale, *Slumdog Millionaire: Music from the Motion Picture*.

Delhi & Agra

India

Located 130 miles (209 km) from each other along the Yamuna River, Delhi and Agra reflect 1,000 years of Indian civilization from medieval times and the glorious Mughal era through the British Raj and modern times.

THE BIG PICTURE

Founded: Delhi, 1206; Agra, 1504

Population: Delhi, 28 million; Agra, 2.1 million

Size: Delhi, 865 square miles (2,240 sq km); Agra, 65 square miles (168 sq km)

Language: Hindi

Currency: Indian rupee

Time Zone: GMT +5:30 (IST)

Global Cost of Living Rank: Delhi, 117; Agra, unlisted

Also Known As: Delhi—City of Rallies; Agra—City of Love, City of the Taj

Although they were roughly the same size 200 years ago, Delhi would grow into the sprawling capital of the entire Indian nation, while Agra would remain a provincial city primarily renowned for hosting the world's most extraordinary building.

DELHI

There really is an old and new Delhi, although the terms now seem obsolete, labels from the colonial past when the British built their Indian Empire capital on the outskirts of the ancient walled town.

Looking back, there have actually been nine different cities in the area that now cradles the Delhi metropolitan area. The sprawling **Mehrauli Archaeological Park** preserves the remains of 11th-century **Lol Kot** and four cities that followed.

The neighboring **Qutb Minar Complex** boasts far more impressive ruins like its namesake victory tower, a miracle of medieval engineering that marks the start of 500 years of Muslim rule over northern India. The tower is surrounded by other Indo-Islamic creations like **Quwwat-ul-Islam Mosque**, the **Ala'i Darwaza** gateway, and the **Tomb of Iltutmish**.

Two centuries later, Shah Jahan mandated the creation of an entirely new city along the south bank of the Yamuna River that would replace Agra as the Mughal capital. He named it **Shahjahanabad**, but it's come to be called **Old Delhi** in modern times.

In much the same way that he left his mark on Agra, Shah Jahan was also responsible for Old Delhi's most astonishing structures. Completed in 1648, the stupendous **Red Fort** is filled with his palaces, gardens, and audience halls. Just outside the citadel, the immense **Jama Masjid** was constructed in the decade that followed. Capable of holding 25,000 worshippers, the mosque's impeccable

Taste the flavors of Old Delhi at the Chandni Chowk Spice Market.

The white marble Taj Mahal was built between 1631 and 1648 to honor Emperor Shah Jahan's favorite wife.

Mughal design had a profound effect on the Anglo-Indian architecture that followed two centuries later.

The Red Fort's main gate empties into **Chandni Chowk**, a name given to both the main street through Old Delhi and a tumultuous market area that surrounds it. As in medieval times, the area is divided into smaller specialty areas hawking gold and silver, food and fabrics, books, religious items, and so much more.

Escape the chaos by strolling the **Raj Ghat**, a riverside park that includes a memorial on the spot where Mahatma Gandhi was cremated after his 1948 assassination. Across the street is the **National Gandhi Museum**.

By the early 20th century, the British had decided they needed a new capital for their Indian Empire, a meticulously planned modern city on the plains beyond the chaotic and overcrowded old town. They christened it **New Delhi** and filled it with extravagant public squares like **Connaught Place** (nowadays one of the city's foremost spots for dining and shopping) and imperious structures meant to reflect their domination of the subcontinent.

URBAN RELIC

Fatehpur Sikri

Although it was occupied for barely 40 years, this ancient Mughal "City of Victory" 25 miles (40 km) west of Agra is a stunning example of medieval urban planning.

Akbar the Great founded **Fatehpur Sikri** in 1571, initially as a religious shrine to honor the prophet who predicted the birth of his son Jehangir. But it quickly grew into a proper city and the capital of his empire.

Fatehpur Sikri is renowned for its red sandstone architecture, superlative structures like the Panch Mahal (Palace of Five Levels), Hiran Minar (Elephant Tower), and a massive gateway called the Buland Darwaza (Victory Gate) that commemorates Akbar's conquest of Gujarat. The city's Jama Masjid is one of the largest mosques in India and a popular pilgrimage site, as is the marble Tomb of Salim Chishti.

Although the truth is lost in history, it's thought that either a severe water shortage or a military campaign in the Punjab compelled Akbar to abandon his purpose-built capital a mere 14 years after it was built. By 1610, Fatehpur Sikri was completely abandoned.

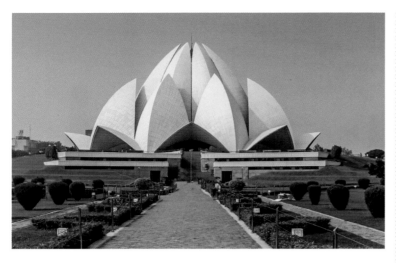

The Lotus Temple, dedicated in 1986, is a prominent Delhi attraction.

Originally the home of the British governor-general, the massive **Rashtrapati Bhavan** is now the official residence of India's president. At the other end of the **Rajpath**, a spacious ceremonial boulevard originally called the Kingsway after the British monarch, is the **India Gate**, a massive arch honoring Indian troops who died fighting for the Union Jack in World War I and other conflicts. The area around the 15th- and 16th-century **Lodi Tombs** was transformed during British times into the lovely **Lodi Gardens**, now one of the city's best outdoor escapes.

New Delhi also boasts many of the city's best museums, including the **National Museum** and its extensive survey of 5,000 years of Indian art and history, the **National Crafts Museum**, the **National Gallery of Modern Art**, and the quirky **Shankar's International Dolls Museum**.

Located on the opposite side of the Yamuna River from the city center, **Swaminarayan Akshardham** is hard to categorize. Opened in 2005, it's the city's largest Hindu temple. But there are definite theme park aspects, like a cultural boat ride through animatronic dioramas reflecting centuries of Indian history, an after-dark **Sahaj Anand Water Show** around a musical fountain, and a video of Indian landscapes played on a giant screen.

AGRA

No amount of images or words prepares you for that first in-person look at the **Taj Mahal**. Unlike so many other world wonders, it always exceeds expectations and preconceived notions formed over a lifetime of hearing and reading about the extraordinary monument on the outskirts of Agra.

For the record, it was created as the mausoleum of Mumtaz Mahal, third wife of Shah Jahan, after she died during childbirth in 1631. When the great Mughal emperor passed away 35 years later, he was laid to rest beside her in the exquisite white marble tomb.

The genius of the design is undeniable. That famous first glimpse is always framed by the Great Gate, followed by a mirror image of the Taj in the reflecting pool. Paths lead through the Persian-style **Taj Mahal Garden** divided into quadrants according to the four gardens of paradise listed in the Koran. Steps lead to a broad marble terrace in front of

---

LOCAL FLAVOR

• **Indian Accent:** Award-winning contemporary Indian cuisine created by fusing global ingredients with traditional Indian flavors and cooking methods; *The Lodhi, Lodhi Road, CGO Complex, Pragati Vihar, New Delhi;* indianaccent.com.

• **Moti Mahal:** The founders of one of Delhi's oldest eateries helped introduce Punjabi dishes like tandoori chicken and butter chicken to India's restaurant scene; *3704 Netaji Subhash Marg, Old Dariya Ganj, Daryaganj, Old Delhi;* motimahal.in.

• **Esphahan:** Live music complements the Mughal cuisine at an upscale restaurant in the Oberoi Amarvilas Hotel with tasting menus for vegetarians and carnivores; *Taj East Gate Road, Agra;* oberoihotels.com.

• **Zorba the Buddha:** Tiny vegetarian restaurant that emphasizes fresh ingredients and safe food practices; *E-19 Shopping Arcade, Gopi Chand Shivare Road, Sadar Bazaar, Agra;* zorbarestaurantagra.com.

the tomb, where it's possible to study the floral motifs and calligraphy that cover the facade. The interior is garnished with intricate *pietra dura* (marble inlaid with semiprecious stones) and delicate *jali* (lattice screens).

The tomb is such a mind-blowing experience that it's easy to forget that it's flanked by two other magnificent structures: **Kau Ban Mosque** and the **Mehmaan Khana** assembly hall, both of which are fashioned from red sandstone. Anywhere else than beside the Taj, they would be astounding landmarks all on their own.

The great challenge that everyone faces with the Taj is *when* to see it. Many swear by early morning, others recommend late afternoon—the light soft and dreamy both of those times. Every lunar cycle, the compound is open five nights for **moonlight viewing**. It's also worth heading across the **Yamuna River**, where the restored **Mehtab Bagh** gardens offer a totally different take on the Taj.

From the back of the Taj compound you can gaze upriver to another significant edifice: the massive **Agra Fort**. Although its construction began under the early 16th-century Lodi dynasty, the fort realized its greatest extent during Mughal times under Akbar the Great, his son Jehangir, and grandson Shah Jahan, who made it the centerpiece of their empire.

Because much of the fort is still an active Indian military base, only certain sections are open to the public. Entry is through the **Amar Singh Gate** to a warren of courtyards, passages, and opulent Mughal-era buildings like **Jehangir's Palace**, the **Khas Mahal** palace of Shah Jahan, and the **Sheesh Mahal** residence called the "Glass Palace" because of its glittering mirrored mosaics.

Opposite the fort are the vintage **Agra Fort Railway Station** (opened in 1874) and its trains to the far corners of India, as well as the magnificent **Jama Masjid** (another Shah Jahan creation, completed in 1648), and a jumble of sights, smells, sounds, and shops that compose the **Kinari Bazaar** market area.

On the western outskirts of town, the **Agra Bear Rescue Facility** and **Elephant Conservation Centre** (both run by the well-respected organization Wildlife SOS) strive to save and rehabilitate exploited and abused captive animals. Tours of both facilities are available. ■

The Agrasen Ki Baoli in New Delhi is a protected monument—a historical stepwell located on Hailey Road.

Jaipur
India

India's most colorful urban area swirls with the vibrant hues of Rajasthani fashion and the dazzling terra-cotta facades that spur people to call it the "Pink City."

THE BIG PICTURE

Founded: 1727

Population: 3.9 million

Size: 195 square miles (505 sq km)

Languages: Hindi, English

Currency: Indian rupee

Time Zone: GMT +5:30 (IST)

Global Cost of Living Rank: Unlisted

Also Known As: Pink City

Despite the chaotic street life that characterizes every Indian urban area, Jaipur is actually one of the world's oldest planned cities. Long before Brasilia, Canberra, and Washington, D.C., Jaipur was created from scratch by order of Jai Singh II, the maharaja of a princely state that would later become Rajasthan.

The nine giant blocks laid out in the 1720s still form the city's core. And at the very center is the magnificent **City Palace**. Although still home to Rajasthan's royal family, much of the palace is open to the public, including the Sarvatobhadra with its massive silver urns, the **Sabha Niwas** throne room, a historic textile collection in the **Mubarak Mahal**, thousands of hand-rendered manuscripts and maps in the **Pothikhana Archives**, and the **Sileh Khana** museum of royal arms and armor.

Just outside the palace walls is the **Jantar Mantar**, an open-air astronomical observatory created by Jai Singh II in the 1730s. The 19 stone or metal instruments were used to track the sun, stars, and planets, to predict eclipses, measure time, and establish calendars. Among the instruments are the world's largest sundial and a massive bronze astrolabe.

Yet the best is still the come. The royal palace and observatory literally pale in comparison to the **Hawa Mahal**—the garish "Palace of the Winds" with its five-story pink facade. Adorned with intricate latticework around more than 900 tiny windows, the structure's unique design allowed the ladies of the royal court to observe street life without being seen by ordinary folk below.

Crowning a hilltop north of the city center is the 18th-century **Nahargarh Fort** ("Abode of Tigers"). In addition to awesome views, the citadel offers several restaurants and the **Jaipur Wax Museum** with lifelike effigies of Indian heroes and celebrities. Remnants of the ramparts that once surrounded Jaipur stretch off to the northeast to the **Gaitore Cenotaphs**, flamboyant marble monuments that honor the memory of the maharajas (including Jai Singh II) cremated on the site.

Perched at the city's northern edge, the **Jal Mahal** ("Water Palace") rises from the waters of an ancient human-made lake. Although not open to the public, it makes an incredible photo op, especially during the twilight hours when the

The Nahargarh Fort, constructed in 1734, stands on the edge of the Aravalli Hills.

Hawa Mahal, or the Palace of the Winds, boasts 953 windows across its latticed facade.

facade is artfully illuminated. On the other hand, the **Galtaji** shrine on the east side of Jaipur is fully open to both the public and the hundred of macaques who give the "Monkey Temple" its more common name.

The incredible **Amer (Amber) Fortress** lies about a 15-minute drive north of the city center. Originally built in the 10th century and considerably expanded and enhanced in later years, the hilltop fort is another exquisite statement of Rajasthan architecture. For years it was traditional to ride an elephant up the steep entrance road, but nowadays visitors are encouraged to hike or hop a taxi.

One of India's best wildlife areas is a three-hour drive south of Jaipur and easily doable as a day trip from the city. **Ranthambore National Park** is world renowned for its tigers, but the massive reserve also safeguards leopards, sloth bears, antelope, monkeys, crocodiles, cobras, and more than 270 bird species. ■

GATHERINGS

• **Jaipur Literature Festival** (JLF): Authors and bookworms flock to the city in January for an event billed as the "greatest literary show on Earth."

• **Kite Festival:** The skies above Jaipur's parks, cricket pitches, and polo grounds are filled with lighter-than-air flying machines on January 14, a date that coincides with a nationwide celebration of the Hindu sun god.

• **Elephant Festival:** Pachyderm polo and a grand procession of elephants decked out in embroidered blankets and silver jewelry are highlights of an event that normally comes right before the Holi festival in February or March.

• **Gangaur:** This two-week festival—which takes place right after Holi—revolves around unmarried women who fast and perform rituals to ensure marital health, wealth, and fidelity. Jaipur celebrates with special foods and markets and colorful processions.

Kuala Lumpur
Malaysia

With a relentless energy similar to its days as a mining boomtown, the Malaysian federal capital has transformed itself into an astonishingly modern city where some of the world's tallest buildings overshadow vibrant ethnic neighborhoods and verdant green spaces.

THE BIG PICTURE

Founded: 1857

Population: 7.8 million

Size: 835 square miles (2,163 sq km)

Language: Malay, English

Currency: Malaysian ringgit

Time Zone: GMT +8 (MST)

Global Cost of Living Rank: 152

Also Known As: KL, Garden City of Lights

In the same manner that gold gave birth to cities like San Francisco and Johannesburg, Kuala Lumpur was born of tin, a late-19th-century rush that transformed it almost overnight from a remote riverside village into a mining boomtown with an exotic mix of ethnic groups and people seeking their fortune.

The same could be said for the present-day Malaysian capital. With one of the world's newest and biggest airports, it's far from remote nowadays. But "KL" (as locals often call it) endures as a melting pot of Malays, Chinese, Indians, and other ethnic groups. And given its phenomenal growth over the past 40 years, it remains a magnet for dreams and aspirations.

As the tin boom took off, Kuala Lumpur arose around an open *padang* (field) used for grazing and the occasional cricket match. It's still there. Only nowadays the huge open space is called **Dataran Merdeka** (Independence Square), the spot where the Malaysian flag was hoisted for the first time in 1957, when the nation gained its independence from Britain. At 328 feet (100 m), the **Merdeka Flagpole** that flies above the square today is one of the world's highest.

During colonial days, the British surrounded the grassy quad with some of the finest architecture in Southeast Asia. Structures like the **Sultan Abdul Samad Building**, **Jamek Mosque**, and the **National Textile Museum** (originally the railroad headquarters) blend Moorish, Mughal, and other exotic influences, while the faux-Tudor **Royal Selangor Club** and neo-Gothic **St. Mary's Cathedral** wouldn't look out of place in the English countryside.

Many of the city's older neighborhoods lie within easy walking distance to Merdeka Square. Old **Chinatown** is just across the **Klang River**, a warren of traditional shops and eating places like the restored art deco–style **Central Market** with its flavorsome food stalls and the **Petaling Street Market** where just about everything under the sun seems to be for sale.

Flanked by early 20th-century shophouses painted a riot of colors,

The Saloma Link pedestrian bridge crosses the Klang River, connecting Kampung Baru and the city center.

The limestone hill of Batu Caves was originally known as Kapal Tanggang from the legend of Si Tanggang.

Jalan Tuanku Abdul Rahman (Jalan TAR) meanders through historic neighborhoods north of the square to the Jalan Masjid India Food Court and the aromatic **Chow Kit Market** and its prepared food, vegetables, meat, and seafood stalls. On Saturday nights, a long stretch of Jalan TAR transforms into a pedestrian-only *pasar malam* (night market).

Heading west from the square, it's just a short walk (or taxi ride) along tree-shaded Jalan Parlimen (Parliament Street) to Kuala Lumpur's premier green space—the gorgeous **Lake Gardens**—or Tun Abdul Razak Heritage Park to use its formal name. Foremost among its many attractions is **Perdana Botanical Gardens**. Established in 1888,

this urban oasis showcases plants and flowers from around the world in a dozen sections, including hibiscus and orchid gardens, a topiary

and herbarium, a grove of national heritage trees, and even a little deer park.

The gardens are enveloped by

URBAN RELIC

Malacca

Founded in the early 1400s as the capital of the Malacca Sultanate, **Malacca** later passed through Portuguese, Dutch, and British hands during 450 years of colonial control.

Each of them left their mark on the city. Although now in ruins, the Portuguese A Famosa Fort (1511) and hilltop St. Paul's Church (1521) are among the oldest European buildings in Southeast Asia.

Down in the flatlands, the bright

red Christ Church (1776) started life as a Dutch Reform chapel before it was converted to an English Anglican church. The nearby Stadthuys (built 1641–60) is probably the oldest Dutch building in East Asia.

But as the city expands outward from the historical core, so does its distance from the colonial past; landmarks like the golden-domed Melaka Straits Mosque and the stunning new Encore Melaka theater reflect modern Malaysia rather than the distant past.

Find beautiful winged species in the Kuala Lumpur Butterfly Park.

other cultural and natural attractions like the **National Museum of Malaysia**, the excellent **Islamic Arts Museum**, the colorful aerial creatures of **Taman Burung Bird Park** and **Kuala Lumpur Butterfly Park**, and the small but thoroughly amusing **Malaysia Cartoon & Comic House**.

The bold modernist design of the **Malaysian Houses of Parliament** punctuates the western side of the gardens, near the **ASEAN Sculpture Garden** with artistry from around Southeast Asia, and the poignant **Tugu Negara** monument dedicated to Malaysia's struggle for freedom.

As much as Merdeka Square and its environs form the heart of the old town, **KLCC Park** and its surroundings reflect the city's 21st-century energy. Named for the adjacent **Kuala Lumpur Convention Centre**, the modern green space revolves around various animated water features including **Lake Symphony**, where two fountains burst into a spectacular sound-and-light show three times each evening.

Looming high above the lake are the **Petronas Twin Towers**, headquarters of Malaysia's national oil and gas company. The world's tallest buildings when they opened in 1999 (1,483 feet/452 m), the towers feature an 86th-floor **Observation Deck** and an acrophobia-inducing **Sky Deck** and **Sky Box** around halfway up.

The towers complex also includes the state-of-the-art **Dewan Filharmonik** concert hall, **Aquaria KLCC** aquarium, and the **Petrosains, the Discovery Centre** science museum, as well as a large shopping mall and several luxury hotels.

The Petronas Towers anchor the northern top of the city's **Golden Triangle,** a boisterous central business district flush with hotels, malls, and an ever expanding forest of high-rises. Newest kid on the block is **Exchange 106**, a 106-story building with an illuminated crown that opened in 2020. But **Merdeka PNB 118**, a 118-story office tower near Chinatown, will soon top it to become the planet's second tallest building (2,113 feet/644 m) upon completion.

The lively **Bukit Bintang** neighborhood along the southern edge of the Golden Triangle boasts a wide range of shopping and entertainment options, including the **Jalan**

LAY YOUR HEAD

• **The Chow Kit:** Overlooking Jalan Tuanku Abdul Rahman in the colorful Chow Kit neighborhood, this new boutique lies within walking distance of KL's historic heart; restaurant, bar, yoga, roof deck; from $63; thechowkit.com.

• **The Majestic:** Set between Chinatown and the Lake Gardens, this fully restored art deco palace was the grande dame of KL hotels when she opened in 1932; restaurants, bar, tearoom, cigar room, swimming pool, barber shop, gymnasium; from $76; majestickl.com.

• **Villa Samadhi:** Contemporary Asia meets tropical chic at this intimate, adult-only hotel just minutes from KLCC Park and the Petronas Towers; restaurant, bar, fitness center, swimming pool; from $143; villasamadhi.com.my.

• **Banyan Tree:** High-rise luxury is the hallmark of this towering hotel in the Golden Triangle, which features a 60th-floor penthouse restaurant, bar, and outdoor terrace; also with spa, fitness center, swimming pool; from $199; banyantree.com.

Alor night food court and the indoor roller coasters and other thrill rides of **Berjaya Times Square Theme Park**.

Though skyscrapers and shopping malls echo modern KL, the fabled **Batu Caves** on the city's northern fringe reflect one of its oldest aspects. Formed around 400 million years ago, the limestone caverns are both a geological wonder and cultural landmark. Roving monkeys and a colossal golden statue of **Lord Murugan** (Hindu god of war) guard the 272 steps to **Temple Cave** with its myriad Hindu shrines. **Ramayana Cave** is embellished with an extravagant depiction of the Hindu saga featuring colored lights and brightly painted statues.

As for the metal that gave birth to Kuala Lumpur, many of its traditional functions were gradually replaced by aluminum. When the global tin market crashed in 1985, Malaysia was forced to shut many of its mines. But it hasn't gone away completely. **Royal Selangor Visitor Centre** in KL features a collection of precious pewter items created from Malaysian tin, as well as craftsmanship demos and workshops during which visitors can make their own pewter accessories. On the city's southern edge, **Mines Wellness City** offers hotels, spas, a shopping mall, and a golf course built around what was once the world's largest open-pit tin mine.

If you're really into tin, consider driving two hours north of Kuala Lumpur to **Tanjung Tualang Tin Dredge No. 5**. The last of its kind in Malaysia, the gargantuan floating machine avoided the scrap heap because of its historic value and now offers guided tours six days a week. ■

HIDDEN TREASURES

• **Taman Tugu:** One of the world's best urban jungles, this 66-acre (27 ha) preserve offers hiking trails through a mix of secondary and old-growth rainforest inhabited by native birds, mammals, reptiles, and insects.

• **KL Forest Eco Park:** A rainforest canopy walk with swinging bridges is the highlight of this pocket park near the Golden Triangle.

• **Krau Wildlife Reserve:** Although most of this large national reserve around 70 miles (113 km) east of KL is closed to the general public to protect the flora and fauna, visitors are welcome at the Kuala Gandah elephant sanctuary and Jenderak Selatan Wildlife Conservation Centre, where wild seladang oxen are bred in captivity.

• **Kanching Rainforest Waterfall:** A steep hiking trail leads to the summit of this seven-level cascade in heavily wooded hills on the city's northern border.

The Sultan Abdul Samad Building was constructed at the end of the 19th century.

China's Cities
Chengdu, Guilin, Xi'an

China's inland cities offer a broad gamut of attractions, from giant pandas and terra-cotta warriors to scenic wonderlands and under-the-radar foodie destinations.

Conservation is the focus of the Chengdu Research Base of Giant Panda Breeding.

CHENGDU

Population: 11.3 million
Size: 706 square miles (1,829 sq km)

The capital of Sichuan Province boasts a long and vivid history, dating back to at least the Shu Kingdom of the fourth century B.C. China's fifth biggest urban area has been a hub of rice and wheat production, silkworm cultivation, and lacquer and filigree production on its way to becoming a modern political, economic, and cultural powerhouse.

Far and away the city's foremost attraction is the **Chengdu Research Base of Giant Panda Breeding**, started in 1987 with six of the black-and-white creatures rescued from distressed situations. Today the facility also breeds equally rare red pandas and other endangered Chinese wildlife in captivity. Wild pandas—as well as 60 other mammals and 300 bird species—inhabit **Wolong National Nature Reserve** 80 miles (130 km) west of Chengdu.

Among the city's other landmarks are **Du Fu Thatched Cottage**, a sprawling park and museum compound dedicated to the beloved Tang dynasty poet, and the large green space that surrounds the third-century **Wuhou Shrine** and museum of Three Kingdoms era. Rounding out the city's trio of popular green spaces, the **Culture Park** (Wenhua Gongyuan) features the Taoist **Qingyang Gong monastery** and **Shufeng Yayun Sichuan Opera House** with regular performances of traditional local music and theater.

Chunxi Road (Hundred Year Gold Street) has been transformed into a neon-dappled pedestrian zone with shops, stalls, cafés, and department stores beneath the skyscrapers of **Jinjiang District**. Meanwhile, age-old lanes like **Kuan Xiangzi** (Wide Alley) and **Zhai Xiangzi** (Narrow Alley) preserve historic teahouses and courtyard homes, many of them now occupied by restaurants and folk-art shops.

Chengdu offers an ideal base for day trips into the Sichuan countryside that might include the third-century B.C. **Dujiangyan irrigation system**, the 233-foot (71 m) **Leshan Giant Buddha**, and **Mount Emei** (aka Emeishan), one of China's four sacred Buddhist mountains.

GUILIN

Population: 950,000
Size: 71 square miles (184 sq km)

An inspiration to centuries of painters and poets, Guilin's often mist-shrouded karst landscape—reflected in the **Li River**, **Shanhu** (Cedar) and **Ronghu** (Banyan) **Lakes**—is considered one of the most beautiful in all China.

A scenic boat cruise or motorized bamboo raft trip along the river between Guilin and Yangshuo affords plenty of time to appreciate the passing rice farmers, fishermen with their trained diving cormorants, and the limestone landmarks like **Elephant Trunk Hill** (that resembles a giant elephant sipping water).

Guilin boasts the Li River and emerald green karst mountains.

Lao Zhai Shan Hill and its celebrated sunsets lie at the summit of 1,100 steps from the riverside, while **Seven Star Park** on the opposite bank is named for seven peaks arranged in a pattern like the Big Dipper constellation. **Hiking** and **biking** are popular along the Seven Star route.

In the city center, **Jingjiang Palace Park** offers a blend of natural and human-made attractions: **Solitary Beauty Hill** with its lofty panoramas and 14th-century **Jingjiang Prince Mansion**, built during the Ming dynasty and older than the Forbidden City in Beijing.

Reed Flute Cave (aka Palace of Natural Arts), a 180-million-year-old limestone cavern, made even more dramatic with multicolored lighting, has attracted curious tourists for more than a thousand years.

One of Guilin's newest attractions, the *Impression of Liu Sanjie* is a nightly outdoor show created by filmmaker Zhang Yimou, who earned international acclaim for directing the spectacular opening ceremony of the 2008 Beijing Olympics. Set against a backdrop of the Li River, the 70-minute show features a cast of 600 and takes place outside of Guilin in **Yangshuo County.**

XI'AN

Population: 6.7 million
Size: 422 square miles (1,093 sq km)

The capital of Shaanxi Province in central China, Xi'an is world renowned for its **Bingmayong**, or Terra-Cotta Army, thousands of life-size figures that accompanied China's first emperor to the afterlife in 221 B.C.

Nowadays they march in silent

The Xi'an halal temple is an ancient site located on Muslim Street.

homage through the **Emperor Qinshihuang's Mausoleum Site Museum.** Amazingly, each figure is unique, with a different face, clothes, and hairstyle. Some of the warriors also inhabit the enormous **Shaanxi History Museum** along with more than 370,000 other relics and artifacts from around the region.

A miniature version of the terra-cotta army inhabits the **Han Yang Ling Mausoleum** of Emperor Jing and Empress Wang. Hidden inside two large earthen mounds, the 8,000-strong throng is arrayed in burial pits viewed through glass floors.

The city's other collections range from Paleolithic to Revolution. **Banpo Museum** offers relics of a prehistoric matriarchal village inhabited between 5000 and 3000 B.C., while the **former residence of General Zhang Xueliang** recalls a Manchurian warlord.

Xi'an's **Old City** is encircled by well-preserved **city walls** with watchtowers, gates, and drawbridges. After biking or walking the 8.7-mile (14 km) ramparts, you'll find quenching refreshment at **Xi'an Brewery** near the **North Gate**.

The modern **metro** reaches into the heart of the Old City and a station near the wooden **Bell Tower** (built in 1384) and **Drum Tower** (erected in 1380) with 24 large drums arrayed around its mezzanine level. Outside the walls, the seven-story **Giant Wild Goose Pagoda**, one of the birthplaces of Buddhism in East Asia, noticeably leans to the west.

A melting pot of regional cuisines, Xi'an is also a foodie paradise. Local favorite *biang biang* noodles are served in restaurants and cafés in the **Muslim Dasi Quarter** and **Yongxing Fang Food Street** in the Old City. ∎

Europe

Traditional houses as well as the Belfry tower line the canal in Old Town Bruges, Belgium (p. 366).

Geneva
Switzerland

Surrounded by France on three sides—and with only a narrow lakeshore connecting it to the rest of Switzerland—this French-speaking metropolis might be the world's most global city. But it's also uniquely Swiss, an urban area where Reformation and particle colliders blend easily with the great outdoors.

THE BIG PICTURE

Founded: first century B.C.

Population: 620,000

Size: 70 square miles (181 sq km)

Language: French

Currency: Swiss franc

Time Zone: GMT +1 and +2 (CET or CEST)

Global Cost of Living Rank: 8

Also Known As: Genève (French), Genf (German), Ginevra (Romansh), Peace Capital

Someone once called Geneva "the city that Switzerland gave to the world." It certainly feels that way while exploring the lakeside burg, given the multitude of languages one comes across and the abundant global institutions. From the Red Cross and World Health Organization to the UN High Commissioner for Refugees and Médecins Sans Frontières, Geneva is said to host more international agencies than any other city (more than 60 total).

Yet beneath that worldly facade is a more conventional city that harkens back to the 16th century when John Calvin established Geneva as a cornerstone of the Protestant Reformation. This is especially true in the **Vieille Ville** (Old Town), where the **Cathédrale de St. Pierre** bore witness to Calvin's radical Reformation. Inside, the flamboyant **Chapel of the Maccabees** has been restored to its full Gothic glory. Calvin and his cohorts are memorialized at the **Reformation Wall** in Parc des Bastions. A detailed scale model in the **Maison Tavel** city museum shows how Geneva looked during Calvin's time.

The Old Town rises above **Lake Geneva** (Lac Léman) and the place where the lake empties into the Rhône at the start of the river's long journey to the Mediterranean. A lengthy waterfront **promenade** stretches five miles (8 km) from the **Conservatoire et Jardin Botanique** with its vintage glass greenhouses to the posh lakeside suburb of Cologny, where **La Belotte** café provides a breezy bar and outdoor terrace for a drink along the lake. On the way are Geneva landmarks like the famous **L'horloge Fleurie** (Flower Clock) in the Jardin Anglais and a fountain called the Jet d'Eau that shoots 130 gallons (492 liters) of water a second as high as 460 feet (140 m) into the Alpine sky.

Those who don't mind chilly liquid (even in summer) can swim at several spots, including Genève-Plage on the **Quai de Cologny** and the year-round Bains des Pâquis public baths (saunas in winter). And there are plenty of ways to mess around in boats: sailboat, powerboat, and peddleboat rentals at **Les Corsaires;** guided kayak tours and paddleboarding on the Rhône with **Rafting-Loisirs;** and white-water rafting on

L'horloge Fleurie, or the Flower Clock, is an installation at the Jardin Anglais.

Lake Geneva, or Lac Léman, is shared between France and Switzerland and is a hub for luxury shops and cobblestone streets.

the nearby Arve and Dranse Rivers with adventure company **Rafting.CH**.

From the lakeside **Pont du Mont-Blanc** bridge, paths run down either side of the Rhône to **Pointe de la Jonction**, where the jade-colored water of the Arve River merges into the sapphire-colored Rhône. Towering high above the multicolored confluence, the lofty **Viaduc de la Jonction** railroad bridge includes a pedestrian footpath to cross (and gaze down on) the meeting of the rivers.

Many of Geneva's institutions open their doors to visitors for guided tours, including the **Palais des Nations**—built as home of the League of Nations and now the United Nations Office at Geneva—and the futuristic campus of **CERN**, where the world's largest physics lab includes the legendary **Large Hadron Collider**.

Geneva's literary heritage unfolds at several spots, including the grave of Argentine author Jorge Luis Borges in the **Cimetière des Rois** and a statue of **Frankenstein's monster** along the Plainpalais that recalls the fact that Mary Shelley was inspired to write the classic horror story after spending the summer of 1816 in Geneva. ■

ALTER EGO

Zurich

As much as Geneva exemplifies French Switzerland, superefficient and hardworking **Zurich** is the epitome of the nation's German side.

Renowned for secret banks, sumptuous chocolate, and exquisite timepieces, the Alpine city sprawls along the **Limmat River** and north shore of **Lake Zurich**.

Unless you're going to open a Swiss bank account, the **Money Museum** is the best place to delve into the world of currency. The private museum offers exhibits on monetary history and the future of money throughout the world. The **Beyer Clock & Watch Museum** covers more than 3,000 years of global timekeeping.

Chocolate requires a far greater commitment: a comprehensive taste of the goodies at **Sprüngli**, **Teuscher**, **Läderach**, and other chocolate shops along Bahnhofstrasse.

Paris
France

It's no accident that Paris offers striking panoramas. Because that's exactly what the kings and presidents of the past 500 years had in mind—transforming the French metropolis into a three-dimensional masterpiece that's even more divine when viewed from above.

THE BIG PICTURE

Founded: third century B.C.

Population: 11 million

Size: 1,098 square miles (2,844 sq km)

Language: French

Currency: Euro

Time Zone: GMT +1 and +2 (CET and CEST)

Global Cost of Living Rank: 33

Also Known As: Lutetia (Roman), La Ville-Lumiere (City of Lights), La Ville de l'Amour (City of Love)

It's not like cities haven't lost iconic structures in the past. After all, only one of the Seven Wonders of the Ancient World survived into the modern age. But the devastation of **Notre-Dame Cathedral** in April 2019 was especially difficult for Parisians and the world at large.

The great cathedral had been around for more than 800 years (construction started in 1163) and survived the Hundred Years' War and the Black Death, the Reign of Terror and Nazi occupation, the Great Flood of 1910 and hordes of tourist kids pretending to be the Hunchback or Esmeralda.

Luckily the cathedral's sumptuous **rose windows**, statue-laden **portals**, and magnificent **western facade** were largely spared, as were many of the ecclesiastical treasures inside the church. However, the spire, the roof, and much of the interior will take years, if not a full decade, to rebuild. Meanwhile, the church and its bell towers, **Parvis de Notre-Dame** square, and the **Archaeological Crypt** will be shuttered.

Yet, Notre-Dame endures in other ways. During the ongoing reconstruction, the orchestra and choir of the **Association Musique Sacrée at Notre-Dame de Paris** is playing historic churches around Paris, including 17th-century **Saint-Sulpice** and 13th-century **Saint-Séverin** in the medieval Latin Quarter along the left (southern) bank of the Seine, and 16th-century **Saint-Eustache** overlooking **Les Halles** park on the right (north) bank.

And there's still plenty of Gothic to explore on the **Île de la Cité**, starting with 13th-century **Sainte-Chapelle.** King Louis IX (aka St. Louis) had the chapel constructed to house the Crown of Thorns and other sacred relics. One of Europe's most exquisite Gothic creations, the church is renowned for its slender flying buttresses, sumptuous woodwork, and 1,300 stained-glass panels, most of them originals.

View the Eiffel Tower from a river cruise along the Seine.

The pyramid structures of the Louvre light up at night, a beautiful display to watch come to life at twilight.

The chapel is surrounded by the massive **Palais de Justice**, originally a royal palace but now the nation's highest court. The complex includes the **Grand Chambre** (Great Hall) where Louis XIV uttered his most famous phrase ("I am the State") and where the Revolutionary Tribunal dispensed a gruesome justice that came to be called the Reign of Terror. Many of the condemned were imprisoned in the palace's **Conciergerie**, where the cells that housed Marie Antoinette, Danton, and Robespierre are preserved as a reminder that even the Age of Enlightenment had its dark moments.

Just across the river is the **Tour Saint-Jacques**, all that remains of a Gothic church that was otherwise demolished during the French Revolution. Visitors can take in the view from the tower summit during 50-minute guided tours. There's another great view from the summit of nearby **Centre Pompidou**, which encases the cutting-edge

BACKGROUND CHECK

• The French metropolis takes its name from the Parisii, an ancient Celtic tribe that settled the area in the third century B.C. and were eventually conquered by Julius Caesar.

• Paris lays claim to the first photograph that featured human beings (Louis Daguerre in 1839) and first public movie screening (by the Lumière Brothers in 1895).

• The Paris coat of arms features a sailing ship because the river traders in medieval times had the most powerful guild and their 13th-century seal was eventually adopted for the entire city.

• Among 300-plus stops along the Paris Métro, a dozen "ghost stations" were constructed but never opened or shut down in later years. Some are now used for operator training, others store unused trains, and still others are used for movie and TV shoots.

• The Flamme de la Liberté—a full-size replica of the flame that tops the Statue of Liberty—has evolved into an unofficial tribute to Princess Diana, who was killed in 1997 in the nearby Alma Tunnel beneath the Seine.

Musée National d'Art Moderne.

Graced with numerous heights both natural and human-made, there are plenty of other places to gaze down at *La Ville-Lumière* (the City of Light), a nom de plume that echoes Paris's status as one of the first cities to mandate street lighting and its more benevolent role in the Enlightenment.

In hindsight, it seems remarkable that the **Eiffel Tower** was planned as a temporary structure. Like many buildings at the 1889 Exposition Universelle, it was earmarked for demolition following the event. It's also extraordinary that so many French artists and architects petitioned to stop construction of a "giddy, ridiculous tower dominating Paris like a gigantic black smokestack."

Creator Gustave Eiffel got the last laugh, because almost at once it was hailed as a wonder of modern engineering. From any of the tower's three levels, visitors can gaze down to the **Champs de Mars**, a grassy esplanade from which the world's first manned hot-air balloon flight took off in 1783 and on which many citizens literally lost their heads during the French Revolution.

Poised at the far end of the "Field of Mars," **Napoleon's tomb** lies beneath the massive dome of the **Hôtel National des Invalides**, a complex that also includes the **Musée de l'Armée** and a small but fascinating **museum of military maps and models**. The tree-filled estate just east of Invalides was once the home and workshop of France's most famous sculpture: Auguste Rodin's "**Le Penseur**" ("The Thinker") is one of the many

Walk through the ornate, statue-filled entry hall of the famous Opéra Garnier.

HIDDEN TREASURES

• **Le Mur des Je t'aime:** In keeping with the "City of Love" meme, the "I Love You Wall" in Jehan-Rictus Square is a large public art installation by Frédéric Baron and Claire Kito that is inscribed with "I Love You" in 311 different languages.

• **Musée des Arts Forains:** The magical collection of carousels and other carnival attractions of Jean-Paul Favand, an actor and antique dealer, is open by reservation or scheduled seasonal tours. Visitors can try out some of the vintage games and rides, which date from 1850 through 1950.

• **La REcyclerie:** Located in an abandoned train station near Porte de Clignancourt, this unique eco-centric café and education center aims to raise public awareness of the three R's (reduce, reuse, and recycle). In addition to fresh and locally sourced cuisine, the center offers gardening and cooking classes, and hosts concerts and seed swaps.

• **Jardin des Plantes:** The city's natural history hub boasts curiosities like the Gallery of Paleontology & Comparative Anatomy, the Room of Endangered & Extinct Species, and the Gloriette de Buffon, one of the world's oldest metal structures.

---

masterpieces displayed in the **Musée Rodin** house and garden.

Looking west from the Eiffel Tower are the River Seine and the **Trocadéro,** an art deco plaza created for the International Exposition of 1937. In addition to terraces that render the absolute best views of the Eiffel Tower, the plaza's curving **Palais de Chaillot** houses the French national **museums of architecture**, **anthropology**, and **maritime history**, as well as the underground **Aquarium de Paris**. Nearby are the chic **Palais Galliera** fashion museum and **Palais de Tokyo**, another remnant of the 1937 expo now used for contemporary art exhibitions and cutting-edge artist creations.

Starting in the 1850s, Emperor Napoleon III commissioned Baron Haussmann to undertake one of history's first large-scale urban renewal projects. One of the legacies of his radical makeover are the broad, tree-lined avenues that dissect central Paris.

Avenue d'Iéna, one of those 19th-century thoroughfares, connects the Eiffel Tower and Trocadéro with another famous vantage point—the **Arc de Triomphe**. Although the giant arch ostensibly honors those who fell in the French Revolution and Napoleonic Wars, it's really a tribute to Napoleon's military prowess and his many battlefield victories.

From the viewing platform atop the Arc de Triomphe, visitors can gaze up the **Avenue de la Grande Armée** to the monumental **Grande Arche de la Défense**—unveiled in 1989 to mark the bicentennial of the French Revolution—and down the celebrated **Champs-Élysées**. A pastiche of luxury boutiques and fast-food outlets, the city's most celebrated boulevard has witnessed its fair share of French history from the guillotine execution of Louis XVI and Marie Antoinette in the **Place de la Concorde** in 1793 to Hitler's goose-stepping victory parade in 1940 and the joyful liberation processions of August 1944. Nowadays it hosts the annual **Bastille Day parade** on July 14.

The Champs-Élysées leads into the heart of royal Paris, an ensemble of palaces and gardens used by French kings and queens as their city residence between the 14th and late 18th centuries. Commissioned by Catherine de' Medici in the 1560s, the **Tuileries Garden** was a royal playground before the revolution converted them into one of the city's first public parks.

Perched at the eastern end of the Tuileries is the incomparable **Louvre**, the world's largest and most visited art museum. Entering via I. M. Pei's sublime glass pyramid, around 10 million people each year trek its lengthy hallways to eternal treasures like the **"Mona Lisa"** and **"Venus de Milo."** The wonderful **Musée d'Orsay** and its Impressionist and post-Impressionist masterpieces are in the old beaux arts train station just across the river from the Louvre, while the old **Bourse de Commerce** reopened in 2021 as a stunning new location for the **Pinault Collection** of contemporary art.

The best of Paris's natural vantage points hovers along the northern edge of the city center. The Romans knew it as the Mount of Mercury, but medieval Parisians called it **Montmartre** ("Hill of Martyrs") because it's where the Romans executed St. Denis. By the late 19th century, it was the city's bohemian quarter, a hangout for the likes of Van Gogh, Renoir, Matisse, Picasso, and Braque, and as such one of the cradles of Impressionism and Cubism.

Montmartre's narrow, twisting cobblestone streets are flanked by bars, restaurants, and small theaters. Crowning the hilltop is **Sacré-Coeur Basilica**, a Romano-Byzantine style church with distinctive domes. The terrace in front of the church affords one of the city's most photographed views, but there's an even better vista for those who scale the 300 steps to the observation area around the central dome and its 360-degree panorama.

Rising above the southern edge of the central city is **Montparnasse**. Once a major artistic hub in its own right between World Wars, it is now a place where visitors can venture to high-end venues and budget-friendly spots.

Much of the neighborhood's boho past was bulldozed to construct the **Tour Montparnasse**, a 689-foot (210 m) office tower that

La Défense is a structurally unique spot to visit and a major business district in Île-de-France.

became the most hated structure in Paris during its 1970s construction. It's still not especially liked, but the view from its **Top of the City** observation deck is arguably the city's best. On the other side of the **Montparnasse Cemetery** is the entrance to the creepy **Catacombs of Paris**, an underground ossuary where an estimated six million people are buried.

The city's most famous burial ground is **Père Lachaise Cemetery** in the 20th arrondissement, where author Oscar Wilde, rock star Jim Morrison, actress Sarah Bernhardt, and many other notable people are buried amid parklike grounds with thousands of elaborate tombs.

Père Lachaise is part of a garland of large green spaces around central Paris that range from the formal **Luxembourg Gardens** and the faux wilderness **Buttes-Chaumont** to the modern **Parc de la Villette**, home to the **Conservatoire de Paris** dance and music school, and the **Cité des Sciences et de l'Industrie** (Europe's largest science museum).

Flanking the city on the east and west are two former royal hunting grounds transformed into huge urban parks. The heavily forested

See Leonardo da Vinci's masterpiece "Mona Lisa" at the Louvre.

Bois de Vincennes provides plenty of space for the **Paris Zoo** and **Parc Floral de Paris** botanical gardens, as well as the **Château de Vincennes**, the classic medieval castle. Out west, the **Bois de Boulogne** wraps itself around **Longchamp** racetrack, **Roland-Garros stadium** (home of the French Open tennis tournament), and **Fondation Louis Vuitton**, a fantastic new modern structure designed by American architect

Frank Gehry that houses contemporary art.

Open-topped **Bateaux Mouches** boats have been the most popular way to cruise the Seine since their introduction in 1949. A more leisurely and thoroughly modern way to get out on the water is renting an electric boat from **Marin d'Eau Douce** for a voyage along the **Canal de l'Ourcq** that can take up to six hours round-trip if you motor all the way to **Sevran** on the eastern edge of Paris. ■

URBAN RELIC

Palace of Versailles

Just a short 12-mile (19 km) jaunt outside of Paris lies the **Palace of Versailles**, a UNESCO World Heritage site. The palace is an ode to opulence and intrigue, in no small part thanks to its most famous residents, Marie Antoinette and Louis XVI, the last royals to call it home.

The remarkable complex was

declared the official royal residence in 1682. The gardens, lined by orange trees and mazelike pathways, took 40 years to complete and are worth a stroll. Inside the palace are stunning displays of luxury and architecture, including King Louis XIV's elegant ballroom, adorned with arched mirrors, gold statues, and sparkling chandeliers. The most recognized space is the Hall of Mirrors, a

baroque-style gallery with 17 floor-to-ceiling arched windows and the same number of large mirrors, composed of more than 350 reflective surfaces.

Outside the palace, the town of Versailles offers Marché Notre-Dame, a covered market with tasty bites; the Hôtel Le Louis Versailles Château, with a cocktail bar fit for royalty; and the Michelin-starred Gordon Ramsay au Trianon.

Stockholm
Sweden

Spread across 14 islands and a warren of waterways, the Swedish capital is one of those rare cities that manages to blend medieval moods and progressive thinking within the guise of a single urban area.

THE BIG PICTURE

Founded: 1252

Population: 1.58 million

Size: 147 square miles (381 sq km)

Language: Swedish

Currency: Swedish krona

Time Zone: GMT +1 and +2 (CET and CEST)

Global Cost of Living Rank: 72

Also Known As: Venice of the North

"It's not a city at all," Ingmar Bergman quipped of his hometown. "It's simply a rather larger village, set in the middle of some forests and some lakes." Stockholm may have seemed that way when the legendary film director was growing up there between the two World Wars. But in the intervening years, the Swedish capital has grown into a full-fledged global city.

Stockholm is renowned for its astonishing modern architecture that ranges from the **Ericsson Globe** (with its SkyView gondolas) and the shimmering **Aula Medica** concert hall to the spiky **Stockholm Waterfront** complex and the glass-wrapped **Scandic Victoria Tower**. With more than 90 stations flaunting various forms of creative expression, the **Tunnelbana** metro

is often called the world's longest art gallery. Among the more striking stations are **Rådhuset**, **Kungsträdgården**, **Stadion**, and **T-Centralen**.

That's not to say that all of Bergman's memories have been swept away by the new wave. Stockholm still swirls around a "village" of sorts—the charming **Gamla Stan** with its cobblestone streets, narrow lanes, and brightly colored houses. Founded in the 13th century, the islandbound old town harbors the **Kungliga Slottet** royal palace (daily guided tours) and its fascinating **Royal Armory**, solemn **Riddarholmskyrkan** church where many of Sweden's monarchs are buried, **Storkyrkan** cathedral with its priceless artworks, and an interactive medieval museum and archaeological dig called the **Medeltidsmuseet**.

Bridges leap the canals around Gamla Stan to other landmarks like the cutting-edge **Moderna Museet** of art on Skeppsholmen Island, the leafy **Kungsträdgården** (King's Garden), and the redbrick **Stockholm City Hall**, a masterpiece of Nordic-style art nouveau completed in 1929. Accessible by stairs or elevator, the **City Hall Tower** (347 feet/

Take in the skyline from the SkyView gondola that encircles the Ericsson Globe.

Colorful facades mark the houses in Stortorget Square.

106 m) renders fabulous views of the city center.

Once a royal hunting ground, the island of **Djurgården** is now dominated by entertainment for the masses, ranging from the historic **Gröna Lund** amusement park (opened 1883) to the new **ABBA: The Museum** dedicated to the celebrated Swedish pop group. Nordic wildlife and vintage buildings from around Sweden are the twin attractions at **Skansen** outdoor museum, while an almost intact 17th-century warship—rediscovered in the 1950s after more than 300 years on the seabed—is the focal point of the popular **Vasa Museum**. The island also boasts an innovative **Museum of Spirits** where visitors can explore the wide world of alcoholic beverages via vodka artwork, a "beer pier," and a four-beverage sampler.

From **water taxis** to **guided canal tours**, Stockholm offers various ways to get out on the water. The vintage ferry **M/S *Emelie*** offers hourly service from Nybroplan in downtown Stockholm to Djurgården and Luma Brygga on the south shore. **Stockholm Nature** organizes guided kayak tours of the central city and kayak camping in the archipelago, while the steamship **M.S. *Stockholm*** (launched in 1931) serves brunch during a three-hour cruise past some of the 24,000 outer islands.

Stockholm also boasts unique culinary experiences. Founded in 1722, **Den Gyldene Freden** in Gamla Stan is one of the world's oldest restaurants still operating from its original location. For a party of 10 or more, **Stadshuskällaren** restaurant in the city hall will reproduce any Nobel Prize banquet menu from 1902 onward. ■

Prague
Czechia

During modern times, the Czech capital has evolved from totalitarian rule and revolutionary response into a vibrant renaissance city with a cherished past and a creative, forward-thinking populace.

THE BIG PICTURE

Founded: ninth century A.D.

Population: 1.37 million

Size: 115 square miles (298 sq km)

Languages: Czech

Currency: Czech koruna

Time Zone: GMT +1 and + 2 (CET and CEST)

Global Cost of Living Rank: 90

Also Known As: Praha, City of a Hundred Spires, Golden City, Heart of Europe

Legend holds that Prague was founded during Roman Empire days by Libuše, a Celtic princess and oracle who predicted that a great city would one day rise beside the Vltava River amid the primeval forest of Bohemia.

Not a shred of historical evidence supports that myth. Yet people can see it with their own eyes, because the Czech capital has grown into one of Europe's most treasured (and visited) urban areas thanks to a remarkably preserved medieval core and its lively and thriving cultural life.

The Vltava River splits the city into **Staré Město** (Old Town) on the right bank and **Malá Strana** (Little Side or Lesser Town) on the left, with the famous **Charles Bridge** linking the two. Despite its name, Malá Strana is actually older and in several respects greater than its cross-river rival. This is where the city was officially born with the founding of hilltop **Prague Castle** (Pražský Hrad) around 880 by the Duke of Bohemia.

Expanded over the millennium that followed into the world's largest coherent castle complex, the vast citadel contains the **Old Royal Palace** and several other regal residences, the large **Royal Garden**, a row of well-preserved 16th-century homes called the **Golden Lane**, and the 14th-century **St. Vitus Cathedral** with its immense bell towers, ecclesiastical **Treasury**, and underground **Royal Crypt** where many a Bohemian king was laid to rest.

Below castle hill, riverside Malá Strana is dominated by another impressive shrine: **St. Nicholas Church**, an 18th-century baroque masterpiece that boasts an organ that Mozart once played, Europe's single largest fresco, 200 classical music performances a year, and a cool World War II resistance story. Among the left bank's more offbeat sights are a small **KGB Museum** and the **Lennon Wall**, an anti-establishment graffiti facade inspired by the ex-Beatle.

Across the river, the Staré Město revolves around **Old Town Square** (Staroměstské Náměstí), just as teeming today as it was when it was first established in the 12th century as a medieval marketplace. The

The Prague Orloj is a medieval astronomical clock, mounted on the Old Town Hall.

See the historic center of Prague and landmarks of Old Town in one sweeping view across the river.

square is flanked by magnificent townhouses like the **Dům U Minuty**, where Franz Kafka lived in the 1890s, as well as the **Church of Our Lady Before Týn** with its dark Gothic spires, and the incredible **Prague Orloj**, a 600-year-old astronomical clock outside the **Old City Hall** that comes to life each hour with a **"Walk of the Apostles"** automaton procession.

Wedged between the Old Town and the river is **Josefov**, the old Jewish Quarter, nearly depopulated by the atrocities of World War II. The **Prague Jewish Museum** offers exhibits, artistic events, and workshops in four historic buildings, including the castle-like **Ceremonial Hall** and the sublime Moorish Revival–style **Spanish Synagogue**.

Another old neighborhood called **Nové Město** owes its name (New Town) to the fact that it was first built outside the city walls in medieval times. The area is renowned for art nouveau masterpieces like **Adam Pharmacy** and the massive **Wenceslas Square**.

The square dominates the area with both its size (nearly as long as eight soccer fields) and significance to Czech history as a gathering place for protests and celebrations like the Velvet Revolution that overthrew the nation's Communist regime in 1989. The colossal **National Museum** (science and natural history) crowns one end of the square, behind a statue of **St. Wenceslas,** patron saint of Czechia. ∎

HIDDEN TREASURES

- **Petřín:** Accessible via a vintage funicular railway and hiking trails, Prague's favorite park looks down on the city from the left-bank hills.

- **Letenský Zámeček:** Quaff some of the city's best suds and drink in the views from this lofty open-air beer garden in Letná Park.

- **Decorative Arts Museum:** From furniture and fashion to pottery and pewter, this fascinating collection in Josefov explores the legacy of Czech creativity.

- **Žižkov Television Tower:** Recently rebranded as Tower Park Praha, this 709-foot (21 m) relic of communist days features steel "pods" with viewing decks, restaurants, and a tiny airborne hotel.

Athens

Greece

For nearly 2,500 years, the Acropolis and Parthenon have been the main attractions. But the Greek metropolis offers plenty of other pickings, from age-old monasteries and world-class museums to cool neighborhoods and hip beach areas.

THE BIG PICTURE

Founded: ca 1400 B.C.

Population: 3.5 million

Size: 225 square miles (583 sq km)

Language: Greek

Currency: Euro

Time Zone: GMT +2 and +3 (EET and EEST)

Global Cost of Living Rank: 115

Also Known As: Athína, City of the Violet Crown

Though the fabled **Acropolis** might seem like the most obvious place to start an exploration of Athens, it's certainly not the best. Instead of diving straight into antiquity, hike the trail or hop the glass-enclosed **funicular railway** to the top of **Lycabettus Hill** and take in a view that looks out across central Athens. The view is especially alluring at dusk as the sunset fades into an evening spangled with the illuminated **Parthenon** and other fabled structures.

Linger for a play or concert at the hillside **Lycabettus Theatre** or ride the funicular back to the **Kolonaki** district for gourmet dining establishments, art gallery hopping, or shopping in the posh boutiques. Kolonaki also flaunts two of the city's best museums: the **Benaki Museum of Greek Culture** with its Greek, Roman, and Byzantine treasures, and precious family heirlooms; and the equally intriguing **Museum of Cycladic Art**, which showcases ancient art and artifacts from the Aegean Islands and Cyprus.

The funicular doesn't start until 9 a.m., but anyone is free to hike or drive to the summit of Lycabettus Hill for sunrise over Athens. Then head straight for the **Acropolis**. The earlier you start, the better in terms of crowds, photographic light, and cooler summer temperatures. Opened in 2009, the new **Acropolis Museum** is remarkable for both its relics—especially the famous **Parthenon Frieze**—and its architecture. A glass floor on the bottom level allows for views of the ruins below while a top-floor glass-enclosed atrium affords sweeping views of the mountaintop temples.

From the museum, it's a short (but sometimes steep) trek to the summit of the Acropolis, where the elegant **Athena Nike Temple** and **Propylaea** gateway offer a prelude to that first dramatic glimpse of the **Parthenon**. Erected in the fifth century B.C. at the height of the city's Golden Age, the temple is one of the globe's most iconic structures. The Parthenon is even more remarkable—given its perfect proportions and built-in optical illusions—when you consider it was built long before modern engineering came along.

Presidential soldiers march in the changing of the guard at the Hellenic Parliament.

The ruins of the Parthenon on the Acropolis are just one of the ancient sites on offer throughout Athens.

That other large structure on the mountaintop is the **Erechtheion.** Dedicated to Athena and Poseidon, the temple is slightly younger than the Parthenon and renowned for its **Porch of the Maidens** supported by six caryatid (female-shaped) columns. Entrance tickets are also good for viewing the **Odeon of Herodes Atticus**, **Theatre of Dionysus**, and other Acropolis ruins.

If you're not totally burned out on antiquities by the time you finish with the Acropolis, meander across the adjacent **Filopappou Hill**, where tree-lined walkways lead to the tiny 16th-century **Church of Agios Dimitrios Loumbardiaris**, the cave-like **Prison of Socrates**, and the **Pnyx**, an open-air meeting place where democracy was born in the fifth century B.C. among the citizen assemblies and great orators that convened there.

Or wander down to the **Agora** and its abundant ruins, an area that served as the political, commercial, and cultural hub of the city's ancient Greek, Roman, and Byzantine eras. Imagine how the area once appeared at the **Stoa of Attalos**, a meticulous reconstruction that now harbors the **Museum of the Ancient Agora**.

The Agora overflows into the **Monastiraki** neighborhood, which takes its name from an early Greek Orthodox monastery where the 10th-century **Church of the**

URBAN RELIC

Sparta

Athens versus **Sparta** was the Ali versus Frazier of Greece's Golden Age. The Spartans may have been superior warriors, but their rigid society—propped up by enslaved laborers—was inherently weak. By the fourth century B.C., the city was in steep decline.

Located on the Peloponnese Peninsula around 135 miles (217 km) from its former rival, modern-day Sparta features a downtown Archaeological Museum with relics discovered in the Sparta Antica archaeological park on the north side of town.

Scattered across the ruined city are the remains of an acropolis, theater, and several temples, as well as a 10th-century Byzantine church.

Sparta's most outstanding attraction is the Museum of the Olive and Greek Olive Oil, which traces 60,000 years of olive history and its significance to Greek culture and cuisine.

ARTBEAT

- **Best Movies:** *Never on Sunday* (1960), *Barefoot in Athens* (1966), *The Burglars* (1971), *Timon of Athens* (1981), *A Touch of Spice* (2003), *10th Day* (2012), *Son of Sofia* (2017).

- **Best Books:** *Walking in Athens* by Constantine Cavafy, *Apartment in Athens* by Glenway Wescott, *Deadline in Athens: An Inspector Costas Haritos Mystery* by Petros Markaris, *Why I Killed My Best Friend* by Amanda Michalopoulou, *Chasing Athens* by Marissa Tejada.

- **Best Plays:** *A Midsummer Night's Dream* and *The Comedy of Errors* by William Shakespeare, *Oedipus Rex* by Sophocles, *The Frogs* by Aristophanes, *The Trojan Women* by Euripides.

- **Best Music:** *Zorba the Greek* by Mikis Theodorakis, *Very Best of Manos Hatzidakis, Rough Guide to Rebetika* by various artists, *Chante la Grèce (Greek Songs)* by Nana Mouskouri, *Sirens* by Astarte.

Pantanassa now stands. Outside the church, cobblestone **Monastiraki Square** is awash in open-air cafés, street performers, and handicraft vendors that continue the length of Ifestou Street to **Avissinias Square** with its antique stalls.

On Sundays, the helter-skelter **Monastiraki Flea Market** unfolds in the same area. The best place to purchase local records is the nearby **Museum of Popular Greek Musical Instruments**, which showcases more than 1,000 pieces from *toum-*

belekia (pottery drums) to *laghouta* (long-necked lutes).

Farther along is the **Plaka**, an unabashedly kitschy but thoroughly charming area of Greek tavernas and souvenir shops catering to foreign visitors. It's not everyone's cup of tea. But one small corner of the Plaka is worth checking out: a tiny neighborhood called **Anafiotika**. Established by Cycladic islanders in the mid-1800s, the area's white-washed houses and tiny churches look like they're straight out of the island of Santorini.

Farther east is the **National Garden**, which offers leafy relief from the masses and sweltering summer heat via playgrounds, ponds, and a couple of outdoor cafés. As main venue for the 1896 Summer Olympics, the park also boasts sporting relics like the elegant neoclassical **Zappeion** (fencing hall) and the

Luxury shops, cafés, and vendors line Ermou Street, the city's most famous commercial thoroughfare.

Kallimarmaro, a stadium with white-marble seating where many of the outdoor events took place. One of the city's newest landmarks lies just a block away: the **Basil & Elise Goulandris Foundation**, an art museum where works by Doménikos Theotokópoulos (aka El Greco), Van Gogh, Gauguin, Chagall, Degas, and Jackson Pollock are on display.

The garden's south end is anchored by the imposing Roman-era **Temple of Olympian Zeus**, while the northwest corner is dominated by the **Hellenic Parliament** building and **Syntagma Square**, where the changing of the guard by **Evzones** troops clad in historical Greek uniforms takes place every Sunday morning.

Like many cities around the globe, Athens is transforming disused industrial areas into exciting new spaces. **Ermou Avenue** provides a stark divide between ancient Monastiraki and upstart **Psirri** with its trendy bars and restaurants—sort of a nouvelle Plaka. Perched at the western end of Ermou, the revamped **Gazi** area is even hipper, a hub for live music, avant-garde theater, and alternative lifestyles. The old gas factory that lends the neighborhood its name has morphed into the **Industrial Gas Museum** with "theatrical"

See scaled replicas of ruins at the Acropolis Museum.

guided tours and a **Technopolis** events space with performances, exhibitions, and educational programs.

Nitty-gritty Piraeus—the staging area for Aegean Island ferries—is gradually revamping its waterfront with new attractions like the **Stavros Niarchos Foundation Cultural Center**. Opened in 2017, the massive steel-and-glass palace is home to the **Greek National Opera** and the **National Library of Greece**, and a vast array of other activities, from

art exhibits to Pilates classes and bike rentals.

Across the harbor, **Mikrolimano** and **Zea Bays** are awash in yachts and surrounded by restaurants serving some of the region's best seafood. Athens's best beaches are scattered along **Attica Peninsula**, a strip of Highway 91 that includes **Voula**, **Lagonisi**, and other popular seaside retreats. Poised at the road's end is the spectacular **Temple of Poseidon** at **Cape Sounion**. ◼

ALTER EGO

Thessaloniki

Greece's second largest city, **Thessaloniki**, carries a much different vibe than the capital. Wrapped around a broad bay, it feels more Mediterranean. And given that it was founded during the Hellenistic era—and named after Alexander the Great's sister—it lacks the classical architecture that epitomizes Athens.

Rather there's a mix of landmarks created by later civilizations that ruled Thessaloniki like the **Roman Forum**, the seventh-century **Church of Agios Dimitrios** and other Byzantine shrines, and the notorious Ottoman-era **White Tower**.

The region's Macedonian heritage comes into focus at the **Archaeological Museum**, but the city's best collection is probably the **Museum of**

Byzantine Culture. The ruins of **Pella**, birthplace of Alexander, are around a 40-minute drive west of the city center.

Fast-forwarding into the 21st century, the **New Waterfront** redevelopment features a long seafront promenade between the White Tower and the avant-garde **Thessaloniki Concert Hall** with its wide array of performing arts.

Oslo
Norway

Tucked at the top of a gorgeous fjord, the Norwegian capital strikes a fascinating balance between old and new, a forward-thinking city that hasn't forgotten its kaleidoscopic past.

If the tormented figure in "The Scream" could somehow come alive today, his look of utter shock would be caused by the dramatic changes in Oslo since Edvard Munch created the iconic painting in 1893.

From a sober and somewhat backward city, Oslo has morphed in modern times into one of the world's most progressive urban areas, a cultural melting pot and architectural superstar that often tops the list of Europe's most livable cities.

Oslo's love affair with modern architecture began with the redbrick **Rådhuset** (City Hall) in 1950 and accelerated with the launch of a massive urban renewal project called Fjordbyen (Fjord City) in the 1980s. The project begat the innovative **Aker Brygge** neighborhood, which mixes residential high-rises with waterfront bars, shops, restaurants, and the **Astrup Fearnley Museum** of contemporary art.

Farther along the waterfront is another modern masterpiece, the **Norwegian National Opera House**, with its year-round slate of opera, ballet, and symphony. Rising nearby is the new **Munch Museum**—a steel-and-glass wedge that houses "The Scream" and other works by the noted Norwegian painter—one of the world's largest collections dedicated to a single artist.

Wedged between all that new are antique structures like the **Nobel Peace Center** (tours, talks, and exhibitions) and **Akershus Fortress**, a hulking medieval castle that hosts summer outdoor concerts and tours of the old royal palace and mausoleum.

Three hundred years before King Haakon V commissioned the castle, Norsemen were masters of Oslo Fjord, an era celebrated at the excellent **Viking Ship Museum**. Located a short ferry ride from downtown on the **Bygdøy Peninsula**, the collection revolves around three ninth-century longships recovered from Norse burial mounds. The peninsula also harbors the **Fram Museum** of polar exploration and the **Kon-Tiki Museum** that showcases the original balsa raft that Thor Heyerdahl skippered across the Pacific, as well as the living history demonstrations and vintage structures of the **Norsk Folkemuseum**.

Named the European Green Capital of 2019 because of its efforts to combat climate change, enhance recycling, and restore urban

THE BIG PICTURE

Founded: 1048 A.D.

Population: 1 million

Size: 185 square miles (480 sq km)

Language: Norwegian

Currency: Norwegian krone

Time Zone: GMT +1 or +2 (CET or CEST)

Global Cost of Living Rank: 55

Also Known As: Kristiania (1624–1925)

Explore Viking history, including watercrafts, at the Viking Museum.

Frogner Park, also known as Vigeland Sculpture Park, on the city's West End, stretches across 79 acres (32 ha).

waterways, Oslo offers dozens of parks, lakes, and islands where residents and visitors can work out, chill, or breathe the city's extraordinarily fresh air (thanks to strict greenhouse gas emission standards).

Frogner Park flaunts more than 200 bronze and marble sculptures created by Gustav Vigeland. **Tøyen Park** hosts the **University of Oslo Botanical Garden**, which includes a section dedicated to plants the Vikings used.

On the city's northern outskirts, **Holmenkollen Olympic Park** offers snow sports during the winter months and rapelling and zipline adventures during the rest of the year—and a chance to view Oslo from the top of the gargantuan steel ski-jump tower.

Holmenkollen is also one of the main gateways for hiking and biking **Nordmarka Forest**. With slopes ranging from black to green, a ski school, gear rentals, and evening floodlights, **Oslo Vinterpark** in the forest is the city's main winter sports area.

Oslo's eclectic after-dark scene ranges from Ibsen plays at the **National Theatre** and concerts at **KulturKirken Jakob** to the trendy neo-Nordic cuisine of **Maaemo** (three Michelin stars) and the boho bars of the revitalized **Grünerløkka** neighborhood. ■

ALTER EGO

Longyearbyen

The world's northernmost town, **Longyearbyen** lies in the Svalbard archipelago between the Arctic Ocean and Greenland Sea. Despite its Norwegian-sounding name, the company was founded by American coal magnate John Longyear in 1906.

Nowadays, Longyearbyen is a hub for polar adventure and scientific research, home port for cruise ships plying the islands, and home base for the **Svalbard Global Seed Vault**.

The small but super **Svalbard Museum** and **North Pole Expedition Museum** offer an excellent orientation to local natural and human history. Guided boat trips visit the historic **Isfjord Radio station** and active Russian coal mine at **Barentsburg**.

But outdoor adventure is the city's forte: **hiking and kayaking**, **dogsled and snowmobile safaris**, and **wildlife-watching** (reindeer and polar bears) in the frozen wilderness surrounding town.

Madrid
Spain

A powerhouse in business and culture, as well as a tourism magnet, Spain's sophisticated capital relishes both its royal days and modern ways, a city of grand plazas, sprawling markets, savory tapas bars, and spirited soccer teams.

THE BIG PICTURE

Founded: ninth century A.D.

Population: 6.4 million

Size: 525 square miles (1,360 sq km)

Language: Spanish

Currency: Euro

Time Zone: GMT +1 and +2 (CET and CEST)

Global Cost of Living Rank: 67

Also Known As: El Foro (The Forum), Villa y Corte (Town and Court)

Of all the world's great cities, Madrid is the one that makes the least sense. The Spanish capital doesn't lie along an especially important river and it's nowhere near the sea. It wasn't a vital market town or mining center, and it had little religious significance.

It all came down to the whim of a single man: A decision in 1561 by King Philip II to relocate his court to a place that wasn't tainted by the regional rivalries or the intrigue and backbiting of other royal enclaves. Madrid has never looked back, growing from a meager 4,000 people in the early 16th century to more than six million today—the European Union's third largest urban area after Paris and London.

Madrid's **Palacio Real**, with more than 3,000 rooms arrayed across six levels, is now the largest functioning royal palace in all of Europe. Several dozen rooms are open to the public, including the magnificent **throne room**, **royal chapel**, and **apothecary**, walls throughout the palace adorned with works by Caravaggio, El Greco, Goya, Rubens, Velázquez, and other great masters.

The palace is surrounded by spacious parks and gardens like the leafy **Campo del Moro,** the awesome

Plaza de Oriente and its **Teatro Real** opera house, and the **Jardines de Sabatini**, graced by the only public monument to Philip II in the city that he essentially kick-started. Nearby is one of his greatest achievements: the fabled **Plaza Mayor.** Over the centuries it's been used for bullfights, jousting, theater performances, and the infamous auto-da-fé public penance (like burning at the stake) of those condemned by the Spanish Inquisition.

"Foodie central" could very well describe the warren of streets around the Plaza Mayor. Erected in 1916, **Mercado San Miguel** features dozens of gourmet food vendors, while the **Calle de la Cava Baja** teems with gourmands questing Madrid's best tapas street. In between is the esteemed **Sobrino de Botín**, declared the world's oldest restaurant by the folks at Guinness. Opened in 1725, it appears in the final scene of Hemingway's *The Sun Also Rises,* when the main characters dine on roast suckling pig.

North of the palace, the **Plaza de España** is celebrated for an **equestrian statue of Don Quixote and Sancho Panza** that honors the great Spanish writer Cervantes. Two of

Sample and buy traditional chocolates at Chocolatería San Ginés.

Traffic passes the Cibeles Fountain, with the city's iconic buildings in the distance.

the city's historic high-rises flank the plaza: **Edificio España** and **Torre de Madrid**. Both neo-baroque masterpieces are now hotels, the latter topped by two fabulous penthouse bars. The dashing twin towers anchor the western end of the **Gran Via**, Madrid's most celebrated street. It's less than a mile (1.6 km) from end to end, but there's so much to see and do along the way that you could spend days wandering the grand thoroughfare, browsing the **El Corte Inglés** department store and the many boutiques, or admiring its blend of art nouveau, art deco, Plateresque, and neo-Mudéjar architecture.

The *paseo por la noche* is a huge

Enjoy dinner, drinks, and skyline views on the terrace of Círculo de Bellas Artes.

part of Madrileños culture, and there's no better place to stroll than the Gran Vía, which comes alive in the evening with people on the hunt for tapas, cocktails, and amusement. Considered Madrid's answer to Broadway, there is usually a blockbuster performance on at the **Teatro Lope de Vega**, **Teatro Coliseum**, or one of the other playhouses, along with flamenco dancing and *zarzuela* (satirical opera accompanied by classical Spanish guitars) at the **Teatro de la Zarzuela**. Before and after, people gather in bars, restaurants, and nightspots of the nearby **Malasaña** neighborhood.

Reaching its eastern extreme, Grand Vía spills out onto another majestic avenue called the **Paseo del Prado**, embellished with extravagant fountains and the over-the-top **Palacio de Cibeles** (Madrid's city hall), as well as its namesake **Museo del Prado**. Founded in 1819, one of the world's foremost art museums fills an immense neoclassical palace and a controversial new cube-shaped building for temporary exhibitions. With more than 5,000 paintings, mainly from the 12th through 19th centuries, the collection includes masterpieces like "Las Meninas" by Diego Velázquez, "The Third of May 1808" by Francisco Goya, and "The Garden of Earthly Delights" by Hieronymus Bosch.

Across the street, the **Museo Nacional Thyssen-Bornemisza** specializes in European art from the 13th through late 20th centuries. Rounding out Madrid's Golden Triangle of Art is the **Museo Nacional Centro de Arte Reina Sofía**. Opened by Queen Sofía in 1986, the glass-and-steel showcase features Picasso's "Guernica," as well as pieces by Joan Miró, Salvador Dalí, and other contemporary Spanish artists.

Miguel de Cervantes, Lope de Vega, and Federico García Lorca are among the Spanish bards who lived

URBAN RELIC

Toledo

The Spanish capital until 1561, **Toledo** occupies a commanding bluff-top location 46 miles (74 km) south of Madrid. A UNESCO World Heritage site since 1986, the city is still entered via medieval gateways that open onto haphazard cobblestone streets teeming with architecture that reflects the city's multicultural heritage.

Toledo's dry, sunny weather is perfect for contemplating landmarks as varied as the Alcázar fortress-palace, the Roman-era Puente de Alcántara bridge, Mezquita del Cristo de la Luz mosque, the Sinagoga del Transito, the Gothic Catedral de Toledo, and the El Greco masterpiece "The Burial of the Count Orgaz" in the Iglesia de Santo Tomé.

The city's museums are also varied, ranging from the Museo del Greco displaying the works of the longtime Toledo resident, Museo Sefardí, and Museo de los Concilios y de la Cultura Visigoda to the cheesy Museo del Queso Manchego, the gruesome Museo de la Tortura (Museum of Torture), and the fascinating Museo de la Espana Magica.

Toledo swords have been esteemed since at least 218 B.C. when Hannibal allegedly used them to defeat the Romans and their inferior bronze weapons. Just one artisan still makes them—Mariano Zamorano—who toils for as long as 50 hours on each blade.

The modern Mercado de San Agustín is spread across five floors with a vertical garden serving tapas, wine, full meals, and more. The city is also celebrated for supersweet marzipan molded into cute little animals, some with an egg-yolk filling, and eaten as snacks.

and wrote in the nearby **Barrio de Las Letras** (Literary Quarter). Quotes from their works are embedded in the sidewalks of a pedestrian precinct that also offers cool street art and tapas bars. The barrio's **Cine Doré** screens classic films and houses the **Filmoteca Española**, which restores and preserves old movies for the Ministry of Culture.

However, not all of Madrid's creativity is found in palaces or museums. The **Matadero,** an ex-slaughterhouse along the **Rio Manzanares**, has converted into an arts center with workshops, galleries, performance spaces, and food stalls. Goya's frescoes inside the **Ermita de San Antonio de la Florida** chapel are considered some of the artist's finest work. Rubens-inspired tapestries complement the ecclesiastical art of the 16th-century **Monasterio de las Descalzas Reales** (Monastery of the Barefoot Royals).

Madrileños also worship two very successful *fútbol* (soccer) teams. **Real Madrid** fans flock to the **Estadio Santiago Bernabéu** to see the team's numerous trophies and artifacts, as well as a stadium tour that includes the locker room. The new 68,000-seat stadium of archrival **Atlético de Madrid** also offers tours.

Speaking of crowds, **Ribera de Curtidores** and adjoining streets transform on Sundays into **El Rastro** market and its 1,000-plus vendors. The nearby **La Latina** neighborhood counters with **Mercado de la Cebada**, which mixes weekday *carnicerías, pescaderías,* and *fruterías* with weekend hipster wine bars and organic eateries.

With more than 40 parks and gardens, it's not hard to find nature in Madrid. A center-city location and rental rowboats make **Parque del Retiro** one of the most popular. The westside **Casa de Campo** is five times larger than New York's Central Park. From there you can hop onto a Teleférico cable car and fly over the city to **Parque del Oeste**.

Madrid also flaunts cutting-edge modern architecture. The gravity-defying **Puerta de Europa** (Gate of Europe) twin towers, poised at 15-degree angles, were the world's first leaning skyscrapers, while the **Ciudad del BBVA** includes a 305-foot (93 m) elliptical tower called **La Vela**. ∎

Retiro Park features the Palacio de Cristal (Crystal Palace), a conservatory built in 1887.

Barcelona
Spain

Larger-than-life personalities of the past and present loom large in Spain's second largest urban area, a colorful Catalonian metropolis that extols an offbeat architect, a cutting-edge artist, and a celebrated soccer player.

THE BIG PICTURE

Founded: first century B.C.

Population: 4.8 million

Size: 415 square miles (1,075 sq km)

Language: Spanish

Currency: Euro

Time Zone: GMT +1 and +2 (CET and CEST)

Global Cost of Living Rank: 84

Also Known As: Ciudad de los Condes (City of Counts), Ciudad de Gaudí (City of Gaudi)

Who's luring you to Barcelona—Gaudí, Miró, or Messi? Because your adoration of the city's legendary residents and their bodies of work will largely determine your bucket list for exploring the effervescent Catalan metropolis.

Antoni Gaudí was born and raised in Reus, about a 90-minute drive west of Barcelona, but moved to the big city in the 1870s to undertake his compulsory military service for the Spanish crown and to study architecture. He would remain in Barcelona for the rest of his life, remaking the cityscape in ways that were previously unimaginable. Seven of his structures make up the **Works of Antoni Gaudí World Heritage site**, one of only a handful of individual people worldwide who have been accorded that honor.

Far and away his most celebrated creation is **Basílica de la Sagrada Família** cathedral, a bizarre blend of Gothic and art nouveau that's the single most visited landmark in Spain. Gaudí allegedly studied how melted wax trickles down candles to devise his radical design. Although construction began in 1882, the church remains unfinished and will likely remain so for years to come.

Park Güell was Gaudí's journey into landscape design, a mind trip of mosaic walls and walkways, unconventional colonnades, and eccentric pavilions that look like something conjured by Dr. Seuss. The park's **Casa-Museu Gaudí**, where the famed architect lived for 20 years, is filled with furnishings and objects that he also designed.

Many observers feel that Gaudí reached his creative peak with residential structures like **Casa Batlló**, its dragon back–shaped roof adorned with wacky towers and chimneys, its facade seemingly fabricated from the skeleton of some giant creature (hence its local name *Casa dels ossos,* or "House of Bones"). Among his other domestic masterpieces are the fortresslike **Palau Güell** with its intricate ironwork and **Casa Milà** (also called La Pedrera) with its breathtaking atrium and ghostly ventilation towers.

Barcelona's **Barri Gòtic** (Gothic Quarter) was the youthful stomping ground of celebrated artist **Joan Miró**, who started off as a Cubist (like fellow Catalan Pablo Picasso) but later drifted toward surrealism and the free-form abstraction for which he's best known.

Catch a *fútbol* (soccer) match at the Camp Nou stadium, home to FC Barcelona.

The world-renowned Park Güell comprises gardens and architectural beauties on Carmel Hill.

Two dozen of his major works are on display at the **Fundació Joan Miró** in the Parc de Montjuïc, including his **"Barcelona Series"** of black-and-white lithographs, "**Solar Bird,**" and the "**Tapestry of the Fundació**" mural. But the artist left his mark all over the city: the much photographed sidewalk mosaic **"Pla de l'Os"** and the towering **"Woman and Bird"** sculpture (his last work of public art) in the **Parc de Joan Miró**, as well as **Miró's birthplace** at No. 4 Passatge del Crèdit in the Barri Gòtic, now occupied by a small but intriguing **Artevistas Gallery** with contemporary works that Miró would have no doubt appreciated.

Lionel Messi is an artist of a much different sort, acclaimed as the best soccer player of the 21st century and one of the greatest of all time by many *fútbol* (soccer) pundits. Although born in Argentina—and an integral member of that country's national team—Messi spent two decades (2001–2021) playing for the city's beloved **FC Barcelona**.

Many of his exploits are featured in the **FC Barcelona Museum**, tucked into the bowels of the massive **Camp Nou stadium** (with a

HIDDEN TREASURES

• **El Recinte Modernista de Sant Pau:** The world's largest art nouveau complex—created by renowned Catalan architect Lluís Domènech i Montaner and declared a UNESCO World Heritage site in 1997—comprises 12 pavilions of Sant Pau Hospital.

• **Museu de la Xocolata:** Created by the city's confectionary guild, the museum features sculptures of the Taj Mahal, the Pietà, Don Quixote, etc., made with a treat introduced to Spain by conquistador Hernán Cortés—a visit made even sweeter by the chocolate

bar everyone receives upon entry.

• **Parc del Laberint d'Horta:** This semisecret storybook hedge maze was created in 1791 as part of the Desvalls family estate that also features a beautiful 14th-century mansion, a waterfall, and strolling-friendly paths. Those who reach the center of the labyrinth discover a statue of Eros (the Greek god of love).

• **MUHBA Turó de la Rovira:** Built in 1938 during the Spanish Civil War, the abandoned "bunkers of Carmel" antiaircraft defenses atop Turó de la Rovira offer awesome city views.

Construction of the Basílica de la Sagrada Família began in 1882.

and a **Barça Store** filled with soccer jerseys and other swag. **Game tickets** are best purchased online at the FC Barcelona website, by telephone, or at some ATMs around the city. Your best chance of snagging a seat at Camp Nou is a weeknight match against a down-the-table La Liga (Spanish League) opponent.

If you're not aligned with any of these larger-than-life personalities, consider starting your Barcelona adventure in the **Plaça de Catalunya**. Night or day, the city's paramount square buzzes with traffic, pedestrians, and pigeons. **Café Zurich** still channels the famous writers and artists who flocked here in the 1930s, but **Farggi 1957 café** is the place to head for ice cream, panna cotta, or other sweet treats.

The streets behind Cafe Zurich plunge into an **El Raval** quarter transitioning from its days as Barcelona's primary red-light district into a haven for trendy restaurants, bars, and galleries clustered around the dazzling **Museu d'Art Contemporani de Barcelona** (MACBA). Deeper into the El Raval are the vast **La Boquería** food market, 13th-century **Sant Pau del Camp** (Barcelona's oldest church), and **El Gat del Raval**, a large bronze cat by Fernando Botero that stalks a palm-lined promenade in the heart of the old neighborhood.

Another web of streets fans out across the Barri Gòtic to the soaring Gothic facade of **Barcelona Cathedral** and the excellent **Museu d'Història de Barcelona** in the **Plaça del Rei**, which spans a cluster of Gothic and Renaissance structures as well as underground Roman ruins. The Gothic Quarter bleeds into an El Born neighborhood that feels even more medieval, its warren of narrow streets home to the

capacity of nearly 99,000, one of the world's largest sporting facilities). But Messi is far from being the club's only superstar. Although founded in 1899 largely by British and Swiss residents, FC Barcelona grew into a global powerhouse on the footwork of Hungarian László

Kubala, Dutchman Johan Cruyff, Argentine Diego Maradona, and other all-time greats.

The stadium offers guided tours—including a **"Players Experience"** that includes the team locker room and a walk on the pitch—as well as the **FCB ice-skating rink**

GATHERINGS

- **Formula One Spanish Grand Prix:** Circuit de Barcelona-Catalunya in Montmeló, an auto race that celebrated its centenary in 2013, roars into action for three days each May; circuitcat.com.

- **Primavera Sound:** This hugely popular indie rock festival in late spring or early summer at the Parc del Forum runs concurrently with side programs at nightclubs and public spaces citywide; primaverasound.com.

- **Grec Festival:** Venues throughout the city showcase theater, dance, music, cinema, and circus events from around the globe during this annual summer (June–July) extravaganza; barcelona.cat/grec/en.

- **La Festa Gràcia:** This sizzling August fiesta in Gràcia barrio features colorful street and balcony display competitions, giant dolls, traditional food and music, and *correfoc* (fire runs) featuring pyromaniacs dressed as devils tossing firecrackers with abandon; festamajordegracia.cat.

- **La Mercè:** Barcelona welcomes autumn with a September celebration that honors the city's patron saint with huge effigies, blazing correfoc parades, and amazing "castellers" (human pyramids); barcelona.cat/lamerce/en.

bustling **Mercat de Santa Caterina** with its chromatic abstract roof, 14th-century **Santa Maria del Mar** church, and a **Museu Picasso** that showcases more than 3,500 of the Catalan artist's earlier works.

Alternatively, you can head straight down **La Rambla**, which is not just Barcelona's most celebrated avenue but also one of Europe's great walking and people-watching spots. If crowds along the boulevard start to overwhelm, simply duck into one of many bars or restaurants, or whatever opera or concert is playing at the **Gran Teatre del Liceu** (opened in 1847).

Although Spanish poet Federico García Lorca famously wished that La Rambla "would never end," the grand avenue terminates at **Mirador de Colom**, a towering tribute to Christopher Columbus on the Barcelona waterfront. The harbor area beyond the monument floats numerous attractions, including the **Museu Marítim** in the old royal shipyard, **Las Golondrinas** harbor cruise boats, and **L'Aquarium Barcelona**. Running north from the docklands is the city's beach strip, a series of sandy strands like **Platja de la Barceloneta** that runs 4.6 miles (7 km) up the Mediterranean coast.

From the harbor, the **Teleférico del Puerto** cable car rises to the crest of **Montjuïc Hill** and its sundry attractions. Hiking paths climb to lofty *miradores* with views across the city and coast. The 17th-century **Castell de Montjuïc** flaunts Catalonia's military history while the open-air **Poble Espanyol** flourishes historic buildings relocated from all around Spain. Montjuïc is also home to ethnology and archaeology collections, as well as the **Museu Nacional d'Art de Catalunya** (MNAC) inside a grandiose domed building erected for Barcelona's 1929 world's fair. ■

An ornate bridge spans buildings in the Barri Gotic quarter.

London
England, United Kingdom

It's hard to imagine London without the Thames, for the river nurtured the English monarchy, sparked the city's rise into an economic power, and nowadays flaunts many of the iconic sights that attract more than 20 million visitors each year to the British capital.

THE BIG PICTURE

Founded: A.D. 47

Population: 11 million

Size: 671 square miles (1,738 sq km)

Language: English

Currency: Pound sterling

Time Zone: GMT +0 and +1 (GMT and BST)

Global Cost of Living Rank: 18

Also Known As: Londinium (Roman), the Old Smoke

"I have seen the Mississippi. That is muddy water. I have seen the St. Lawrence. That is crystal water. But the Thames is liquid history," quipped 19th-century British politician John Burns.

No more eloquent words have ever been written about the waterway that cleaves London. And perhaps no other river on the planet has such a fascinating story to tell. From the ancient Romans and Norman conquerors to Henry VIII, Shakespeare, Sir Francis Drake, and the Rolling Stones, the river has been woven into the stories of many who have shaped British history and global culture.

After a 147-mile (237 km) journey from its source in western England—via **Oxford**, **Reading**, and **Windsor**—the River Thames finally enters London near Heathrow Airport. The spot where the river crosses into the city is marked by **Garrick's Temple to Shakespeare**—a famous 18th-century actors' tribute to the Bard—and the houseboat *Astoria*, a floating music studio where Pink Floyd recorded several of its albums.

The forest along the left bank is **Bushy Park**. London's second largest green space started life as a royal hunting reserve for Henry VIII but is most remembered as the place where Gen. Dwight D. Eisenhower and his staff planned the D-Day landings of 1944. Henry VIII also features at nearby **Hampton Court**, the most magnificent of all the city's royal palaces and home to Britain's oldest surviving hedge maze (planted in the 1690s).

Between the districts of Hampton and Kew, the Thames is still more of a bucolic river than big-city waterway, flanked by residential areas that feel more like villages than London's outer suburbs. Kayaks and eights, small sailboats and holiday barges dominate this stretch of river—a floating cavalcade best observed from the two **Teddington Lock Footbridges** (erected in the 1880s).

The first big suburb on the left bank, **Twickenham** fosters two beloved British institutions: rugby and cinema. Massive **Twickenham Stadium** (82,000-person capacity) is

Soldiers march in the Changing of the Guard at Buckingham Palace.

Take in London's skyline, including the historic Tower Bridge, from the southwest side of the Thames River.

where England's national rugby team plays its home games. Meanwhile, some of the most famous films in British history were shot at **Twickenham Studios**, which opened in 1913.

Hammertons Ferry—the last remaining privately owned ferry on the tidal Thames—provides passage across the river between Twickenham's **Marble Hill Park** and the 17th-century **Ham House and Garden**. Nearby **Eel Pie Island** was a hotbed for London's jazz and blues scene of the 1950s and '60s; among the young musicians who appeared there were Eric Clapton, Rod Stewart, and Jeff Beck. Nowadays the narrow isle is home to the **Eel Pie Island artists collective**, two dozen studios that open their doors to the public on June and December weekends.

Rising above the right bank is **Richmond-upon-Thames**,

renowned for historic pubs like **The Cricketers** (opened in 1770) and **The Old Ship** (1735), as well as the huge leafy expanse of **Richmond Park**. The largest of

London's royal parks is also the city's most important wildlife reserve, with an amazing array of creatures, including more than 600 red and fallow deer.

Nature continues in riverside **Kew** and its famed **Royal Botanic Gardens**, a UNESCO World Heritage site that nurtures more than 50,000 trees and other plants, many lodged in stunning structures like the **Temperate House**, opened in 1862 and still the world's largest Victorian glass house. The river meanders through **Chiswick** and **Hammersmith** to the **London Wetland Centre**, a watery nature reserve with birding hides and a visitors center that features exhibits, lectures, and guided tours.

Reaching Victorian-era **Putney Bridge**, the Thames finally starts to feel urban. Especially when you see the giant smokestacks of **Battersea Power Station** looming ahead. Once a hallmark of London's industrial might, the vintage plant and its surroundings are transitioning into a mixed-use residential, entertainment, and cultural area.

Chelsea Bridge—perhaps the prettiest of the 30-plus bridges that span the Thames in London—leaps the river to **Chelsea** and its posh **King's Road** shopping area. One of the polestars of the psychedelic sixties, Chelsea was home to the Beatles and the Rolling Stones. Founded in 1682, the **Royal Hospital Chelsea** is both an active retirement home for old soldiers and a venue for the wildly popular **Chelsea Flower Show** each spring.

Around another big bend in the river is the architectural **Big Ben**, the famous clock tower that crowns the **Houses of Parliament**. The **Tate Britain** national gallery of art squats along the same bank, and from **Victoria Tower Gardens** you can see

Westminster Abbey has hosted 17 royal weddings and is the resting place of many British luminaries.

the twin spires of **Westminster Abbey**, where many a luminary is buried.

Not far from the river are the **10 Downing Street** home of the British prime minister and the **Churchill War Rooms**, a secret underground bunker where the government took refuge during World War II. From **Parliament Square**, Great George Street and Birdcage Walk lead inland to **St. James Park** and **Buckingham Palace**.

Westminster Pier offers an array of boating experiences from slow-going sightseeing cruises to the superfast **ThamesJet** speedboat experience. Hovering high above the floating dock is the **London Eye**, a giant observation wheel that offers one of the best bird's-eye views of the central city. Several theme park attractions—**Sea Life Aquarium**, **Shrek's Adventure**, and the **London Dungeon**—have made their home near the historic **County Hall** building beneath the Eye.

Suspended from either side of a Victoria railroad crossing, the **Golden Jubilee Bridges** offer pedestrians a scenic passage from the **South Bank** performance halls to **Trafalgar Square**. London's most celebrated square features **Nelson's Column**, the four black **Landseer Lions**, and hundreds of pigeons, but its real star is the **National Gallery** of art and adjoining **National Portrait Gallery**.

Trafalgar Square is only a short walk from the movie houses of **Leicester Square**, neon-spangled **Piccadilly Circus**, restaurant-packed **Chinatown**, the **West End** theater district, and the sundry attractions of **Covent Garden**. A broad avenue called **The Strand**—a name that betrays the fact this was once a sandy riverside—connects Trafalgar and the old **City of London**.

The heart of Roman and medieval London, and now the central business district, the old city of London was almost completely destroyed by the Great Fire of 1666. **St. Paul's Cathedral** arose right after the blaze as a symbol of London's resurgence. Nowadays the Millennium Bridge, an innovative pedestrian suspension span, connects the cathedral with the avant-garde **Tate Modern** art museum, the moody Gothic interior of **Southwark Cathedral** (founded in 1106), and **Shakespeare's Globe**, a re-creation of the playhouse the playwright opened on the same site in 1599.

St. Paul's dominated the city skyline until the late 20th century when radical head-turning skyscrapers shot up. Topped by a glass-enclosed atrium with bars, restaurants, and viewing decks, the **Sky Garden** bills itself as London's highest public garden. Across the river, the 95-story **Shard**—Britain's highest building at 1,016 feet (310 m)—features a lofty observation area simply called **The View**.

The Shard literally overshadows the **H.M.S. *Belfast***, a floating naval museum permanently moored along the Thames. A little farther downstream are two of England's most recognizable landmarks: the **Tower of London** and **Tower Bridge**, monumental constructions that mark the transition from the city to the storied **East End**, a working-class area famed for gangsters, entertainers, and Jack the Ripper. Among the old pubs along this stretch of river is the **Mayflower**, opened around 1550 and later renamed for the ship that took the Pilgrims from **Rotherhithe** to America.

Limehouse Reach borough takes the Thames around another great bend to the **London Docklands.** The historic hub of the city's maritime trade (and by extension the British Empire) found itself obsolete with the advent of container shipping in the 1960s and has been redeveloped since then into a business, residential, and recreation area that centers around **Canary Wharf** and the **Isle of Dogs**. Almost lost amid all the modern is the **Virginia Quay Settlers Monument** in

The world's largest museum, the Victoria and Albert Museum (V&A) boasts more than 2.27 million items.

Poplar, which marks the spot where the Jamestown colonists departed for the New World in 1606. Nearby **Trinity Buoy Wharf** now supports a much different kind of colony: an arts quarter with multiple galleries, studios, and event spaces.

From the southern end of the Isle of Dogs, **Greenwich Foot Tunnel** dips beneath the river to old **Greenwich** town. Out of all the riverside communities that line the Thames in London, Greenwich has the deepest connection with the sea. For more than 500 years, it was the mainstay of British sea power as home to both the **Royal Naval College** and **Royal Dockyards**. The hilltop **Royal Observatory** was established to "find out of longitude of places," and perform "navigation and astronomy." Its heritage endures through the **National Maritime Museum** and clipper ship *Cutty Sark*.

The giant dome that punctuates the **Greenwich Peninsula** is an entertainment and sports complex called the **O2,** which hosted several sports at the 2012 London Olympics. Centered around a 20,000-seat arena and a retail area called the Avenue, the futuristic tentlike structure includes the **Cineworld** movie complex, a bowling alley, a virtual reality experience, and more than two dozen bars and restaurants.

From the peninsula's **Emirates Aviation museum**, visitors can "fly" across the Thames on cable cars that come back to earth beside the **London Royal Docks** and two unique water sports facilities: the **Love Open Water** swimming area (your chance to take a dip in the Thames) and the **WakeUp Docklands** wakeboarding park.

Downstream from the docklands are the shiny silver **Thames Barrier**, a sophisticated moveable barrage that

See pieces dedicated to the fictional detective at the Sherlock Holmes Museum.

closes whenever the city is threatened by high tides or storm surges, and the free **Woolwich Ferry** for passengers and vehicles (launched in 1889).

Of course, many London landmarks are not along the river. World-renowned museums like the **British Museum** and the **Victoria and Albert Musem** (V&A); incredible green spaces like **Hyde Park**, **Regent's Park**, and **Hampstead Heath**; and the shop-till-you-drop ethos of **Oxford Street**, **Portobello Road**, and **Camden Market**.

The Thames finally departs London between **Rainham Marshes Nature Reserve** and the town of **Dartford**, where Mick Jagger was born and raised (his boyhood home is on Denver Road). By now it's in full estuary mode, a waterway that smells, feels, and acts more like the ocean than a river as it merges with the North Sea. ∎

LOCAL FLAVOR

• **Mr. Ji:** This low-frills Soho establishment offers high-flavored Taiwanese dishes served on red plastic plates or in paper cartons. It's worth trying the sit-down menu, which boasts spring rolls, prawn toast, the Sichuan burger, and spicy double-cooked chicken thigh bites; *72 Old Compton St.*; mrji.co.uk.

• **Bong Bong's Manila Kanteen:** This stall at Seven Dials Market is worth the trip to the indoor food emporium. Dishes are modern takes on traditional Filipino fare, including crispy pata pancakes, adobo-glazed chicken wings, and a selection of ice creams and sorbets for dessert; *Unit 6 Kerb, Seven Dials Market, Earlham Street*; bongbongs.co.uk.

• **Delhi Grill:** Indian street food comes to the Islington neighborhood at Delhi Grill. Curries, butter chicken, and murgh makhana should not be missed. Nor should the lamb chops, samosas, or chaats; *21 Chapel Market, Islington*; delhigrill.com.

Munich
Germany

Munich's lively beer-hall culture and remarkable medieval facades disguise the fact that Bavaria's largest city has long been a mold breaker and global trendsetter in everything from motorcars and movies to museums and social mores.

THE BIG PICTURE

Founded: 1175

Population: 2 million

Size: 180 square miles (466 sq km)

Language: German

Currency: Euro

Time Zone: GMT +1 and +2 (CET, CEST)

Global Cost of Living Rank: 52

Also Known As: München, Millionendorf, World City with Heart

Munich has long marched to the beat of a different drummer than the rest of Germany. And not just to the rhythm of the *oom-pah-pah* bands that enliven the city's famous beer halls. Despite a passion for local traditions—like lederhosen, terrific beer, and Bayern München winning the Bundesliga nearly every year—Munich residents are also renowned for their creativity, innovation, and progressive social attitudes.

Hometown film directors like Rainer Werner Fassbinder and Werner Herzog turned Munich into Germany's modern movie capital, a heritage revealed during guided tours of the city's **Bavaria Filmstadt** (Film City) and daily screenings at the **Munich Film Museum**.

Innovation of a much different kind is on display at **BMW Welt & Museum**, where more than a century of Bayerische Motoren Werke cars, motorcycles, and other machines are on display. Guided tours are offered at the museum, adjacent car factory, and **BMW Group Recycling and Dismantling Centre** in suburban Unterschleißheim.

Once upon a time, **Schwabing** was the epicenter of Munich nightlife, especially **Kurfürstenstraße** boulevard with its many cafés and nightclubs.

Munich's reverence for modern art includes the pop-art-focused **Museum Brandhorst**—part of a museum cluster that also includes **Alte Pinakothek**, **Pinakothek der Moderne**, and **Munich Egyptian Museum**—and the groundbreaking **Museum of Urban and Contemporary Art** (MUCA), which showcases Banksy as well as graffiti artists whose work is normally found on abandoned buildings and railroad cars.

Although civic authorities have tolerated nude sunbathing for nearly a century, Munich made it official in 2014 by establishing six **Urban Naked Zones**, including three spots in the **Englischer Garten** and two in the **Flaucher** with several of these clothing-optional areas on beaches along the **River Isar**. The English Garden also features one of the city's most unique sports venues: a human-made surfing and white-water kayaking run called the **Eisbach.**

There's still plenty of tradition, especially in the circular **Altstadt** (Old Town). Crowded around the medieval **Marienplatz** are the

Surf in the city at Eisbach, a unique sporting venue.

Nymphenburg Palace served as the summer residence for Max Emanuel, heir to the throne, who was born in 1662.

14th-century **Altes Rathaus** (Old Town Hall), and 19th-century **Neues Rathaus** (New City Hall) where the famous **Glockenspiel** clock unleashes a parade of mechanical knights, dancers, and other medieval figures two or three times each day.

Marienplatz is surrounded by a web of cobblestone pedestrian streets leading to the **Frauenkirche** cathedral with its iconic double bell towers, the **Viktualienmarkt** gourmet food market and beer garden, and the dazzling **Münchner Residenz** where Bavaria's rulers lived between 1509 and 1918. Among the neighborhood's many shops is **Steindl Trachten**, which sells traditional and modern versions of lederhosen, dirndls, and other Bavarian clothing.

Beyond Old Town, tradition endures at **Schloss Nymphenburg**, a sprawling summer palace and gardens that easily rival Versailles in size and splendor. Munich's other larger-than-life attraction, the vast **Deutsches Museum** of science and technology features thousands of exhibits from an underground mine and model Cave of Altamira to a virtual reality experience and high-voltage Faraday cage. ∎

GATHERINGS

• **Maidult:** The first of three "Auer-Dult" festivals throughout the year, the April–May version is a combination flea market and carnival in the plaza that fronts Mariahilfkirche; auerdult.de.

• **Tollwood Summer Fest:** Spread over four weekends in June and July, this smorgasbord of cultural and environmental activities includes music, food, and handicrafts; tollwood.de/en.

• **Christopher Street Day**: Munich's version of Pride Week culminates in a grand parade through the Altstadt and street festival in the Marienplatz; csdmuenchen.de.

• **Kaltenberger Ritterturnier:** Jesters, troubadours, and jousting knights suffuse Europe's largest medieval festival, which unfolds each July at Kaltenberg Castle near Munich; ritterturnier.de/en.

• **Oktoberfest:** More than just beer, Munich's biggest bash also features Bavarian foods, music, dance, a costume parade, and carnival rides at Theresienwiese fairgrounds, most of it in September; oktoberfest.de.

Florence
Italy

Although Florence is world renowned for its art and architecture, the Tuscan city can also be explored through the lives of the men and woman who have left an indelible mark on the urban area while it was raising the bar on the Renaissance.

THE BIG PICTURE

Founded: 59 B.C.

Population: 835,000

Size: 85 square miles (220 sq km)

Language: Italian

Currency: Euro

Time Zone: GMT +1 and +2 (CET, CEST)

Global Cost of Living Rank: Unlisted

Also Known As: Florentia (Roman), Firenze, City of Lilies

Galileo, Machiavelli, Michelangelo, and the Medici—names that defined the Renaissance and whose work in art, science, literature, and politics helped shape both Italy and the wider Western world. Their achievements continue to reverberate through Florence.

The celebrated astronomer and physicist Galileo Galilei was born in Pisa (1564) but exiled to Florence after the pope accused him of heresy. He died while under house arrest in 1642 and was buried in the **Basilica di Santa Croce**. The **Museo Galileo** preserves his scientific instruments and his finger in a glass bell. Visits to his **Villa Il Gioiello** in suburban Arcetri can be arranged through the **Museo di Storia Naturale di Firenze**.

Renowned as author of *The Prince*—the seminal work of political manipulation—Niccolò Machiavelli was a lifelong Florentine. As one of the top officials in the short-lived republican government, his work revolved around the **Piazza della Signoria** with its sculpture-filled **Loggia dei Lanzi** and **Palazzo Vecchio**, which now doubles as the city hall and civic museum. After his death in 1527, Machiavelli was also interred at Santa Croce.

Michelangelo was born in Caprice, a village 70 miles (113 km) east of Florence but spent much of his life in the city creating artistic masterpieces for both the republic and the Medici. Some of his most famous works—like "David" and the unfinished "Prigioni"—are among the highlights of the **Accademia**.

Other works are on display at the **Casa Buonarroti** where Michelangelo's family once lived, the **Palazzo del Bargello** museum, the **Medici Chapels** in the **Basilica di San Lorenzo**, and the **Opera del Duomo Museum**, which showcases sculptures created for the adjacent **Duomo** cathedral. After his death in Rome in 1564, Michelangelo was buried at Santa Croce like so many other prominent Florentines.

The House of Medici, a powerful banking clan renowned for both their ruthlessness and patronage of the arts, ruled Florence for much of the 15th and early 16th centuries. The massive family art collection

The statue "David" stands in the Galleria dell'Accademia.

At dusk, the Piazzale Michelangelo and the Duomo are illuminated above the city streets.

was bequeathed to the city of Florence by an 18th-century Medici heiress and subsequently divided between the **Uffizi Galleries** and **Palazzo Pitti**. But scores of other places in Florence have a Medici connection, like the once secret **Vasari Corridor** that crosses the Arno River on top of the **Ponte Vecchio** bridge, the **Hall of Maps** and **Studiolo Francesco 1** inside the Palazzo Vecchio, and even the **Boboli Gardens** with its abundant statues and fountains.

Dante Alighieri, another famous Florentine, was born in the late Middle Ages rather than the Renaissance. His epic poem *Divine Comedy* continues to define our perceptions of heaven and hell. **Casa di Dante** museum is built over the ruins of the place where Dante was allegedly born around 1265. However, his elaborate marble tomb in Santa Croce is empty; Dante's remains are buried in Ravenna, where he died in exile in 1321. ■

LAY YOUR HEAD

• **Antica Torre di Via Tornabuoni:** The rooftop terrace with its romantic views across Florence is reason enough to book a room in this boutique hotel set inside a 13th-century medieval tower; restaurant, bar, sun terrace, luxury boutiques; from $199; tornabuoni1.com/en.

• **Hotel Bernini Palace:** Located right behind the Palazzo Vecchio, the palace and its frescoed Sala Parlamento was once a meeting place for top government officials; restaurant, bar, fitness center, sun terrace, bike rental; from $203; hotelbernini .duetorrihotels.com.

• **Baglioni Relais Santa Croce:** Step back in time at this elegant hotel and its renowned Tuscan restaurants, the city's only Relais & Châteaux property and the host of musical and theater events in the sumptuous Sala della Musica; restaurants, bar, fitness center; from $310; baglionihotels.com.

• **Il Salviatino:** This bucolic villa-style hotel with Italian gardens and outdoor swimming pool crowns a hilltop northeast of the city; restaurant, bar, spa, fitness center, bike rental; from $663; salviatino.com.

Budapest
Hungary

Divided by the Danube River, the Hungarian capital is a fusion of ancient Buda in the west bank and upstart Pest on the east, a shotgun marriage of sorts that compelled two very different urban areas to merge into one of Europe's most intriguing cities.

THE BIG PICTURE

Founded: 1873

Population: 2.5 million

Size: 375 square miles (971 sq km)

Language: Hungarian

Currency: Hungarian forint

Time Zone: GMT +1 and +2 (CET, CEST)

Global Cost of Living Rank: 162

Also Known As: Aquincum (Roman), Queen of the Danube, City of Festivals

Budapest was born of revolution—an 1840s uprising against the Hapsburgs that resulted in twin capitals for the Austro-Hungarian Empire—a heritage that spilled over into the 20th century when the city rose against the Nazis and Soviet Communist rule. That rebellious streak persists in a metropolis that continues to nurture artistic brilliance and contentious politics.

If ever a city had a split personality, it's Budapest. With its hilltop citadels and bosky forest reserves, Buda feels more like something from the distant past, all at once medieval and primeval. Across the water, Pest sprawls across flatlands sculpted with elegant squares and broad boulevards, born of the days when it was on the cutting edge of art, architecture, commerce, and many other aspects of modern civilization.

Rising high above the left bank, the **Várnegyed castle quarter** (also called Buda Hill) boasts many of the city's oldest and most esteemed institutions. Visitors can bus, taxi, or hike to the summit of the 560-foot (170 m) hill, or climb aboard the **Sikló**, a historic funicular railway operating since 1870.

A pastiche of architectural styles from medieval to art nouveau, **Buda Castle** (Budavári Palota) is home to the **Hungarian National Gallery** of art, as well as the **Budapest History Museum** and **National Széchényi Library** with its rare books, maps, and medieval manuscripts. Farther north are the flamboyant **Matthias Church** (founded in the 11th century) and the famous **Fisherman's Bastion** (Halászbástya) with its bird's-eye views of Budapest. A network of caves inside the hill safeguards all sorts of weird and wonderful things, from the **Panoptikum Labirintus** where Vlad the Impaler (Vlad Tepes) was allegedly imprisoned in the 15th century to the **Hospital in the Rock Nuclear Bunker Museum**.

The undulating terrain along the city's western edge is where the Budapesti hike, bike, and picnic in green spaces like the **Határnyereg** and **János-hegy** nature reserves. **Trails of Budapest** offers guided hikes in the western hills while the

Once a factory, the Szimpla Garden was converted into a pub and open-air cinema.

The statue of King Stephen I and the Parliament Building stand in front of the famous Fisherman's Bastion.

Zugliget Chairlift whisks visitors to **Elizabeth Lookout** at the city's highest point (1,730 feet/527 m). Gracing the west bank are the ruins of **Aquincum**, an ancient Roman city preserved within an archaeological park that features villas, public baths, and a well-preserved amphitheater.

There are numerous ways to cross the Danube, including the historic **Széchenyi Chain Bridge** (opened in 1849). Although it's been settled for nearly as long as the west bank, Pest is largely a product of the city's late 19th- and early 20th-century heyday, when iconic structures like the exquisite **Hungarian Parliament Building**, the enormous **Dohány Street Synagogue** (largest in Europe), and **St. Stephen's Basilica** were constructed.

The crowded **Inner City** that surrounds these landmarks doubles as downtown Budapest, the city's financial and commercial hub, and creative heart via institutions like the **Hungarian State Opera** and **Liszt Academy** of music. The ghosts of the Hapsburg-era café society linger at the elegant **New York Café**. The boho **Jewish Quarter** around **Klauzál tér** square is ground zero for the city's modern after-dark scene, home to electronic dance clubs, trendy restaurants, and hip "ruin bars" like the rambling **Szimpla Kert**. ■

GATHERINGS

- **Budapesti Tavaszi:** Bartók, Liszt, and other classical masters are the headliners at this spring festival that also stages ballet, opera, and other traditional performing arts for two weeks in April; btf.hu.

- **Floralia:** This revival of an ancient Roman festival at the Aquincum archaeological park in May features food, wine, song, dance, ancient stories, and gladiatorial combat; www.aquincum.hu/en.

- **Sziget Festival:** Old Buda Island provides a bucolic venue for this massive outdoor music fest in August that features more than 1,000 performances spanning rock, pop, rap, reggae, and EDM; szigetfestival.com/en.

- **Christmas Markets:** Great for traditional Hungarian handicrafts and foods, the city offers huge yuletide markets in Vorosmarty Square and Szent Istvan square from late November through New Year's Day; budapestchristmas.com.

Vienna
Austria

One can hike the Vienna Woods or trip the light fantastic at a Viennese ball. But what Austria's capital really excels at is sitting—in one of its many concert halls, coffeehouses, cavernous churches, or along the banks of the "Blue Danube" on the city's eastern edge.

THE BIG PICTURE

Founded: ca first century B.C.

Population: 1.85 million

Size: 130 square miles (337 sq km)

Language: German

Currency: Euro

Time Zone: GMT +1 and +2 (CET or CEST)

Global Cost of Living Rank: 37

Also Known As: Wien (German), Vindobona (ancient Roman), City of Music, City of Dreams

Many urban areas lend themselves to walking. And with features like the **Vienna Woods** and multiple paths along the **Danube River**, Vienna offers plenty of options for peregrination. But few can match the Austrian metropolis when it comes to the pure joy of sitting, the wide variety of places that you can perch yourself, and the things you can do while seated.

Like sipping some of the world's best coffee, a local obsession since the 17th century, when the Viennese began brewing beans looted from the Turkish army. The first **coffeehouse** appeared in the 1680s as a venue for gossip, political discourse, and intellectual thought by the likes of Freud, Klimt, Mahler, and Trotsky. Vienna's coffeehouse culture is so unique that UNESCO named it an Intangible Cultural Heritage of Humanity.

Esteemed coffeehouses like **Café Landtmann** (opened in 1873) opposite the city hall, **Café Sperl** (opened in 1880) in the Mariahilf neighborhood, and **Café Central** (opened in 1876) in the Palais Ferstel are still among the favorite places for locals and visitors to sit, chat, and sip a traditional Kleiner Schwarzer, Wiener Melange, or Einspänner.

Sitting takes a much different form at iconic performing arts venues like the sumptuous **State Opera**, the city's foremost showcase for opera and ballet, as well as longtime host of the legendary Vienna Opera Ball. Music lovers can also sink into comfortable seats at the **Musikverein**, the handsome red-and-gold home of the **Vienna Philharmonic** and a music hall that ranks among the world's best for acoustics. Gothic **Stephansdom** cathedral and baroque **Peterskirche** are among the many churches that offer classical music concerts, while the immense **Karlskirche** presents *The Four Seasons* by Vivaldi in the very church where the composer is buried. Learn more about sound at the city's **Haus der Musik**, a unique interactive museum.

Vienna became the music center of the Western world in the early 18th century, when the Hapsburgs flooded their capital city with intellectuals and artists from around their empire. A child

Dine out at the Viennese Café Central, linked to Freud and Trotsky.

The Schloss Belvedere was constructed by Johann Lukas von Hildebrandt as a summer residence for Prince Eugene of Savoy.

prodigy who often performed for the royal court, Mozart wrote 21 piano concertos, six symphonies, and four of his most celebrated operas while living in Vienna. Beethoven lived most of his adult life in the city and composed most of his major works there. Meanwhile, "Waltz King" Johann Strauss was blending folk tunes from the countryside around Vienna with classical music to create 19th-century pop hits like "The Blue Danube." Striking up a waltz was definitely one way to get the Viennese out of their comfortable seats and onto the dance floor.

The city honors its maestros in various ways. Guided tours and musical events are offered at the **Mozarthaus,** which preserves the apartment where the celebrated Austrian composer lived in the 1780s.

Burgtheater is remembered as the place where Mozart premiered three operas and where Beethoven's Symphony No. 1 was performed in public for the first time.

Beethoven's statue commands a small park across the street from

Sand in the City urban beach, **Mozart** tops a cherub-covered monument in the **Burggarten,** while a golden **Johann Strauss** plays the violin in the **Stadtpark.** Beethoven and Strauss, as well as Schubert, Brahms, and Salieri, are buried in

BACKGROUND CHECK

• With around 51% of the city comprising parks and gardens, Vienna boasts a higher percentage of green space than any other major world city.

• Vienna topped Mercer's list of world cities with the best quality of living for 10 years in a row between 2010 and 2019.

• In the same manner as Berlin, Vienna was divided between the Allied powers after World War II

(1945–1955), with separate British, French, American, and Soviet sections.

• With more than 1,700 acres (688 ha) of vineyards, Vienna is both the world's largest wine-growing city and the only global capital that makes its own wine.

• Among the items invented in Vienna were the the croissant (18th century), the snow globe (1900), and the PEZ candy dispenser (1927)

A gilded statue of Austrian composer Johann Strauss stands, violin in hand, at the Stadtpark.

the city's immense **Zentralfriedhof** cemetery along with more than 300,000 other past residents.

Erected between the 13th and early 20th centuries and featuring architecture from all those eras, the sprawling **Hofburg** palace was once the primary residence of the Hapsburg rulers. Much of the sprawling structure is now open to the public, including the crown jewels of the **Kaiserliche Schatzkammer**, the opulent **Kaiserappartements** royal residence, and the **Kunsthistorisches Museum** filled with priceless art collected by the Hapsburgs.

But the palace also offers cool places to sit, like watching performances by the celebrated Lipizzaner stallions of the **Spanish Riding School**. Founded in 1572 with white steeds imported from Iberia, the stable carries on a 450-year-old tradition of equestrian grace and skill. Protected like priceless artworks during warfare and civil strife, the horses are another Vienna institution honored by UNESCO as an Intangible Cultural Heritage to Humanity. Headquartered at **Palais Augarten** on the outskirts of the central city, the famed **Vienna Boys' Choir** was created in 1498 for the exclusive entertainment of the royal court. Today anyone visiting Vienna can hear their heavenly voices at

---

ARTBEAT

- **Best Movies:** *Waltzes From Vienna* (1934), *The Third Man* (1949), *The Night Porter* (1974), *Amadeus* (1984), *Before Sunrise* (1995), *A Dangerous Method* (2011), *Museum Hours* (2012), *Woman in Gold* (2015).

- **Best Books:** *The Piano Teacher* by Elfriede Jelinek, *Auto-da-Fé* (*The Blinding*) by Elias Canetti, *The Star of Kazan* by Eva Ibbotson, *Letter From an Unknown Woman* by Stefan Zweig, *The Hotel New Hampshire* by John Irving, *Palace of Treason* by Jason Matthews.

- **Best TV Shows:** *Assignment Vienna* (1972–73), *Freud: the Life of a Dream* miniseries (1984), *Kommissar Rex* (1994–2004), *Vienna Blood* (2019–present).

- **Best Art:** The luminous gold-suffused paintings of Gustav Klimt and the often disturbing Expressionist portraits of Egon Schiele.

concerts in the **Hofburg Chapel** and other venues around the city.

At the opposite end of the Viennese sitting spectrum—out in the fresh air and blue skies—are the city's numerous parks and gardens, many with distinct personalities that set them apart from the others.

The **Burggarten**, for example, is a favorite haunt of students and young people who relish the tropical heat of the old royal **Palmenhaus** in winter or who sun themselves on the lawns of what was once the emperor's private garden. Farther up the **Burgring** boulevard, the **Volksgarten** is popular with foot-weary tourists who recover on green folding chairs around the **Rose Garden** fountain. Long gone are the days when Strauss hosted waltzing in the garden, but after all these years there's still dancing—at the after-dark **Volksgarten Club Disco** on Thursday through Saturday nights.

One of Vienna's newest sitting places is the **Maria-Theresien-Platz**, a once lackluster courtyard in the **MuseumsQuartier** that underwent an extreme makeover in the early 21st century with breezy outdoor cafés and colorful abstract benches. The plaza is surrounded by more than half a dozen museums, including the **Leopold** and **Mumok** museums of modern art, **Architekturzentrum Wien**, and an interactive children's museum called **Zoom**.

The immaculate **Belvedere Gardens** are sandwiched between a pair of baroque palaces of the same name commissioned by Prince Eugene of Savoy, a hero of Austria's last war against the Turks. The Upper Palace now houses the **Österreichische Galerie Belvedere** and a collection that includes "The Kiss" and "Judith and the Head of Holofernes" by Gustav Klimt.

ALTER EGO

Innsbruck

With around 200,000 people in the metro area, **Innsbruck** is about one-tenth Vienna's size. But the big difference is topography: Located amid the Austrian Alps, the Tyrolean city is surrounded by 8,000-foot (2,438 m) peaks.

In addition to staging the Winter Olympics twice (1964 and 1976), Innsbruck is a year-round hub for snow sports and summer hiking, biking, and mountain climbing.

Among its sporting landmarks are the **Bergisel Ski Jump** and the **OlympiaWorld** bobsled run, while a modern funicular whisks passengers to the excellent **Alpenzoo** and lofty **Hungerburg peak**.

Beyond the great outdoors, Innsbruck also boasts an **Altstadt** (Old City) spangled with castles, palaces, and churches, as well as the **Swarovski Crystal Worlds** museum, factory, and gardens.

Even more inspiring is **Schloss Schönbrunn**, the 1,441-room rococo palace and garden complex commissioned by Queen Maria Theresa in the 18th century as her summer home away from crowded Hofburg. Her husband Franz Joseph was so impressed that he transformed it into a year-round royal abode. It almost goes without saying that both places offer superb sitting, especially the benches on **Gloriette Hill** with their panoramic views of Schönbrunn.

Vienna also affords an opportunity to sit and move at the same time. The city boasts one of the world's oldest municipal tram systems, launched in 1865. And it's still one of the world's largest. The electric trains of the **Wiener Linien** zip along more than 109 miles (175 km) of track. Over on east side of the Danube Canal, the vintage **Prater** amusement park (founded in 1766) features the **Wiener Riesenrad** (Vienna Giant Wheel), which was the world's tallest Ferris wheel for much of the 20th century. ∎

Street cafés line the pedestrian-friendly Innsbruck town center.

Berlin
Germany

Berlin may have grown into Germany's largest and most intriguing city via the mandates of kings and tyrants, but the reunited capital has come to exemplify how human beings can set aside their animosity and antagonism for the greater good.

THE BIG PICTURE

Founded: 1237

Population: 4.1 million

Size: 520 square miles (1,347 sq km)

Language: German

Currency: Euro

Time Zone: GMT +1 and +2 (CET and CEST)

Global Cost of Living Rank: 60

Also Known As: Die Hauptstadt, the Gray City

"*Ich bin ein Berliner,*" declared President Kennedy in 1963, standing before the infamous wall. Words that still ring true. Because all of us truly are Berliners, our minds suffused with preconceived notions about a place that so much of the world considers the city of spies and informants, espionage agencies and secret prisons, broken hearts and lost dreams.

There's no denying that for much of the 20th century, Berlin was all that and more. But like one of those many-layered German cakes, the city offers many different and sometimes contrasting flavors.

There's the 18th-century Prussian city of Frederick the Great, the powerhouse urban area that emerged under Otto von Bismarck after German unification, the anything-goes metropolis of the Roaring Twenties, and the goose-stepping capital of the Third Reich, followed by the Cold War that's still yielding consequences (Berlin is where young Vladimir Putin cut his teeth as a young KGB officer).

It wasn't until after the **Berlin Wall** came down in 1989 that all these Berlins reemerged, along with new sides that make the capital of a once-again-united Germany one of the world's most diverse and fascinating cities.

Nearly every era of Berlin history has contributed something along the **Unter den Linden**, a tree-lined boulevard that starts beneath the majestic **Brandenburg Gate**, an enduring symbol of the divided Cold War city but in reality a monument to Prussian military might.

Farther down the avenue are the luxurious **Hotel Adlon** (originally opened in 1907), a shop selling **Ampelmännchen** swag based on the two entirely different styles of pedestrian traffic lights used in East and West Berlin before unification, and a fancy **Volkswagen showroom** that reflects Germany's postwar industrial rise.

The eastern end of Unter den Linden is dominated by imperial-era structures like the **Staatsoper** (opened in 1742), **St. Hedwig's Cathedral** with its prominent green dome, and the campus of **Humboldt University** (opened in 1810). Having traded

Explore the beautiful architecture of the old warehouse district at sunset.

Brandenburg Gate was opened in 1791 under King Frederick William II during the Batavian Revolution.

rifles for relics, a 17th-century arsenal called the Zeughaus now houses the **Deutsches Historisches Museum**.

The statue-adorned **Schlossbrücke** (Palace Bridge) spans the **Kupfergraben** canal to the city's extraordinary **Museum Island**. Since the early 1990s, collections of art and antiquities from both sides of the wall and reconstruction of museum buildings destroyed during World War II have created a critical mass of treasures easily equal to the Louvre or British Museum.

Choosing a favorite depends on the historical era that most floats your boat. The sprawling **Pergamonmuseum** safeguards treasures from ancient Greece, Rome, and the Middle East, including the indigo blue

Ishtar Gate from Babylon. From a Neanderthal skull to an incomparable bust of **Queen Nefertiti**, the recently resurrected **Neues Museum** flaunts artifacts from prehistoric times and ancient Egypt. The temple-like **Alte Nationalgalerie** showcases European art from the 18th through early 20th centuries, while sculpture takes precedence at the **Bode-Museum**.

The island's tallest structure is the **Berliner Dom,** a massive neo-Renaissance cathedral finished in 1905 and renowned nowadays for its choir, organ concerts, and annual Bach Project. Across the street is the new **Humboldt Forum**, a rebirth of the colossal **Stadtschloss** where Prussian kings and German emperors once resided. Among the massive structures' new tenants are the **Ethnological Museum** and **Museum of Asian Art**, relocated from their old premises in West Berlin.

The extinct German Democratic Republic (aka East Germany) comes into focus on the other side of the **River Spree**, where a bronze statue in the **Marx-Engels-Forum** honors the authors of *The Communist Manifesto*. The adjacent **DDR Museum** offers a glimpse of everyday life in East Berlin under the repressive Communist regime.

Looming high above the **Alexanderplatz** is another symbol of those days—the 666-foot (203 m) **Fernsehturm** tower—an observation deck and revolving restaurant inside its shiny silver sphere. A few blocks away is **Karl-Marx-Allee**, a monumental boulevard flanked by Stalinist-era apartments (hyped as palaces for the working class) and a

The Bode-Museum overlooks the River Spree, where tourists can take sightseeing cruises.

ARTBEAT

• **Best Movies:** *Metropolis* (1927), *The Blue Angel* (1930), *One, Two, Three* (1961), *Cabaret* (1972), *Wings of Desire* (1987), *Run Lola Run* (1998), *Good Bye, Lenin!* (2003), *The Lives of Others* (2006).

• **Best Books:** *Berlin Alexanderplatz* by Alfred Döblin, *Goodbye to Berlin* by Christopher Isherwood, *Alone in Berlin* by Hans Fallada, *The Spy Who Came in From the Cold* by John le Carré, *The Innocent* by Ian McEwan, *Stasiland* by Anna Funder, *Midsummer Night* by Uwe Timm.

• **Best TV Shows:** *Deutschland 83, 86,* and *89* (2015–2020), *Homeland* season 5 (2015), *Berlin Station* (2016–19), *The Same Sky* (2017), *Counterpart* (2017–19), *Berlin Babylon* (2017–present).

• **Best Music:** *Twilight of the Gods: The Essential Wagner Collection* by Richard Wagner, *100 Berlin Cabaret Classics of the '30s & '40s* by Marlene Dietrich and others, "Winds of Change" by the Scorpions.

---

longtime venue for East Berlin's annual May Day military parade.

Checkpoint Charlie was the legendary interface between American and Russian forces during the Iron Curtain days. Only three sections of the infamous Berlin Wall remain in their original locations. A 660-foot (201 m) stretch survives along Niederkirchner Strasse a block west of Checkpoint Charlie, covered in graffiti and largely chipped away, but moving all the same.

This area was also the hub of Nazi Berlin. The gripping **Topographie des Terrors** museum behind the

wall details Nazi atrocities on the site of Hitler's **Gestapo headquarters** and SS secret police nerve center. The gray government office block on the north side of Niederkirchner was Hermann Göring's **Ministry of Aviation**. An information board on Gertrud-Kolmar Strasse marks the site of **Hitler's Bunker**, buried beneath tons of rubble and covered by a parking lot. It's no coincidence that Berlin's **Memorial to the Murdered Jews of Europe** lies just a block away.

Painted by more than 100 artists from around the world, the **East Side Gallery** along Mühlen Strasse is the longest preserved stretch of the wall, as well as one of the globe's largest open-air art galleries. Another original section along **Bernauer Strasse** features display panels and photos of those who died trying to escape across the wall. The adjacent **Gedenkstätte Berliner Mauer** museum and memorial features old newsreel footage of people crawling through barbed wire and jumping from windows to flee East Germany.

Exploring the **Potsdamer Platz** today, it's hard to believe that three decades ago it was virtually a desert, a no-man's-land strewn with barbed wire and antitank barriers; booby-trapped with trip wire, machine guns, and mines; and patrolled by the notorious Stasi border guards. As soon as the wall came down, the neighborhood transformed into the largest construction project of 1990s Europe. The end result was an astonishing postmodern precinct designed by some of the world's leading architects.

Europe's fastest elevator speeds visitors to the top of redbrick **Potsdamer Platz No. 1** and the penthouse **Panoramapunkt** café and observation deck. **Bluemax Theater**

GATHERINGS

- **Festival of Lights:** More than 100 buildings, monuments, and landmarks around the city transform into an outdoor gallery of creative light displays over 10 days in mid-September; festival-of-lights.de/en.

- **Berlin Oktoberfest:** Copious amounts of beer, wurst, and music are the cornerstones of a September/October festival at venues throughout the city, including the Zentraler Festplatz near Tegel Airport, the Alexanderplatz, and Die Wilden Wiesn near the East Side Gallery; berlin.de/en/events /oktoberfest.

- **German Unity Day:** Brandenburg Gate is one of many places that Berliners gather each year on October 3 to celebrate the reunification of East and West in 1990.

- **Christmas Markets:** The yuletide season is a magical time in Berlin, which hosts more than 70 holiday markets between late November and the end of December. The Spandau and Gendarmenmarkt bazaars are the best known; berlin.de/en/christmas-markets.

provides a full-time home for the **Blue Man Group**, while the billion-dollar **Sony Center** accommodates dozens of attractions beneath its soaring tentlike roof, including the **Deutsche Kinemathek** film and television museum and **Legoland Discovery Centre**.

The huge green space beyond Potsdamer Platz is the **Tiergarten**, Berlin's largest inner-city park and a favorite escape for walking, running, biking, and other outdoor pursuits. Established as a 16th-century royal hunting ground, the park features monuments to **Wagner**, **Goethe**, **the Soviet army**, and even **Michael Jackson**.

Tucked down in Tiergarten's southwest corner are the **Berlin Zoo** and **Aquarium Berlin**, while the 18th-century **Schloss Bellevue** home of Germany's president graces the north side. The wonderfully rebuilt **Reichstag**, where Germany's parliament convenes, looms above the park's northeastern corner. Paths meander through the lovely **Rosen-garten**, **Englischer Garten** (with its popular teahouse), and along the banks of the Spree.

Berlin was never a city that was especially interested in the River Spree. But that changed after German reunification via a multi-decade renewal project that's made the waterway clean enough to fish and swim in again, its banks lined with alfresco cafés and trendy outdoor *strandbars* like **Badeschiff** and **Sage Beach**, where Berliners clad in board shorts and bikinis sip cocktails on sand trucked in from the Baltic Sea.

Along the north bank of the Spree opposite Museum Island, **Spandauer Vorstadt** and adjoining **Scheunenviertel** are among the city's oldest and most atmospheric neighborhoods. Rather than the monumental structures that so much of Berlin is known for, these areas' big draws are art nouveau–era courtyard complexes like the **Hack-esche Höfe** that mix residential, office, entertainment, and creative spaces. The heart and soul of **Jewish Berlin** before the Holocaust, the two neighborhoods were largely derelict during communist days. Artists and other creative types moved in after reunification, followed by **gallery owners and fashion designers** as the area became a showpiece of ad hoc urban renewal, rebirth initiated by individuals rather than government bodies. The latest arrivals are **trendy bars**, **outdoor cafés**, **boutique hotels,** and the risqué nightlife

A total of 2,711 concrete slabs make up the Memorial to the Murdered Jews of Europe, also known as the Holocaust Memorial.

The Ishtar Gate of Babylon, constructed in 575 B.C., can be seen at the Pergamon Museum.

that was strictly verboten during the Cold War.

Moses Mendelssohn School beside the historic cemetery on Grosse Hamburger Strasse—used as a deportation center for Berlin's Jews during the Holocaust—and the **Blindenwerkstatt** factory where Otto Weidt safeguarded his disabled Jewish workers until the end of the war are reminders of the area's dark past. But there's quirky modern stuff too, like the **Magicum** magic museum, **Puppentheater Firlefanz,** and the freaky robots that inhabit the **Monsterkabinett** underground museum.

The German capital is surrounded by a cordon of imperial playgrounds included in the **Palaces and Parks of Potsdam and Berlin World Heritage site**. Closest to the city center is the expansive **Schloss Charlottenburg**, a vast residence and garden completed in the early 18th century as a summer palace for Prussia's first queen consort (Sophia Charlotte). In the three centuries since it was built, **Schloss Schön-hausen** in northern Berlin has gone from imperial garden parties to hosting the likes of Khrushchev, Castro, and Ho Chi Minh during the communist days.

But the most precious of all is **Sanssouci**, a flamboyant rococo-style summer palace on the edge of **Potsdam** constructed in the 1740s at the behest of Frederick the Great. It's not especially large, but the decoration and decor are exquisite and reflective of a ruler who actually had good taste according to one observer. Instead of other kings and potentates, Frederick used his "carefree" castle to host the philosopher Voltaire, the composer Johann Sebastian Bach, and other cultural icons of the Age of Enlightenment. From the majestic **Weinberg Terrace** that spills down from the palace to the oddball **Chinese House**, the palace gardens offer an entirely different treat. ◼

Venice
Italy

Although people tend to think of Venice as the epitome of a Renaissance city, the Queen of the Adriatic flaunts a vast range of moods and styles, from outlying islands settled by Roman citizens fleeing the barbarian hordes to a modern museum created by a 20th-century American socialite.

THE BIG PICTURE

Founded: sixth century A.D.

Population: 430,000

Size: 50 square miles (130 sq km)

Language: Italian

Currency: Euro

Time Zone: GMT +1 and +2 (CET and CEST)

Global Cost of Living Rank: Unlisted

Also Known As: Venezia, Queen of the Adriatic, City of Bridges, City of Masks, La Serenissima (The Most Serene)

Like a celebrity stalked by paparazzi, the Queen of the Adriatic just can't seem to stay out of the headlines: stories about Venice under threat by climate change, overrun by tourists, and barring giant cruise ships from coming too close. The city protests yet endures these trials and tribulations with an aged elegance. And people continue to flock here: to make movies, write books, paint landscapes, or merely sip espressos in the illustrious **Caffè Florian** (founded in 1720) in the **Piazza San Marco**.

Those who call on Venice by winter often have vast parts of the city to themselves and local residents. On the other hand, going in the summer often requires the expertise of a jewel heist: Sneak in, grab what you can, and get out fast. This snatch-and-grab approach normally entails a sprint from Santa Lucia

Festivalgoers don traditional masks and costumes for Venice Carnival festivities in San Marco Square.

railway station to the **Rialto Bridge** and across the **Grand Canal** to the **Basilica San Marco**, the **Doge's Palace**, and crowded cathedral square with its towering **Campanile di San Marco**.

For those staying overnight or longer, many guidebooks recommend setting aside a day for each of the *sestieres* (districts) that compose the city: **San Marco**, **San Polo**, **Dorsoduro**, **Cannaregio**, **Castello**, and **Giudecca**. Others advocate themes: churches one day, museums the next, and so on.

A more novel approach is exploring Venice chronologically, starting with the oldest parts and working your way up to the 20th century—which means hopping a *vaporetto* water taxi to **Torcello**, a tiny island on the north side of the lagoon where Italians seeking refuge from the barbarian invasions first settled in the fifth century A.D. The red-brick **Santa Maria Assunta** church (founded in 639) is the city's oldest structure of any kind, a confluence of Byzantine and Romanesque architecture that safeguards saints' relics and ancient frescoes. Archaeological relics ranging from Paleolithic

St. Mark's Basilica was completed in 1092 and consecrated in 1117. It lies on the eastern end of the Piazza San Marco.

stone axes and Etruscan pottery to Greek sculptures and Roman oil lamps are on display at the **Torcello Museum**.

Two other lagoon isles have most definitely been discovered by tourists but still give a feel for what life was like before Venice evolved into a global power.

Nearby **Burano Island** is long celebrated for its *punto in aria* ("stitch in air") technique of making lace with needles rather than looms. The **Museo del Merletto** offers equisite examples of Renaissance and baroque lacemaking, and there are plenty of places to purchase Burano embroidery in shops scattered among the island's brightly painted buildings.

Another one of Venice's oldest surviving structures—the Romanesque-style **Santa Maria e San Donato**—rises above **Murano Island**. Far more than its mosaics, the church is renowned for the large fossilized bones that hang behind the altar, allegedly all that remains of the dragon slain by St. Donatus,

---

HIDDEN TREASURES

• **Libreria Acqua Alta:** Cats lounging on the counters, a staircase made from old flood-damaged tomes, and books stored in a gondola are all part of the fun at this eccentric "high-water" bookstore down a back street in Castello.

• **John Cabot House:** A marble memorial topped with a Venetian lion marks the Renaissance manse on Via Garibaldi where famed 15th-century navigator Giovanni Caboto lived when he wasn't discovering the North American coast for the king of England.

• **Drogheria Màscari:** A throwback to days when Venice was the western end of the Silk Road from Asia, this vintage shop on the Ruga del Spezier (Spice Alley) renders herbs and spices, wines and liqueurs, gourmet foods, and delicious Italian sweets.

• **Piscina di Sacca Fisola:** Swimming in the canals may not be wise, but pack your bikini or board shorts for a dip in this popular community pool on an island all its own in Guidecca.

• **Enoteca Millevini:** Oenophiles flock to this tiny wine-tasting shop near the Rialto Bridge to get the lowdown on its incredible array of French and Italian wines and spirits.

Take a gondola cruise through Venice's famous canals, under bridges, and past colorful homes.

but more likely the remains of an ancient mammoth.

But the island is most known for its glassblowing, a tradition that started when the government required the highly profitable Venetian glass industry to relocate to Murano to more closely guard its secrets. Precious glassworks from medieval through modern times are on display at the island's **Museo del Vetro**, while master glassblowers continue the tradition at the **Vivarini glass factory**.

By the 11th century, the **Piazza San Marco** had become the heart of Venice and its architectural innovation. The city's most famous building, the **Basilica San Marco** combines aspects of Romanesque and Byzantine design in a dazzling structure that features bulbous

domes, winged lions, and copious golden mosaics. Although it's been reconstructed numerous times following fires and earthquakes, the 323-foot (98 m) campanile opposite the church also combines aspects of Romanesque and Byzantine design. The city boasts other examples from that era—like the eerily flooded **Crypt of San Zaccaria**—but nothing rises to the brilliance of the beloved basilica.

Venice would have to wait for the Gothic era for the rest of the city to catch up with its mother church in terms of sheer flamboyance. The 14th-century **Doge's Palace**, home to the city's magistrate, offers the most dramatic vision of Venetian Gothic. Although much of the interior decoration and lovely **Bridge of Sighs** date from the Renaissance,

older sections like the **Piombi** dungeon (where Casanova was imprisoned) hark back to medieval days.

Venetian Gothic reaches a fever pitch along the Grand Canal in structures like the **Fondaco dei Tedeschi** shopping emporium, the blood red **Palazzo Bembo**, and **Ca' d'Oro** (House of Gold) art museum. The most Gothic of the city's many churches is the **Basilica di Santa Maria Gloriosa dei Frari** with its immense nave. Venice's version of Westminster Abbey, **Santi Giovanni e Paolo** (San Zanipolo) church is the last resting place of 25 doges and other famous Venetians. Decorated with works by Titian, Giovanni Bellini, Tintoretto, Donatello, and Veronese, either church could easily double as an art museum.

By the Renaissance, Venice was

pouring much of its creative energy into painting and sculpture rather than startling new architecture. As central Venice was already built out, the city's classic Renaissance-style churches are across the water in Giudecca: **San Giorgio Maggiore** (which sports its own lofty campanile) and **Il Redentore** with its classic Palladian facade. Rialto Bridge, the **Torre dell'Orologio** clock tower, and the **Il Ghetto** Jewish Quarter were also born of the Renaissance.

But the most exquisite Renaissance structures are the richly decorated *scuole grandi* (great schools) that served as headquarters for Catholic charitable organizations. Most astounding of these are the **Scuola Grande di San Rocco** and its marvelous Tintoretto mosaics, and **Scuola Grande di San Marco** with its intricate white marble facade. During the Napoleonic occupation, two of the old scuole were combined into a new home for the **Gallerie dell'Accademia**, one of the world's foremost art museums.

The baroque era yielded structures like the sublime **Ca' Rezzonico**, now

Stunning examples of glass artwork are on display at the Murano Glass Museum.

a museum dedicated to the decadence of 18th-century Venice, the **Punta della Dogana** customhouse transformed into a tantalizing space for modern art, and the hulking **Santa Maria della Salute** church, constructed in the mid-1600s to thank the Virgin Mary for suppressing an outbreak of bubonic plague that had killed a third of the city's residents.

Despite its antique image, Venice also has its modern landmarks. Starting in 1949, an American socialite transformed an 18th-century palace into a modern home that eventually became the **Peggy Guggenheim Collection** of modern art. And then there's **Harry's Bar**, a celebrated drinking spot since 1931 and the place where the peach Bellini was invented. ∎

LOCAL FLAVOR

• **Caffè del Doge:** Founded in 1645, Europe's first coffeehouse is still the place to sip a freshly roasted arabica espresso or a hazelnut Giacometta; *Calle dei Cinque 609, San Polo;* caffedeldoge.com.

• **Terrazza Danieli:** If you can tear yourself away from the skyline view at this rooftop restaurant atop the Hotel Danieli, the menu features sumptuous regional dishes like Acquerello risotto, gnocchi made with Biancoperla polenta, and veal fillet with pistachios;

Riva degli Schiavoni 4196, Castello; terrazzadanieli.com.

• **Cantina Do Spade:** This neighborhood café is renowned for *cicchetti, a* Venetian version of tapas that features small plates of calamari, mozzarella, pumpkin blossom stuffed with cod, and tuna rissoles; *Sotoportego de le Do Spade, San Polo;* cantina dospade.com.

• **Gelateria Nico:** A local treat since 1935, this casual waterfront café offers flavored coffees, ice

cream cones, cassatas, coppas, and classic banana splits; *Fondamenta delle Zattere, Dorsoduro;* gelateria nico.com.

• **Trattoria alla Madonna:** Decorated with paintings by many of Italy's most celebrated modern artists, this sophisticated eatery near Rialto Bridge features Adriatic delights like seafood risotto, vermicelli with cuttlefish ink, and fish soup; *Calle della Madonna 594, San Polo;* ristorante allamadonna.com.

Kraków
Poland

Poland's medieval capital blends dungeons and dragons with a young, contemporary vibe that has transformed it into one of Eastern Europe's most alluring cities since the Iron Curtain came down.

Legend holds that Poland's royal city was founded after the warrior sons of King Krakus slew Smok Wawelski—a fire-breathing, man-eating dragon that lived in a cave beneath Wawel Hill—although a second version of the story claims that a humble shoemaker named Skuba actually did the deed (and subsequently married the king's daughter).

Either way, Kraków has had a fiery history. Poland's royal capital for more than 500 years (1038–1596) endured the Black Death and numerous invasions, the Nazi atrocities of World War II, and nearly half a century of harsh Soviet rule. It wasn't until the advent of Solidarność (Solidarity, the self-governing trade union) and the fall of communism that Kraków resumed its role as a guiding light of Polish culture and education.

Wawel Royal Castle crowns the hilltop where the fabled dragon once nested. Converted into a museum after World War II, its stout walls now safeguard the **Crown Treasury and Armoury**, the **Royal State Rooms and private apartments**, and several fine art collections. And yes, there's a **Dragon's Den**, a natural limestone cavern that exits beside a menacing metallic statue of Smok Wawelski that actually breathes fire.

THE BIG PICTURE

Founded: fourth century A.D.

Population: 770,000

Size: 84 square miles (220 sq km)

Languages: Polish

Currency: Polish złoty

Time Zone: GMT +1 or +2 (CET, CEST)

Global Cost of Living Rank: Unlisted

Also Known As: Stołeczne Królewskie Miasto Kraków (Royal Capital City of Kraków)

The legendary hill overlooks the **Vistula River** at the southern end of the **Stare Miasto**—the Polish name for Kraków's Old Town. Demarcated by circular **Planty Park** (where the city walls and moat once stood), the entire Old Town is a UNESCO World Heritage site packed with museums and churches, bars and restaurants, university buildings, and performing arts venues.

The Old Town revolves around the sprawling **Rynek Główny**, the largest town square in Europe. Arrayed around the square are the 14th-century **St. Mary's Basilica**, a Renaissance trade emporium called the **Sukiennice** (Cloth Hall) and the **Town Hall Tower**. Visitors can scale the tower for a wide-angle city view or descend into a basement once used as a dungeon.

A warren of passages beneath the square—the remains of medieval streets and market stalls—has been converted into a high-tech museum called **Rynek Underground** that spins local history via artifacts, holograms, and videos. Medieval implements of torture are the focus of the subterranean **Muzeum Kata** (Executioner Museum) on the square's western side.

Visit the sausage stalls at the Easter Market in the Rynek Glowny square.

Kraków Main Square is lined with eateries, cafés, and historic buildings.

With dozens of bars and restaurants—and thousands of university students—the Old Town doesn't die after dark. Renowned for its political cabaret during the communist era, **Piwnica Pod Baranami** is now a subterranean pub and music club on the edge of the Rynek. **Pod Aniołami** serves classic Polish dishes in a 13th-century Gothic cellar, while the **Wodka Bar** behind the cathedral offers cozy digs to sample Poland's homegrown libation.

Kraków's once substantial Jewish population was decimated by the Holocaust. But there are plenty of reminders, from the historic **Kazimierz Jewish Quarter** and the **Ghetto Heroes Square** to **Oskar Schindler's factory** and **Auschwitz-Birkenau** concentration and extermination camp. Kazimierz has

rebounded into a small but thriving Jewish community with services at historic synagogues and restaurants like **Klezmer Hois**, which offers kosher dishes and performances of traditional Jewish music and poetry.

The subterranean theme continues

on the outskirts of town. **Wieliczka Salt Mine** is a veritable underground city chiseled from salt, while **Nowa Huta Underground,** opened as a museum in 2019, offers guided tours of half a dozen Soviet-era fallout shelters. ■

ALTER EGO

Warsaw

Despite its own ancient history, Poland's current capital city feels much more like a huge modern city than Kraków. Having nearly two million residents contributes to that vibe. But so does the fact that **Warsaw** has been substantially rebuilt since its extensive destruction during World War II.

Skyscrapers now ring the central

city, with brand-new **Varso Place** (opened in 2020) topping out at 1,017 feet (310 m), including a restaurant and viewing decks near the summit.

Still, much of the Old Town, a neighborhood around the **Rynek Starego Miasta** square that harbors the cathedral and royal castle, has been restored. Legendary Poles are recalled at the **Copernicus Science Centre, Chopin Museum**, and ̇abiński Villa at the **Warsaw Zoo**.

Frankfurt & Stuttgart

Germany

Two of the cornerstones of modern life—finance and luxury driving—trace their roots to a duo of urban areas in southwest Germany that followed very different paths to success.

THE BIG PICTURE

Founded: Frankfurt, first century A.D.; Stuttgart, ca A.D. 90

Population: Frankfurt, 1.9 million; Stuttgart, 1.35 million

Size: Frankfurt, 255 square miles (660 sq km); Stuttgart, 184 square miles (477 sq km)

Language: German

Currency: Euro

Time Zone: GMT +1 and +2 (CET, CEST)

Global Cost of Living Rank: Frankfurt, 54; Stuttgart, 102

Also Known As: Frankfurt—Bankfurt, Mainhattan; Stuttgart—Schwaben Metropole

FRANKFURT AM MAIN

Poised at the confluence of two great rivers, Frankfurt has long been a place where people from all over Europe have come together to trade goods and ideas. Financial dealings at the city's medieval trade fairs are considered the genesis of the modern stock market, while the 18th-century Rothschild clan is credited with the invention of modern banking.

Fast-forward several centuries, and Frankfurt still thrives on trade and finance. The metropolis hosts several of the world's paramount trade shows, as well as Germany's biggest inland port and aviation hub. More than 200 banks and hundreds of other financial institutions anchor a high-rise downtown that includes Germany's 10 tallest buildings.

An open-air observation platform and restaurant at the summit of the 56-story **Main Tower** offers the city's best vantage point, a panorama that takes in the modern **banking district** and vintage Altstadt, and the **River Main** flowing west toward its rendezvous with the **Rhine**.

Frankfurt's favorite shopping street for more than a century, the **Zeil** stretches about a mile (1.6 km) beneath the skyscrapers. Anchoring the western end is the **Hauptwache**, a sprawling plaza with outdoor cafés, flower stalls, and buskers who perform in the shadow of the **Galeria** shopping mall and the 17th-century baroque **Katharinenkirche**, rebuilt after the Allied bombing of World War II that destroyed nearly all of central Frankfurt.

Legendary author **Johann Wolfgang von Goethe** grew up in a house just south of the plaza that's now a museum devoted to his Romantic-era life and times. Meanwhile, the high-end shopping continues along the tree-lined **Goethestrasse**, which heads west to the **Alte Oper** (Old Opera), an elegant Italian Renaissance theater that stages more than 300 annual events from opera, ballet, and

The Frankfurter Römer served as Frankfurt's city hall for over 600 years.

The Deutsche Bank AG twin tower headquarters (right) and the Frankfurter Sparkasse Tower (center) highlight the financial district skyline.

symphony to Broadway musicals and rock concerts.

The **Altstadt** (Old City) has also risen from the ashes, with many of the half-timbered townhouses and stone palaces and churches resurrected in remarkable detail. The neighborhood revolves around the **Römerberg**, a cobblestone square that once hosted everything from jousting tournaments to public executions and that's still the focus of the city's raucous **Fastnacht** carnival before Lent.

Looking like something from a German fairy tale, the **Frankfurter Römer** city hall overlooks the square. Behind its medieval façade is the **Kaisersaal**, where the coronation banquets of the Holy Roman Emperors took place. The new emperors were chosen by seven electors from around the realm and then crowned in the nearby **Kaiserdom Sankt**

Bartholomäus, a pink-stone Gothic cathedral with a 311-foot (95 m) bell tower that remained Frankfurt's tallest structure until the 1960s.

The **Eiserner Steg** iron pedestrian bridge jumps the River Main to the south bank and the **Museumsufer**, an area boasting 38 world-class museums, ranging from the old master and Impressionist prodigies of the **Städel** to 4,000 years of sculpture at the **Liebieghaus** to a national **Deutsches Filmmuseum** that examines German moviemaking from Fritz Lang's *Metropolis* to the present day.

The **Sachsenhausen** neighborhood beyond museum row is renowned for *apfelwein* (apple wine) taverns like the **Adolf Wagner** and **Fichtekränzi** that also serve traditional dishes like schnitzel, wurst sausages with sauerkraut, and pork knuckles.

Frankfurt's iconic green spaces are arrayed around the edge of the inner city. The expansive **Palmengarten** harbors more than 13,000 tropical and subtropical plant species in outdoor beds and glass conservatories such as the **Tropicarium**. A leader in global wildlife conservation since the 1950s, the **Frankfurt Zoo** shelters more than 4,500 animals and excels at the captive breeding of

endangered species. Frankfurt is also a convenient jumping-off spot for tours of the nearby **Rheingau** wine region and cruises through the **Rhine Valley**.

STUTTGART

Stuttgart is a tale of two cities, at once an innovative commercial dynamo that nurtures two of the globe's most celebrated carmakers and an overgrown country town with folksy traditions and an age-old Swabian dialect that even other Germans struggle to comprehend.

Boosting the city's rustic image is the fact that more than 60 percent of the metro area is green space and nearly 40 percent is protected parkland—higher rates than any other German city. Among its leafy confines are the **Höhenpark Killesberg**, with its wacky spiral observation

tower, and the riverside **Rosenstein-park**, home to the interactive and kinder-friendly **State Museum of Natural History Stuttgart**, the **Wilhelma Zoo and Botanical Garden**, and the expansive English gardens of **Rosenstein Palace**.

Meanwhile, Stuttgart is the only German metropolis with extensive vineyards inside its city limits. Local winemakers produce eight different varieties but are especially known for their fruity red **Trollinger** and white **Riesling** wines. Among the cellars that open their doors to visitors are the **Collegium Wirtemberg** in Rotenberg and the **Alte Kelter** in Uhlbach, which also flaunts a wine museum. Vinophiles can easily combine wine and workout by hiking the seven-mile (11.5 km) **Stuttgarter Weinwanderweg** wine trail.

The first grapes were planted

Stuttgart's baroque Neues Schloss (New Castle) features three main wings.

around 2,000 years ago, when Stuttgart was a far-flung outpost of the Roman Empire. Cultivating luxury autos is a much more recent pastime. Local inventor **Gottlieb Daimler** invented the world's first motorcycle and motorized carriage in the 1880s, around the same time that **Karl Benz** produced the first patented motorcar—a three-wheeler with a top speed of 10 miles (16 km) an hour. **Volkswagen** was founded in Stuttgart in the 1930s with the invention of its first vehicle, the "bug." The companies would grow to be among the world's largest sports-car manufacturers. That heritage is the focus of the **Mercedes-Benz Museum** in Bad Cannstatt and a **Porsche Museum** in Zuffenhausen in cutting-edge structures that are just as innovative as the cars within.

Stuttgart's fondness for avant-garde architecture continues in the gleaming glass-cube **Kunstmuseum** modern art collection and the **Staatsgalerie**, a fusion of metal and stone that showcases Germany's foremost collection of contemporary art. Out in the burbs, the revolutionary **Weissenhof Estate** harbors residential units designed in the 1920s by architectural masters Le Corbusier, Walter Gropius, and Mies van der Rohe.

Yet it's not all brave new world. For more than 900 years, Stuttgart was home base for the House of Württemberg, a dynasty that ruled much of southwestern Germany. Like royals everywhere, they built fabulous royal palaces. The hulking **Altes Schloss** (Old Castle) now houses the excellent **Landesmuseum**, where the **Württemberg crown jewels** are among the many treasures of local culture and history. Erected in the 18th century, the adjacent **Neues**

A Mercedes-Benz 300 SL on display at the Mercedes-Benz Museum in Stuttgart

Schloss (New Castle) is a massive baroque-style palace modeled after Versailles.

Stretching more than two miles (3.2 km) between the castles and the **Neckar River**, the sprawling **Schlossgarten** (Castle Garden) features lakes; leafy lanes; an alfresco beer garden; the **Staatstheater**, home of the Stuttgart Opera, Ballet, and Orchestra; and the state-of-the-art **Planetarium Stuttgart**. Farther up the Neckar is something truly unique: a former slaughterhouse that now houses the amazing **SchweineMuseum** with more than 5,000 piggy banks and other porcine relics. ■

GATHERINGS

- **Fastnacht Carnival:** Crazy costume parades through Frankfurt's old town and nearby Mainz highlight five days of endless parties, music, and feasting before Lent; visitors can rent garb from local costume stores.

- **Stuttgarter Weindorf:** The city's annual wine festival takes over three historic plazas in the city center with wine from 500 regional vineyards and epicurean feasts at 30 outdoor eateries; August–September.

- **Cannstatter Volksfest:** Stuttgart's version of Oktoberfest unfolds each fall as thousands gather at a fairground beside the Neckar River for carnival rides, live music, and copious amounts of local beer; September–October.

- **Stuttgart German Masters:** The city named for a royal stud farm (*Stuotgarten*) hosts one of the globe's premier equestrian competitions with horses and riders from more than two dozen countries; November.

- **Frankfurt Christmas Market:** A tradition stretching back to the Middle Ages, this yuletide fair and its bounty of food, drink, decorations, and handicrafts stretches all the way across the old town from the Hauptwache to the Römerberg; November–December.

Copenhagen
Denmark

The largest city in Scandinavia, Denmark's progressive capital is a template for blending old and new into a seamless urban experience best explored by foot, bike, or boat.

Separate from her roles as national icon, literary landmark, and never ending photo op, **The Little Mermaid** statue is a constant reminder that Copenhagen is a city of the sea. It's a husky port city and island metropolis that owes its fame and fortune to a strategic position on the Øresund, the waterway that connects the Baltic Sea and Atlantic Ocean.

Much of what makes the Danish capital such a delightful place to explore clusters along the waterfront. With its gabled facades and vintage sailing craft, **Nyhavn** canal is like going back in time to the era (1845–1864) when **Hans Christian Andersen** lived and wrote at Nyhavn 67. Restaurants and bars occupy many of the old houses, and the canal is now a haven for water taxis and harbor tours rather than fishing boats.

Nyhavn flows into the much larger **Inderhavnen** (Inner Harbor), the mainstay of Copenhagen's maritime life for centuries, although you would hardly know it these days. The old shipyards and naval base have given way to some of the city's most astounding modern architecture: the **Black Diamond** annex to the Royal Danish Library (with its dark granite cladding), the **Royal Danish Playhouse** (crowned by a copper-clad tower), and the neo-futuristic **Copenhagen Opera House** on Holmen Island (one of the most advanced and expensive theaters ever built).

But history endures farther down Inderhavnen. The elegant **Amalienborg** complex has housed the Danish royal family since the 18th century. One of its four sections—**Christian VIII's Palace**—has been transformed into a museum of royal life. Like wooden soldiers from *Babes in Toyland* come to life, the **Royal Life Guards** change shift in Amalienborg's cobblestone plaza daily at noon. That massive patina-domed structured looming behind the palace is the Lutheran **Frederiks Kirke** (or the Marble Church), a rococo masterpiece that offers a year-round classical music series.

The famous water nymph lounges on a rock beside the **Kastellet**, a classic star-shaped citadel originally constructed in the 17th century to protect the Inner Harbor and its royal residents from seaborne attacks. Although part of the castle is still an

A statue of the Little Mermaid perches on a rock at Langelinie.

Colorful buildings, sailing ships, and cafés line the Nyhavn pier in Old Town Copenhagen.

active military base, much of the grounds are open to the public for free concerts and tree-shaded strolls along the moat. Beyond the fort is **Nordhavnen**, the city's mixed-use modern port area and terminal for ferries to Norway.

Away from the water, Copenhagen's medieval old town is laced by vehicle-free pedestrian lanes like the sinuous **Strøget**, one of Europe's longest shopping and eating streets. And given that Copenhagen is one of the world's most bike-friendly cities, pedal power is also welcome. Visitors can rent their own wheels through **Bycyklen** (City Bike) or join guided tours offered by **Bike Mike**, **Nova Fairy Tales**, and other outfitters.

Gracing Strøget's western end is **Tivoli**. Proof that Danes rather than Disney pioneered the theme park, the gardens opened in 1843 and now feature music and stage shows, numerous restaurants, an aquarium, an amusement park, and three dozen carnival rides. Nearby are two of Copenhagen's finest museums: the **National Museum of Denmark** (priceless Viking artifacts) and **Ny Carlsberg Glyptotek** (European sculpture and painting).

Beyond Tivoli, the **Vesterbro** ("West Bridge") neighborhood has morphed in recent years from the city's gritty meatpacking district into a trendy restaurant and nightlife hub, especially the stretch along **Istedgade**. Punctuating its repute as a great place to imbibe, the district also boasts the **Carlsberg Brewery** and the **Vodka Museum**. ∎

Rome
Italy

Like the Forum of old, Rome's piazzas provide plenty of scope for the "street opera" and sweet life that make the Eternal City an everlasting place for romance, adventure, and just plain fun.

People have been trying to pigeonhole Rome for nearly as long as the city has been around. To some, the Italian capital is a living history book that flaunts 2,000 years of civilization from ancient Rome through the Renaissance and baroque. To others, it's the pinnacle of a faith that revolves around the pope and his Vatican home. And for some, it's the epitome of romance: three coins in the fountain, living *la dolce vita,* that's *amore.*

Rome is all of these things and more, an urban matrix that's at once decrepit and dynamic, hectic and easygoing, incredibly friendly and unbelievably frustrating at times. An alluring blur of old and new, fast and slow, ebb and flow—that's Rome. Especially when it comes to the city's many piazzas, public squares where residents and visitors alike gather from dawn to well after dusk to eat, drink, and make merry.

But in grand fashion, of course. Romans are "naturally inclined toward arranging a spectacle, acting a character, staging a drama. Opera in the streets on a daily basis," wrote prominent 20th-century Italian author Luigi Barzini. "The show

THE BIG PICTURE

Founded: ca 753 B.C.

Population: 4 million

Size: 430 square miles (1,114 sq km)

Language: Italian

Currency: Euro

Time Zone: GMT +1 and +2 (CET and CEST)

Global Cost of Living Rank: 47

Also Known As: Roma, the Eternal City, City of Seven Hills, City of Fountains

is . . . many times more important than the reality."

That daily drama is especially intense in four legendary plazas—the Piazza Navona, Piazza Venezia, Piazza di Spagna, and Piazza San Pietro—which are attractions on their own and launchpads for exploring the surrounding neighborhoods.

The oblong shape of the popular **Piazza Navona** betrays its birth as an ancient Roman Hippodrome where chariot races once played out. In the 17th century, the derelict stadium was transformed (by order of the reigning pope) into a magnificent public square graced by Bernini's **Fontana dei Quattro Fiumi** and Borromini's **Chiesa di Sant'Agnese in Agone church**.

Nowadays, the spacious piazza is a bastion of street performers and sidewalk cafés as well as the **Gladiator Museum** with its reproduction weapons and armor, and luxurious residences like the **Palazzo Pamphilj** and adjacent **Palazzo Braschi**, which houses the **Museo di Roma** art collection. Piazza Navona is also renowned for hosting the city's largest **Christmas market**.

A fountain stands at the bottom of the Spanish Steps.

See old Rome from atop St. Peter's Basilica in Vatican City.

The **Parione** neighborhood around the piazza is home to the colorful **Campo de' Fiori** food market and the oddball **Pasquino**, one of several "talking statues" around Rome where residents have posted anonymous satirical poems, notes, and other political expression since the 16th century.

Heading east from Navona, a zigzag of narrow streets leads to three other well-known landmarks. Erected in the first century A.D. by the Emperor Trajan, the domed **Pantheon** is one of the pinnacles of ancient Roman engineering. Farther along are **Santa Maria sopra Minerva** (Rome's only unblemished Gothic church) and the beloved **Trevi Fountain**, a rococo tour de force that features in several

romantic films. The same area also boasts one of the city's newest attractions: **Time Elevator Roma**, which hosts the 40-minute, multimedia *Story of Rome* on three giant screens.

Even though it's dominated by the massive **Vittoriano** (Victor Emmanuel Monument)—which honors the man who united Italy in the 19th century—the **Piazza Venezia** is really more of a prelude to the wonders of ancient Rome that lay beyond.

Starting beside the great white monument, a broad staircase ascends to the **Campidoglio**, a small perfectly proportioned plaza designed by none other than Michelangelo and flanked by palaces that now house the **Capitoline Museums**. Founded in 1471 as the world's first public museums, the collection ranges from Renaissance paintings to classical statues like **"The Dying Gaul," "Esquiline Venus,"** and **"Lupa Capitolina,"** which depicts Rome's mythical founders Romulus and Remus suckling at the breast of the she-wolf that raised them.

Just beyond the museums are the ruins of the **Roman Forum**, arrayed on either side of the **Via Sacra** (Sacred Road) that leads down to the Arch of Titus and the towering Colosseum (soon to be outfitted with a new "gladiator view" floor). But ancient Rome doesn't stop there, because the archaeological park also includes extensive ruins on the **Palatine Hill**, Nero's **Domus Aurea** palace, and the **Arch of Constantine**.

The past spills over into the adjoining **Ripa** and **San Saba**

A must-see on any visit to Rome: the stunning ceiling of the Sistine Chapel in the Vatican Museums.

neighborhoods, where you'll find the immense **Circus Maximus**, the truth-or-dare **Bocca della Verità** (Mouth of Truth), and the ruined **Baths of Caracalla**. Just beyond the baths is the start of the **Via Appia** (Appian Way), the immortal Roman road that stretches into the countryside south of Rome, with medieval churches and underground catacombs along the way.

There's a much different vibe at the **Piazza di Spagna**, which assumed its current shape in the early 1700s. Almost at once, the triangular plaza became the darling of foreigners visiting or living in Rome who gathered around the boat-shaped **Fontana della Barcaccia** or lounged on the **Spanish Steps** ascending to **Trinitá dei Monti** church.

Among those who lived beside the piazza were German composer Felix Mendelssohn and English poet John Keats, who passed away in 1821 in the **Keats-Shelley House** at the bottom of the steps. His former home is now a museum dedicated to the

early 19th-century Romantic poets and writers who frequented Rome.

There's more literary and musical history on **Via dei Condotti**, where the old **Caffè Greco** (opened in 1760) hosted Dickens, Goethe, Byron, Hans Christian Andersen, Wagner, Berlioz, and many other distinguished artists. Spangled with many of the world's best known luxury brands, Condotti is also the city's poshest shopping street.

For something completely different, walk a few blocks south from the piazza along the **Via Sistina** to the leaping dolphins of the **Fontana del Tritone** and **Chiesa di Santa Maria della Concezione dei Cappuccini**. The 17th-century baroque church isn't much to look at from the outside, but its crypt safeguards the remains of more than 4,000 Capuchin friars arranged as macabre artworks in the Crypt of Skulls, Crypt of Pelvises, Crypt of Leg and Thigh Bones, and other underground vaults.

From the top of the Spanish Steps it's around a 10-minute walk to the

LAY YOUR HEAD

- **Hotel Eden:** The epitome of Roman elegance since 1889, the Eden flaunts a sophisticated aesthetic throughout its 98 rooms and suites; restaurants, bars, spa; from $760; dorchestercollection.com/en/rome/hotel-eden.

- **Palazzo Manfredi:** A 17th-century palace provides a romantic setting for this Relais & Châteaux boutique hotel tucked between the Colosseum and Domus Aurea; restaurant, bar, massage, personal shopper, luxury car rental; from $356; palazzomanfredi.com/en.

- **Hotel Pulitzer Roma:** Chic, monochromatic decor distinguishes the 83 rooms of this quiet, modern hotel near the Nuvola Congress Center and Marconi Metro Station in the Esposizione Universale Roma (EUR district) south of the city center; restaurant, bar, garden, gym, roof terrace with hot tub; from $76; hotelpulitzer.it/en.

- **Hotel Ariston:** Centrally located near the Basilica of Santa Maria Maggiore and Termini train station, this budget option offers 83 modern and comfortable rooms; restaurant, bar, lounge, fitness center; from $67; hotelariston.it/en.

Villa Borghese, Rome's biggest and best park. Founded in 1605 as the private garden of Cardinal Scipione Borghese, the garden is crisscrossed by leafy avenues and gravel trails leading to lofty overlooks of Rome, **San Carlino puppet theater**, and the excellent **Bioparco** zoo.

There's also art among the trees. **Borghese Gallery** in the cardinal's opulent mansion showcases works by **Caravaggio**, **Titian**, **Raphael**, and other Renaissance masters. The park's equally sumptuous **Villa Medici**—where Galileo was under house arrest for three years during his heresy trial—offers guided tours and temporary art exhibits.

The western side of the **Tiber River** is one of Rome's most revered squares: **Piazza San Pietro**. Technically it's beyond the city limits, part of an independent nation called **Vatican City**. But St. Peter's Square is just as evocative of Roman life as the others, splashed with Renaissance fountains, punctuated with an ancient Egyptian obelisk, flanked by Bernini's elegant colonnades, and literally overshadowed by the world's largest church.

Dine at Campo de' Fiori, a traditional Roman restaurant.

The ruins of the Roman Forum, viewed from Capitoline Hill, are just one of many iconic ancient structures throughout the city.

One of the ultimate statements of Renaissance architecture and engineering, **St. Peter's Basilica** was consecrated in 1626 after more than a century of construction and embellishment with masterpieces like Michelangelo's **"Pietà,"** Giotto's **"Navicella"** ("Bark of St. Peter") mosaic, and Bernini's twisting bronze **Baldachin (**canopy) above the high altar. There are more treasures in the basilica's **Museo Storico Artistico**, and visitors can peer down on the piazza (and the rest of Rome) from Michelangelo's superlative dome.

Among the other parts of Vatican City open to the public are the **Sistine Chapel**, the expansive **Vatican Gardens** (guided tour only), and the seemingly endless **Vatican Museums**, where 20,000 items are on display at any given time. Set aside an entire day for the latter, which offers more than two dozen different collections, ranging from ancient Egyptian mummies and Roman sculpture to carriages, maps, and Renaissance paintings. The Holy See offers several ways to experience the museums and Sistine Chapel, including **"skip the line" tickets**, **after-hours twilight tours**, and **early morning breakfast tours** that can all be booked on the Vatican website.

As with the other great squares, St. Peter's offers a great jumping-off point for exploring elsewhere in the same area—in this case the Tiber's west bank. Overlooking the river, **Castel Sant'Angelo** was erected in the second century A.D. as a tomb for Emperor Hadrian. It was converted into a papal fortress during medieval times and nowadays shelters the **Museo Nazionale di Castel Sant'Angelo** and its assemblage of medieval weapons, Renaissance ceramics, and other relics.

Farther downstream, the **Trastevere** neighborhood is a throwback to Rome of centuries long ago or more, a warren of narrow streets, tiny squares, and age-old churches like **Santa Maria in Trastevere** (founded in the fourth century) and **Santa Cecilia in Trastevere** (founded in the fifth century). Rising behind the neighborhood, the **Parco del Gianicolo** on Janiculum Hill offers panoramic views across the city and a daily cannon shot at noon from a vintage howitzer manned by Italian Army troops. ■

Lisbon
Portugal

Rays aren't the only thing that visitors can catch in Lisbon. Beyond the blue skies and abundant sunshine, Portugal's capital also shines when it comes to food and wine, song and dance, and a historic waterfront that launched the age of discovery.

THE BIG PICTURE

Founded: ca 1200 B.C.

Population: 2.7 million

Size: 370 square miles (958 sq km)

Language: Portuguese

Currency: Euro

Time Zone: GMT +0 and +1 (WET and WEST)

Global Cost of Living Rank: 83

Also Known As: Lisboa, City of Seven Hills

Lisbon makes it easy to fall in love. Not just with another human being—it's without a doubt one of Europe's most romantic places—but also with the city itself. What's not to like about a city blessed with an almost perpetually sunny Mediterranean climate, amazing fresh-from-the-sea meals (complemented by a bottle of local *vinho do Alentejo*), a sunset stroll along the waterfront, and then an evening listening to sentimental fado in a local bodega?

But get ready to walk, because Lisbon is also rich in hills. The highest is crowned by the imposing **Castelo de São Jorge** with its sweeping views of the red-roofed old town, the waterfront, and **Tagus (Tejo) Estuary**. Started by the Visigoths and expanded by the Moors and medieval Christians, the castle and its 11 towers served as Portugal's royal palace from the 14th to 16th centuries.

If it's just the view that lures—perhaps a romantic spot to pop the question—skip the castle and head straight for nearby vistas like the **Miradouro das Portas do Sol** and the bougainvillea-covered terrace at **Miradouro de Santa Luzia**.

Cobblestone streets tumble down to the **Alfama** district, an old Moorish and Jewish Quarter that evolved into the crossroads of Christian Lisbon after the Reconquista. Erected between the 11th and 13th centuries, **Sé de Lisboa** cathedral looks as Gothic as the day it was consecrated, while the **Mosteiro de São Vicente de Fora** is gloriously Renaissance. Rounding out the area's spiritual trifecta is the **Panteão Nacional**, a baroque masterpiece that harbors the tombs or funeral monuments of many celebrated Portuguese, including Prince Henry the Navigator, explorer Vasco da Gama, and conquistador Afonso de Albuquerque.

Its narrow street lined with numerous bars and restaurants, the Alfama is also the place to catch fado music in its natural habitat and dine on traditional dishes like *bacalhau* (salted codfish), *caldo verde* soup, and *sardinhas assadas* (grilled sardines). Among the moody spots to catch live music are **Casa de Linhares** in a Renaissance mansion, the popular **Clube de Fado,** and the cave-like **Mesa de Frades** in an old

Guitarists play traditional fado music.

São Jorge Castle stands tall above the skyline and colorful buildings of Lisbon.

chapel. One of Portugal's national treasures, the melancholic music and its 200-year history is celebrated at **Museu do Fado**, where interactive listening posts feature dozens of fado songs.

Largely destroyed by the 1755 Great Lisbon earthquake and tsunami, the nearby **Baixa** district was rebuilt along neoclassical lines around the grand square called the **Praça do Comércio**. The plaza's **Lisboa Story Centre** offers a one-hour interactive journey through city history while the **Arco de la Rua Augusta** commemorates the city's rebirth after the deadly quake.

From there, the **Rua Augusta** pedestrian zone runs eight blocks through the middle of Baixa to the **Elevador de Santa Justa.**

The magnificent neo-Gothic elevator (opened in 1899) lifts passengers to the hillside Chiado neighborhood graced by the **Museu Arqueológico do Carmo** in the ruins of a 14th-century convent and church destroyed by the quake, the rampantly rococo **Basílica de Nossa Senhora dos Mártires**, and **Café A Brasileira**, an art nouveau hangout that's been around since 1905.

Farther along is **Bairro Alto** (Upper Town), another slice of medieval Lisbon that survived the great quake. By daylight, the neighborhood is refreshingly quiet, a great place to wander along cobbled lanes flanked by ancient facades. But after dark it's a different story. Bairro Alto comes alive with the sound of fado played in its myriad cafés, bars, restaurants, and nightclubs.

The parklike **Avenida da Liberdade** provides a pleasant stroll from the Bairro Alto to the **Casa-Museu Medeiros e Almeida**, which renders real insight into the home of Lisbon's uber rich circa 1896. Its 25 rooms are tastefully decorated with furniture, porcelain, silverware, and fine art. Farther up the avenue is the **Praça do Marquês de Pombal**, a monument square that honors the prime minister largely responsible for rebuilding Lisbon after the 1755 catastrophe. Dead ahead is Lisbon's beloved **Parque Eduardo VII** with its undulating lawns, box hedging, and marvelous **Estufa Fria** greenhouse.

Many of Lisbon's top museums are located along the Tagus waterfront between Baixa and Belém, on either side of the rust-colored **Ponte 25 de Abril** bridge. Housed inside a 17th-century palace in the **Lapa** district, the **Museu Nacional de Arte Antiga** (MNAA) safeguards national treasures like the 16th-century "Belém Monstrance" and 15th-century "Saint Vincent Panels." Asian art amassed during 500 years of Portuguese trade and colonialism in the Far East is the forte of the nearby **Museu do Oriente**, housed in a restored codfish warehouse.

Unveiled in 2016, the shockingly modern **Museu de Arte, Arquitetura e Tecnologia** (MAAT) building on the Belém waterfront is an artwork unto itself, with a design that fully reflects the cutting-edge invention and ideas inside. On the other hand, the **Museu Nacional dos Coches** is a total throwback, a collection of extravagant 16th- through 19th-century horse-drawn carriages, including a coach that belonged to Philip II of Spain.

Belém's **Mosteiro dos Jerónimos** offers an outstanding example of the grandiloquent Manueline style that arose in Portugal during the age of discovery, a late Gothic vibe that continues along the waterfront at the much photographed **Torre de Belém**. The nearby **Museu Coleção Berardo** is all about 20th-century style as expressed by Lichtenstein, Picasso,

A streetcar passes in front of Sé de Lisboa Cathedral.

See historic royal carriages at the National Coach Museum.

Bacon, and other modern masters.

Lisbon's northern waterfront also flaunts stark contrasts. The **Museu Nacional do Azulejo** traces the history of Portugal's azulejo (blue ceramic) tiles from the art form's introduction by Arab artisans through masterpieces by living artists. The delicate tiles stand in stark contrast to bold modern creations like the **Ponte Vasco Da Gama** bridge and the **Estação do Oriente** rail terminus.

Speaking of trains, Lisbon's historic and utterly charming **Tram 28** offers an excellent way to navigate the city's hilly terrain without leaving your seat. Operated with 1930s Remodelado trams, the route meanders through Alfama, Baixa, Bairro Alto, and other vintage neighborhoods between **Praça Martim Moniz** near Castelo de São Jorge and the **Campo de Ourique** on the heights above Lapa. A nonstop one-way journey takes 45 to 60 minutes, with plenty of places to hop off and on. ■

LOCAL FLAVOR

• **The Food Temple:** Bohemian-chic vegetarian restaurant with simple indoor/outdoor seating and a menu of soups, tapas, and main dishes that changes frequently; *Beco do Jasmim 18, Graça;* facebook.com /FoodTemple.

• **Pharmacia:** Housed in the same mansion as the Museu da Farmácia, this quirky pharmacy-themed restaurant has decorated medical vials, elixir bottles, and first aid kits, with the bill presented on a prescription pad; *Rua Mal. Saldanha 1, Santa Catarina;* facebook.com /restaurantepharmacia.

• **Belcanto:** Chef José Avillez's creative interpretations of classic Portuguese fine dining has earned this elegant eatery two Michelin stars; *Rua Serpa Pinto 10A, Chiado;* belcanto.pt.

• **Time Out Market Lisbon:** Vintage food hall transformed into a modern gourmet food court with a variety of inventive eats and new restaurant concepts from Lisbon's best chefs; *Mercado da Ribeira, Avenida 24 de Julho 49, São Paulo;* timeoutmarket .com/lisboa/en.

Córdoba, Granada & Sevilla

Spain

Iberia's sweltering south is home to three fabled cities— Córdoba, Granada, and Sevilla—that share common ancestry and a commitment to preserving the urban landscapes and rich traditions that have come to epitomize the whole of Spain.

If geography determines destiny, then Spain's deep south is a sterling example. Andalusia is closer to Africa than any other part of Europe—a mere 8.1 miles (13 km) across the Strait of Gibraltar. Thus, it comes as no surprise that the region's scorching weather and arid landscapes are comparable to Morocco . . . or that Islamic invaders used this close proximity to start their invasion of the Iberian Peninsula in A.D. 711.

Although dynasties came and went, the Moors ruled Andalusia for nearly 800 years. From guitars, gazpacho, algebra, and incredible architecture to contributing more than 400 words to the Spanish language, their legacy endures.

Yet the Moors biggest impact was urbanization. Taking their cue from the great cities of the Middle East, they transformed Sevilla, Córdoba, and Granada from backwater burgs originally established by the Romans or Visigoths into flourishing urban areas that remain among Europe's most exotic and alluring cities.

CÓRDOBA

Founded: ca 152 B.C.
Population: 345,000
Size: 484 square miles (1,254 sq km)
Language: Spanish
Currency: Euro
Time Zone: GMT +1 and +2 (CET, CEST)
Global Cost of Living Rank: NA
Also Known As: Colonia Patricia (Roman)

Wrapped around a big bend in the **Guadalquivir River**, Córdoba has preserved more of its Moorish past than any other Andalusian city. Nowadays it doesn't rank among Spain's 10 largest urban areas. Yet 1,000 years ago, it was Europe's richest and most advanced city, capital of the Umayyad Caliphate—which ruled nearly all of the Iberian Peninsula—as well as a great center of science, learning, art, and trade. As the world's largest metropolis of the early Middle Ages, Córdoba may have sheltered as many as one million residents, a blend of Muslims, Christians, and Jews.

A warren of alleys, gardens, plazas, and courtyards, the entire **Casco Histórico** (Historic Center) is a UNESCO World Heritage site. Restaurants, taverns, handicraft shops, and small hotels occupy many of the ancient townhouses and villas surrounding the neighborhood's historic landmarks.

Foremost among these is the

One of the 12 magnificent courtyards at the Palacio de Viana

This nighttime view shows the Puente Romano, which crosses the Guadalquivir River to the Mezquita-Catedral de Córdoba.

incomparable **Mezquita-Catedral de Córdoba**, an eighth-century mosque converted into a cathedral after the reconquest of Andalusia by Spain's Catholic monarchs. A masterpiece of Moorish architecture, the building revolves around a central prayer hall with more than 850 marble, granite, and jasper columns supporting majestic red-and-white arches. The **mihrab** and its elaborate Islamic mosaics and the shady **Patio de los Naranjos** survived the Christian conversion, while the minaret morphed into the **Torre Campanario**, a bell tower that visitors can scale for a panoramic view of the Old Town.

The belfry of the **Cathedral of Córdoba** and several chapels rise abruptly from the middle of the Mezquita. A fusion of Gothic, Renaissance, Mannerism, and the Moorish/European Mudéjar style adopted after the reconquest, the church offers a fascinating contrast in design and faith to the grand mosque. Even more stunning is the **Capilla de Villaviciosa**, a chapel with a wondrous blend of Moorish and Gothic arches.

The mosque-cathedral compound is engulfed in a tightly packed Old Town that also offers the excellent **Museo Arqueológico**; the **Alcázar de los Reyes Cristianos** fortified palace and plush gardens, which once housed the city's Catholic rulers; and the **Centro Flamenco Fosforito**, which pays homage to Andalusia's iconic song and dance via films, exhibits, and live performances.

The **Judería** (Jewish Quarter) occupies a significant part of the Old Town, testament to the fact that Córdoba supported a sizable Jewish population until the reconquest and the Spanish Inquisition. Among the landmarks along the super-narrow **Calle Judíos** are the 14th-century **Sinagoga de Córdoba**; the **Casa de Sefarad**, dedicated to the history, culture, and traditions of Sephardic Jews; and a statue of **Maimonides**, the city's celebrated 12th-century Jewish philosopher, physician, and astronomer.

Córdoba eventually outgrew its Moorish confines, expanding outward into neighborhoods that were quintessentially Spanish. Built between 1492 and 1704, the antique-and-art-filled **Palacio de Viana** and its 12 courtyards showcase the city's aristocratic life during the Renaissance and Age of Discovery. Commissioned by King Philip II in 1570 to breed Andalusian horses, the **Caballerizas Reales** (Royal Stables) opens for daytime training

sessions and evening equestrian shows. The nearby **Patios de San Basilio** offers a journey through five flower-filled courtyards and vintage houses still occupied by local residents.

Down along the Guadalquivir, the first-century **Puente Romano** (Roman Bridge) leaps the river to the **Torre de la Calahorra**, a Moorish gate and guard tower that now harbors the **Museo Vivo De Al-Andalus**, an outstanding city museum with multilingual audio guides. A few blocks east of the tower, the **Centro de Creación Contemporánea de Andalucía** (C3A) stages a variety of exhibitions and events that showcase the city's artists, musicians, dancers and even circus performers.

GRANADA

Founded: fifth century B.C.
Population: 230,000
Size: 34 square miles (88 sq km)
Language: Spanish
Currency: Euro
Time Zone: GMT +1 and +2 (CET, CEST)
Global Cost of Living Rank: NA
Also Known As: Elibyrge (ancient), Iliberis (Roman), Gharnata (Arabic), Elvira

Swaddled in the foothills of the snowcapped **Sierra Nevada** range, Granada boasts a milder climate than its counterparts in the Guadalquivir Valley. But that's not the only thing that sets the city apart from its Andalusian cousins.

Granada may not have Sevilla's big-city swagger or Córdoba's enchanting Old Town. But it does have two unique lures: the sprawling Alhambra palace and a gitano heritage renowned for flamenco song

Moorish architecture adorns the Patio de los Leones at the Alhambra in Granada.

A flamenco dancer performs at the Zambra María la Canastera in Granada.

and dance. It's also a lively university town, home to more than 80,000 college students plus the lively bars, clubs, and cafés where they hang between classes.

Founded as an ancient Iberian fortified town, Granada later fell under the sway of the Romans and Visigoths before the eighth-century Moorish invasion. It remained fairly obscure until the 1200s, when the ruler of the newly established

Emirate of Granada started construction of a massive hilltop fortified palace complex the Moors called **Alhambra** (the Red One) because of its ruddy walls.

One of Spain's most visited attractions, the Alhambra has five main sections, including a military bastion called the **Alcazaba**; the **Generalife gardens**, where the emirs cooled off in summer; and the **Parador de San Francisco**, a luxury hotel tucked inside a 16th-century monastery. Dating from the Renaissance, the massive **Palacio de Carlos V** harbors the **Museo de la Alhambra** and its priceless Moorish artifacts, as well as the **Museo de Bellas Artes** and its collection of post-reconquest Spanish art. The climax of any Alhambra visit is the **Palacios Nazaríes**, the ultimate expression of Moorish art and architecture. This cluster of 14th-century royal residences includes masterpieces like the **Patio de los Leones** (Lion Courtyard), the **Salón de Embajadores** with its incredible golden dome, and the lavishly decorated **Sala de los Abencerrajes**.

Granada doesn't have a historic center so much as a **downtown** area that meshes historic landmarks and

---

GATHERINGS

• **Feria de Abril:** Originally a post-Easter livestock show, Sevilla's April fair has evolved into an exuberant showcase of Andalusian dance, song, food, and equestrian skill. Many attendees don historic garb.

• **Córdoba Guitar Festival:** Strummers from around the globe gather each July in the City of the Guitar for 10 days of acoustic tunes at historic venues in the Old Town.

• **Festival de Granada:** From its 19th-century roots as symphony concerts in the Alhambra, this month-long summer fest grew into a wide range of international song and dance—and a huge dose of flamenco—in the city's churches, gardens, and palaces.

• **Fiesta de los Patios:** Since 1921, Córdoba residents have battled one another in this May competition to see who can create the most gorgeously decorated traditional and modern courtyards.

Boaters navigate the canal running through the beautiful Plaza de España in Sevilla.

modern city functions. The jumble of architecture runs a broad gamut from the Renaissance-baroque style of the **Catedral de Granada** to stylish art nouveau and late 20th-century boring. The highlight is the cathedral's opulent **Capilla Real**, where Columbus-commissioning monarchs Ferdinand and Isabella are entombed. The **Basílica San Juan de Dios** with its mighty dome and the tranquil **Monasterio de la Cartuja** are the city's other sacred gems.

Unlike the city center, the **Albaicín** and **Sacromonte** neighborhoods retain much of their ancient ambience. The latter is renowned for its gitano residents. Also called Romany, the gitano people were indispensable to the evolution of

flamenco dance and song. The hillside **Centro de Interpretación del Sacromonte** visitors center offers exhibits on gitano contributions to Spanish culture and guides tours of the nearby caves where many gitanos once lived. **Zambra María la Canastera** is one of several Sacromonte taverns that stages gitano-style zambra flamenco shows.

SEVILLA

Founded: eighth century B.C.
Population: 1.1 million
Size: 105 square miles (272 sq km)
Language: Spanish
Currency: Euro
Time Zone: GMT +1 and +2 (CET, CEST)
Global Cost of Living Rank: Unlisted
Also Known As: Seville, Hispalis (Roman), Ishbiliyah (Arabic)

Sprawling along the Guadalquivir not far from the Atlantic coast, Sevilla rose to fame during the Age of Discovery as the primary port that dispatched Spain's explorers and conquistadors to the far corners of the world. Enriching itself on the spoils of the New World, it ascended to the heights of Europe's greatest cities, with a larger-than-life reputation that has attracted writers and composers for centuries.

The fictitious libertine Don Juan was immortalized by the 1630 stage play *El burlador de Sevilla (The Trickster of Sevilla)*. Georges Bizet's Spanish seductress Carmen lived and loved in Sevilla, while Mozart depicted Figaro and Susanna getting married here. Rossini's classic *The Barber of Seville* further enshrined the city's romantic credentials.

Unlike Córdoba and Granada, which somewhat slumber in the past, Sevilla marched boldly into the future by hosting two international expositions (1929 and 1992), erecting cutting-edge modern buildings, and cultivating southern Spain's most diverse and robust economy.

Yet there are still plenty of oldies but goodies, starting with the **Catedral de Sevilla**, the world's largest Gothic church. Rather than occupy a former mosque à la Córdoba, Sevilla's 15th-century Catholic luminaries decided to demolish the city's Muslim house of worship and start from scratch. The only thing they didn't dismantle was the massive minaret, converted into the 342-foot (104 m) **Giralda** bell tower with its bird's-eye views of the surrounding city. Several Spanish monarchs and **Christopher Columbus** are buried in the cathedral.

South of the cathedral is the elegant Renaissance-style **Archivo General de Indias**, where centuries of maps, documents, drawings, and other artifacts relating to Spanish exploration, conquest, and colonization are stored. Treasures on display here include **logbooks from Ferdinand Magellan's circumnavigation of the globe**.

A few blocks farther south lies the **Real Alcázar**, a fortress-palace renowned for its expansive gardens and elaborate Mudéjar architecture, a mashup of Moorish and European elements developed after the reconquest. Among its many highlights are the **Palacio de Don Pedro** and its **Patio de las Doncellas**.

Wedged between the cathedral and the Alcázar, **Barrio Santa Cruz** was the Jewish Quarter prior to the reconquest. Nowadays it's Sevilla's most charming neighborhood, a maze of narrow, cobblestone lanes flush with cafés, taverns, souvenir shops, and cultural institutions like the **Hospital los Venerables** art museum with works by Velázquez, Murillo, and other Spanish masters. The barrio's **Tablao Flamenco Museo** in the **Casa de la Guitarra** presents dance and acoustic music performances, while the **Casa de Pilatos** offers a glimpse of patrician life during the Age of Discovery in a ducal mansion still owned and lived in by the aristocratic Medinaceli family.

The Ibero-American Exposition of 1929 was Sevilla's introduction to the modern world. Its architectural legacy includes the leafy riverside **Parque de María Luisa** and the amazing **Plaza de España**, a tour de force of Moorish, Renaissance, and Baroque Revival design graced by exquisite tilework, a bridge-spanned canal, and grand fountain. Farther upstream, the 13th-century **Torre del Oro** has transitioned from medieval watchtower into maritime museum.

Fast-forwarding to the 21st century, the incredible **Metropol Parasol** on the north edge of the city center is reputedly the world's largest wooden sculpture. Tucked beneath the massive mushroom-shaped "umbrellas" are several bars and restaurants, a live event space, and the **Museo Antiquarium**, which features Roman and Moorish ruins and artifacts uncovered on the site.

Across the water, the **Isla de La Cartuja** personifies modern Sevilla via the **Isla Mágica** amusement park, big-name concerts and soccer matches at **Estadio La Cartuja**, and the 592-foot (180 m) **Torre Sevilla** skyscraper, the tallest building in southern Spain. ■

One of the best panoramic views of Sevilla is from the top of the Metropol Parasol.

Helsinki
Finland

As one of the hubs of global design, Finland's cutting-edge capital is renowned for its futuristic buildings, innovative furnishings, and vibrant fashion, but not at the expense of age-old traditions like pickled herring, classical music, and the beloved birch-bark sauna.

THE BIG PICTURE

Founded: 1550

Population: 1.27 million

Size: 259 square miles (641 sq km)

Language: Finnish

Currency: Euro

Time Zone: GMT +2 and +3 (EET, EEST)

Global Cost of Living Rank: 56

Also Known As: White City of the North, Helsinge fors (medieval), Hesa (slang), Stadi (slang)

The Finns have been global leaders in design since the 1920s and '30s, when Eliel Saarinen emerged as one of the world's first truly modern architects and Alvar Aalto started creating innovative furniture like the Paimio Chair and Model 60 Stool. Inspired by nature, Finland's contemporary design tends to be simple yet elegant and very striking. Nowhere is this more evident than Helsinki, the nation's design-savvy capital city.

With more than 75,000 objects and 170,000 images and drawings, the city's **Design Museum** traces the roots of Finnish creativity back to the 1870s. Scale models, blueprints, photographs, and renderings fill the nearby **Museum of Finnish Architecture**, the world's second oldest collection devoted exclusively to building design.

The museums are wrapped inside Helsinki's vibrant **Design District**, which harbors more than 180 design shops, workshops, and galleries interspersed with hip bars, restaurants, and boutique hotels. Many of the top design houses have showrooms along the chic **Esplanadi**, including **Artek** (furniture),

Marimekko (fabrics and fashion), and **iittalla** (glassware and ceramics). While you're there, snap a selfie in front of the **"My Helsinki"** sign or the art nouveau–style **Havis Amanda** mermaid statue—Helsinki's unofficial mascot—on the esplanade.

Helsinki's odes to modern architecture include the **Temppeliaukio rock church** with its incredible glass-and-copper dome and the oval **Kamppi Chapel of Silence**, an extraordinary wooden structure built in 2012 as part of Helsinki's designation that year as World Design Capital. **Finlandia-Talo Hall**—hailed as a paragon of modern design when it opened in 1971 on a site overlooking Töölönlahti Bay—is now complemented by the **Musiikkitalo**, new home of the **Helsinki Philharmonic** and **Radio Symphony Orchestra**. Another eye-catching structure is the cutting-edge **Oodi Central Library**, which features music studios, handicraft workshops, gaming stations, a rooftop sauna, and robots shuttling books around.

Before Saarinen and Aalto, the maestro of Finnish design was Carl Ludvig Engel. As Helsinki's chief

The deck at the Allas Sea Pool urban spa offers views of the Presidential Palace.

The Helsinki library, called Oodi, was inaugurated in December 2018 and is one of 37 branches of the Helsinki City Library.

architect during Russia's 19th-century rule over the Grand Duchy of Finland, Engel created **Senate Square** in the city center and then enveloped it with majestic neoclassical buildings like **Helsinki Cathedral** (completed in 1852), the **National Library** (1844), the **Prime Minister's Office** (1822), and the **University of Helsinki** (1832).

On the square's south side, the **Helsinki City Museum** (Helsingin Kaupunginmuseo) is housed in a cluster of historic buildings, including the **Sederholmin talo**, the city's oldest structure of any kind (erected in 1757). A block farther south is harborside **Kauppatori** market square with its fresh produce, aromatic *silli* (herring), and souvenir stalls. The new **SkyWheel** lifts riders 131 feet (40 m) above the waterfront and includes a wood-paneled sauna gondola.

Ferries flit from the Kauppatori docks to outer islands like **Suomenlinna.** A UNESCO World Heritage site, the "Fortress of Finland" was originally built by Swedes during the mid-18th century and expanded over the years. Among its many attractions are the old **citadel walls and tunnels**, **history museum**, **toy museum**, a former **Russian Orthodox church** (with a dome that doubles as a lighthouse), and a vintage World War II submarine called the *Vesikko,* as well as several food and beverage outlets like the **Suomenlinnan Panimo** microbrewery. ■

HIDDEN TREASURES

• **Café Ekberg:** Opened in 1852, Helsinki's oldest bakery is renowned for its fresh bread, cakes, and pastries. Open from 7:30 a.m. weekdays and 9 a.m. Saturdays.

• **Hietaranta:** For those who want to take a dip in the Baltic, this urban beach is accessed via a huge cemetery. There's also miniature golf, a wakeboard course, volleyball, and a beachside bistro.

• **Löyly:** Located on the Munkkisaari Peninsula, this seaside hangout offers public and private saunas, yoga sessions, food, beverages, and a marvelous sundeck.

• **Yrjönkatu Swimming Hall:** Helsinki's oldest public pool (€5.50 per visit) hides inside a handsome Nordic Classicism building opened in 1928.

• **Alexander Theatre:** This opulent Russian-era opera house (opened in 1879) now hosts a year-round menu of music, dance, and drama.

Reykjavik
Iceland

Iceland's seaside capital merges Scandinavian sophistication with an end-of-the-world vibe that endures from the days when outcast Vikings settled the North Atlantic island.

THE BIG PICTURE

Founded: ca A.D. 870

Population: 220,000

Size: 31 square miles (80 sq km)

Languages: Icelandic

Currency: Icelandic króna

Time Zone: GMT + 0 or +1 (UTC)

Global Cost of Living Rank: Unlisted

Also Known As: Bæinn (the Town)

You would never know it today—given the affable nature of most Icelanders—but Reykjavik was founded by the medieval equivalent of a motorcycle gang: Norsemen banished from their native Norway because they were too rowdy, even by Viking standards.

More than a thousand years of isolation has taken the edge off. Today, Reykjavik isn't just the world's most northerly national capital, but it's also one of the world's more mellow. Icelandic winters can be pretty harsh, but Reykjavik's climate is moderated by its position on the island's western side near the top of the Gulf Stream.

Although many visitors rush through Reykjavik on their way to the island's celebrated glaciers, geysers, and waterfalls, the city itself should not be missed. Most of the major sights are within easy walking distance from one another.

Towering above even the highest skyscrapers, the steeple of **Hallgrímskirkja** church offers a lofty view of the city's brightly colored houses. Out front, the **Statue of Leif Eriksson** pays tribute to the native son who sailed to Greenland and North America around A.D. 1000. Across the square, the **Einar Jónsson Museum** presents

the works of Iceland's most noted sculptor.

Running downhill from the church, **Skólavörðustígur** is a great shopping street (art galleries, design shops, handicraft outlets) and the quickest way to descend into the **Miðborg** city center. **Culture House** highlights Icelandic art and culture from Viking times through the present day, including medieval illuminated sagas like the *Codex Regius of the Elder Edda*. Nearby **Arnarhóll** (Eagle's Hill) offers rolling lawns for picnics, reading, or sunning yourself on a bright summer day.

West of busy Lækjargata boulevard, the area around **Ingólfur** and **Austurvöllur Squares** is ground zero for Reykjavik nightlife, with cocktail bars and music clubs that have earned the city a global reputation for all-night partying. There's also history: The interactive **"The Settlement Exhibition—871±2"** preserves the remains of a 10th-century Norse village.

The popular **Sculpture and Shoreline Trail** meanders 2.5 miles (4 km) along the waterfront past the **Reykjavik Maritime Museum**, various **whale-watching** outfits, cruise ship piers, and the ultramodern

Take a rejuvenating dip in the geothermal waters of the Blue Lagoon.

A statue of Leif Eriksson guards the entrance to the Hallgrimskirkja church, one of the tallest structures in the country.

Harpa concert hall, which stages everything from symphony, opera, and ballet to long-running comedy shows like *Icelandic Sagas: The Greatest Hits* and *How to Become Icelandic in 60 Minutes.*

Another great urban walk is the path around **Tjörnin lake**, a stroll that easily combines with visits to the **National Gallery** of art, the **National Museum**, and the marshy **Vatnsmýrin Nature Reserve**, home to many native plants and birds.

Reykjavik makes a great base for exploring southwest Iceland. Among the popular day trips are the thermal waters of the **Blue Lagoon**, geysers and waterfalls along the **Golden Circle** drive, and **Thingvellir National Park**, where the world's oldest parliament (the Althing) met for more than 800 years and was the venue for a rocky fissure that marks the divide between the Eurasian and North American tectonic plates.

With its hiking trails, nesting birds, Viking ruins, outdoor art, and views of the volcanic Snæfellsnes Peninsula—where Jules Verne set *Journey to the Center of the Earth*—**Viðey Island** offers another cool day trip from the capital. ■

HIDDEN TREASURES

• **Icelandic Punk Museum:** An old underground public toilet serves as a deliberately ironic venue for this ode to 1980s music.

• **Kolaportid Flea Market:** Vintage clothing, music, and books complement the local foods and people-watching at this indoor emporium near Harpa concert hall.

• **Höfði House:** Famed as the place where Reagan and Gorbachev staged their 1986 summit, the government guesthouse has also hosted Winston Churchill, Marlene Dietrich, and other luminaries.

• **Laugardalur:** "Hot Spring Valley" park offers a geothermal swimming pool, botanical garden, polar animal zoo, and the arena that hosted the 1972 chess showdown between Bobby Fischer and Boris Spassky.

• **Keilir Golf Club:** Play a round of 18 under the midnight sun on a course hewn from a black lava field (summer only, of course).

Amsterdam
Netherlands

More than a century before Adam Smith endorsed the wealth of nations, capitalism was born along the canals of Amsterdam. The resulting riches went into creating the incredible art, architecture, and atmosphere that make the Dutch city such an awesome place to visit today.

THE BIG PICTURE

Founded: 1275

Population: 1.7 million

Size: 195 square miles (505 sq km)

Language: Dutch

Currency: Euro

Time Zone: GMT +1 and +2 (CET and CEST)

Global Cost of Living Rank: 44

Also Known As: Mokum ("Safe Haven" in Hebrew), Venice of the North

The world economy as we know it today wasn't born in London banks or the New York Stock Exchange, but instead in a wine trader's house in an area that later became Amsterdam's famous Red Light District. There, in 1594, a group of local merchants met to figure out a way around the Spanish and Portuguese monopoly on global trade.

Their confab led to both the first Dutch voyage to obtain Asian spices and the creation of history's first corporation—the Dutch East India Company (VOC). The founders didn't have enough cash to float the VOC on their own, so they devised a radical idea: Anyone could invest money in the company, including tradesmen and household servants.

People began trading their paper "shares" on the **Nieuwe Brug**, which nowadays spans the **Damrak** canal fronting Amsterdam's central train station. Inclement weather often forced them indoors at nearby **Sint Olofskapel**, a 15th-century church that accidentally served as the world's first paper stock exchange.

Once the riches started flowing from Asia, Amsterdam became Europe's biggest boomtown in a golden age that created many of the city's canals, grand buildings, and wealthy patrons for the likes of Rembrandt, Vermeer, and other legendary artists.

One of the first things that the fathers of Amsterdam built with their newfound wealth was Europe's largest city hall, an extravagant structure that became the **Royal Palace**. When the Dutch monarchy isn't using it for functions, the place is open to the public on guided tours that include the grand **Burgerzaal** (Citizens' Hall), where the marble floor features two large maps showing the extent of Holland's 16th-century trading empire.

Royal weddings, coronations, funerals, and the like take place in the adjacent **Nieuwe Kerk**, a 14th-century Gothic cathedral that predates the Golden Age. Between royal spectacle, the church is used for cultural exhibitions and recitals on the nation's largest organ. Older still is the nearby **Begijnhof**, a medieval women's religious retreat that endured until the 1970s. Set around a tranquil courtyard, the complex includes the **Houten Huis**, erected in the 1420s and now Amsterdam's oldest remaining wooden house.

The Van Gogh Museum honors the painter and the work of his contemporaries.

Amsterdam is famous for its canal system, as well as being a biker- and pedestrian-friendly city.

A former orphanage beside the Begijnhof has been converted into the **Amsterdam Museum**, with numerous exhibits on the city's history. The museum's best exhibits are an installation highlighting Amsterdam as the first global city, a replica of a legendary Red Light District café, and an interactive area where modern-day kids can discover what it was like to live in a 17th-century orphanage. Just down the road is the erudite **Spui**, a square surrounded almost entirely by bookstores—such as the **Athenaeum Boekhandel**—and which serves as the venue for the Friday **Spui Book Market** for old, rare, and out-of-print tomes.

The notorious **Red Light District** still tenders plenty of vice. But it's much tamer than days gone by, packed with curious out-of-towners wondering what all the fuss is about. **Red Light Secrets Museum**, the **Hash Marihuana & Hemp Museum**, and others have parlayed the area's debauchery into tourist attractions, while the **Prostitution Information Center** (PIC) offers guided tours of the district. Proving that devotion preceded decadence, the district also boasts the 14th-century **Oude Kerk** and the well-hidden **Ons' Lieve Heer op Solder** (Our Lord in the Attic), a chapel built into the top floor of a canal house in the 17th century after the Dutch Reformed Church banned Catholic churches.

The old town wraps itself in the **Grachtengordel**, a ring of four human-made, crescent-shaped waterways otherwise known as the **Canal Belt**. Basically the wealthy suburbs of the golden age, the area is Amsterdam at its picture-postcard best: canal boats, bridges, and lovely gabled homes. In 2010, UNESCO

GATHERINGS

- **Amsterdam Light Festival:** This winter celebration has lit up the canals of the city for more than a decade with illuminated art best seen on a walking or boat tour.

- **Taste of Amsterdam:** For four days surrounding the first weekend of June, foodies take over Amstelpark. During the festival, attendees can sample bites from restaurants from around the city and take part in judging a live cook-off among top chefs.

- **Bacchus Winefestival:** Wine enthusiasts from all corners of the world flock to Amsterdam for this festival, which spans two weekends in June, to enjoy tastings, food pairings, and live music.

declared this "homogeneous urban ensemble" and longtime "model of large-scale town planning" a World Heritage site.

The waterways are easily explored on canal cruises with the **Blue Boat Company** and other outfits. But there's also plenty to discover on foot. Like the fragrant **Bloemenmarkt** where tulips and other blooms are sold from floating stalls, the cutting-edge **Foam** photography gallery, and the offbeat **Kattenkabinet** museum of feline art.

Several of the canal zone's patrician mansions—the **Willet-Holthuysen** and **Van Loon**—now house museums with fascinating insights into the golden age. Another canal-side residence is the famous **Anne Frank House**, just down the canal from the Renaissance-style **Westerkerk**, where Rembrandt was buried in an unmarked grave. The artistic

superstar is also recalled in the **Rembrandtplein**, a rowdy nightlife area packed with pubs, karaoke bars, and the superclub **AIR Amsterdam**.

For many years (1639 to 1656), Rembrandt lived and painted in the **Jodenbuurt** (Jewish Quarter) on the other side of the Amstel River from the square that bears his name. The **Rembrandt House Museum** features many of his etchings, drawings, and personal mementos. Founded by Sephardic Jews who fled Iberia during the Inquisition, the quarter supported one of western Europe's largest Jewish communities until World War II. Their heritage is preserved by the **Portuguese Synagogue**, the **Huis De Pinto** mansion, and the **Joods Historisch Museum**.

The Amstel River is strung with numerous landmarks, including the new **Stadhuis** (City Hall) and the

state-of-the-art **Dutch National Opera & Ballet**. Farther along are the **Hermitage Amsterdam** (a satellite of the celebrated St. Petersburg museum), concerts and dance performances at the **Royal Theatre Carré**, and the delightful afternoon tea at the historic **Amstel Hotel**.

Sometimes called the city's second golden age, Amsterdam underwent another renaissance during the 19th century, growth that pushed the urban area beyond the Canal Belt and created many of the city's most endearing institutions.

The **Plantage** (Plantation) area was already home to **Hortus Botanicus**, a botanical garden founded in 1682 to cultivate spices and other plants the VOC was collecting around the globe. Then along came **ARTIS Amsterdam Royal Zoo**, one of the world's oldest municipal menageries. Elsewhere in Plantage,

Vondelpark is a favorite spot for a rest or a walk around the beautifully maintained green space.

The Rijksmuseum is a national museum dedicated to history and the arts.

Dutch resistance to the Nazi occupation comes into focus at the **Verzetsmuseum Amsterdam**.

During the 1880s, marshlands and pastures beyond the western edge of the Canal Belt were transformed into a **Museum Quarter** that began with the launch of the vast **Rijksmuseum** and its million-strong collection of artworks and artifacts and the **Stedelijk Museum** of modern art. One of those who visited the Rijksmuseum during its first year was a struggling 32-year-old artist by the name of Vincent van Gogh. Over the next five years he produced nearly everything in the quarter's **Van Gogh Museum**. The **Moco Museum** of contemporary art and the legendary **Concertgebouw** music hall are also worth a visit.

Nearby **Leidseplein** is another lively nightlife area with numerous restaurants, bars, and entertainment outlets ranging from the vintage Paradiso rock club to the performing arts at the **Internationaal Theater Amsterdam** (ITA). Across the Leidsekade

canal is **Vondelpark**, the city's largest and most diverse green space.

Amsterdam's latest golden age revolves around the **IJ**, a shipping channel and waterfront area that's undergone a radical remake in recent times.

Shaped like a giant green ship cruising down the waterway, the **NEMO Science Museum** offers five floors of interactive exhibits, including a giant bubble machine and a Rube Goldberg–like chain-reaction circuit. Across the **Oosterdok Bridge**, the innovative **OBA Oosterdok** library building is crowned by the restaurant **Babel** with sweeping harbor views. Another foot/bike bridge crosses a channel to the new **Muziekgebouw** concert hall with its dramatic glass facade and black box annex called the **Bimhuis** that specializes in live jazz.

IJ Ferries flit across the harbor from the Central Station to the **Amsterdam Noord** neighborhood. That giant, white building along the waterfront is the **EYE Film Institute Netherlands**, which houses a movie museum and four cinemas that screen current and classic films. Perched at the top of the adjacent **A'DAM Toren** are an open-air observation deck and **Moon Amsterdam**, a revolving restaurant where diners can enjoy a 360-degree view of the world's first truly global city. ■

HIDDEN TREASURES

• **Amsterdam Music Project:** A nexus for diverse musicians and musical styles, AMP creates "secret" and *gezellig* (cozy) pop-up music gatherings like "Rooftop Jazz Club" and "Hidden Concert in a Houseboat" at unique entertainment venues around the city.

• **Micropia:** Proof that petri dish cultures can be beautiful *and* scientifically significant, this unique, ultramodern museum near ARTIS focuses on the microscopic bacteria and microbes that live on and among us every day.

• **De Poezenboot** (The Catboat): Founded by Henriette van Weelde in

1966, this floating cat sanctuary houses about 50 needy felines at any one time. Visitors and their donations are welcome.

• **Nieuwe Spiegelstraat:** Renowned for its many gabled canal houses, this incredibly well-preserved neighborhood flaunts antique stores (many offering the sought-after traditional blue Delft tiles) as well as oddball items like the medical, scientific, and pharmaceutical instruments sold at Thom & Lenny Nelis *(Keizersgracht 541)* and the globes, sextants, telluriums, and other nautical aids of Staetshuys Antiquairs *(Nieuwe Spiegelstraat 45a)*.

Bruges
Belgium

From medieval towers and cobblestone squares to french fries, waffles, and some of the world's best beer, the old and venerable city of Bruges excels at many of the things that make Belgium such a marvelous place to visit.

THE BIG PICTURE

Founded: ninth century A.D.

Population: 120,000

Size: 53 square miles (138 sq km)

Language: Flemish, French

Currency: Euro

Time Zone: GMT +1 and +2 (CET, CEST)

Global Cost of Living Rank: Unlisted

Also Known As: Brugge, Venice of the North, City of Swans

Like Barcelona and Venice, Bruges is a victim of its own success: More visitors than residents flood the cobblestone streets during summer and winter holiday seasons. The municipal council took the extraordinary step in 2019 of canceling the annual ice sculpture festival to avoid overtourism.

But that doesn't mean you should ignore one of the world's most authentic medieval cities. Just time it right. Visit in the spring when the courtyard of the 13th-century **Begijnhof** convent is carpeted in yellow daffodils, or fall, when the ivy-covered bridges and buildings along the **Groenerei Canal** are ablaze with autumn color. Better still, spend a few midweek days wandering Bruges in January or February, when you just might have the towering **Belfort** belfry (366 steps to the top) or the tiny **Hof Arents** park and its old stone **Lovers' Bridge** all to yourself (and your significant other).

For more than 900 years, civic and commercial life has revolved around the cobblestone **Markt.** Each week the cobblestone square is consumed by the **Wednesday**

Locals dress for a Procession of the Holy Blood.

Market, a colorful ensemble of fruit, flowers, fish, handicrafts, and the occasional waffle stall. Many of the city's classic buildings surround the square, including the aforementioned Belfry, the neo-Gothic **Provinciaal Hof**, and 15th-century **Huis Boechoute,** renowned for the huge compass that crowns its facade. A virtual reality experience inside the **Historium** brings to life the golden age of 15th-century Bruges.

Nearby **De Burg Square** is another cache of classic architecture, including the **Stadhuis** (City Hall) and its exquisite **Gothic Hall** (completed in 1421)—testament to just how rich Bruges became during its golden age. That dark structure with the golden statues crouching in a corner of the square is the **Basilica of the Holy Blood**. The city's oldest building (completed in 1157) takes its name from the **Relic of the Holy Blood**, a gold-and-jewel-encrusted vial that contains a piece of cloth stained with the blood of Christ.

Bruges offers plenty of ways to sample Belgium's iconic dishes. Quaff the city's legendary beer at craft breweries like **De Halve Maan**

The Gothic-style City Hall is adorned with 19th-century murals and ornate architecture.

and **Bourgogne des Flandres**, as well as old-fashioned pubs **L'Estaminet**, **Cambrinus**, and the truly ancient **Café Vlissinghe** (opened in 1515). A tablet guides visitors through the **Bruges Beer Experience**, an interactive museum that also offers local suds and brewery swag.

The wonderful world of french fries is the focus of the unique **Frietmuseum**, stuffed with potato art, artifacts, and recipes from around the globe. And yes, they do serve delicious *pommes frites*. The **Choco-Story** chocolate museum highlights another Belgian specialty via displays, demonstrations, and a tasting room. Bruges also boasts nearly two dozen chocolate shops, including **Pralinette** on Wollestraat.

Those who visit Bruges during the high seasons can escape the throng by heading for the outskirts of the old town. Starting from **Koeleweimolen** windmill (built in the 1760s), a 3.5-mile (5.6 km) footpath runs beside remnants of the moat (now a canal) that surrounded the medieval city. Along the way are the old city gates and tranquil **Minnewaterpark**. Or you can head down to the shore and the broad sandy beaches of **Zeebrugge** on the North Sea, around a 20-minute drive from the Markt. ■

GATHERINGS

• **Bruges Beer Festival:** More than 300 types of Belgian brew are available at this February event, which includes rare beers and new beer launches; brugsbierfestival.be.

• **Procession of the Holy Blood:** Listed as a UNESCO Intangible Cultural Heritage of Humanity, the annual Ascension Day pilgrimage features thousands of costumed participants parading and reenacting stories from the Bible and the life of Christ.

• **Musica Antiqua Bruges:** Harpsichord, organ, and recorder competitions color this August celebration of medieval, baroque, and other early music; mafestival.be.

• **December Dance:** Modern dance from around the globe is the focus of this cutting-edge fest at the Concertgebouw.

Sofia
Bulgaria

One of Europe's oldest urban areas, Sofia blends its ancient past with a modern vibe and a laid-back attitude that makes the Bulgarian capital the most surprising city in the Balkans.

THE BIG PICTURE

Founded: ca 7000 B.C.

Population: 1.28 million

Size: 80 square miles (207 sq km)

Languages: Bulgarian

Currency: Bulgarian lev

Time Zone: GMT +2 or +3 (EET or EEST)

Global Cost of Living Rank: 165

Also Known As: Serdica (Roman), Sredecheski (medieval)

Sofia isn't often mentioned among the must-see places of eastern Europe. But that's actually a good thing for those who cherish cities where they don't have to line up for every church or museum or wade through crowds to cross a famous square.

That's not to say that Bulgaria's capital doesn't deserve attention. With a history that harkens back to 7000 B.C. and a parade of cultures that left behind Roman ruins and Stalinist architecture, Ottoman mosques and Orthodox cathedrals, Sofia makes for a most intriguing sojourn.

Many of the city's landmarks are arranged along busy **Boulevard Tsar Osvoboditel** between **Plostad Nezavisimost** (Independence Square) in the city center and **Knyazheska Garden**. Looming above the square is the **Largo**, a trio of three Stalinist-style buildings from which the Communist Party once ruled Bulgaria. Their imposing facades literally overshadow medieval **St. Petka Church** and an archaeological dig that reveals the ruins of Roman-era **Serdica**.

Right behind the Largo is the **National Archaeological Museum**. Housed inside a 15th-century Ottoman mosque, the collection includes Greek, Roman, and medieval artifacts from around Bulgaria, as well as the golden **Valchitran Treasure** from ancient Thrace (ca 1300 B.C.).

Next along the boulevard is the former **Royal Palace**, constructed in the 1880s but converted into the **National Art Gallery** when the monarchy was abolished after World War II. The palace is surrounded by the leafy **Royal Garden**, beside an alleyway that leads to the subterranean remains of third-century **Serdica Amphitheatre**, discovered in 2004 during construction of a hotel that now surrounds the ruin.

Opposite the royal palace is the **City Garden,** known for its outdoor cafés and cake shops, as well as **Sofia Christmas Market** during the holiday season. **Ivan Vazov National Theatre**—Bulgaria's premier stage for dance, drama, opera, and music—rises on one side of the square. A little farther along the boulevard is the small but dazzling **St. Nikolas Russian Church**, crowned by five golden domes and a green majolica spire.

An equestrian statue honoring **Alexander II**—the Russian tsar who liberated Bulgaria from the Ottoman Turks in the 1870s—graces the half-circle plaza in front of the **National Assembly**. Rising behind

A fountain leads the way to the entrance of the Ivan Vazov National Theatre.

St. Alexander Nevsky Cathedral is believed to be one of the world's 50 largest Christian church buildings.

the assembly is the city's most beloved structure: **St. Alexander Nevsky Cathedral**. The mother church of the Bulgarian Orthodox faith was completed in 1912, a neo-Byzantine building with a massive dome that ranks among the largest churches in the Balkans. Try to catch a service in the cavernous interior (8 a.m. and 5 p.m. on weekdays).

Elsewhere in the central city, a nine-block stretch of **Vitosha Boulevard** has transformed into a pedestrian precinct and after-dark entertainment zone with dozens of bars, restaurants, and coffee shops. Along the way is **Greenwich Book Center & Café**, a hub of local literary life and a great place to pick up foreign-language books on Bulgaria.

The walking street takes its name from **Vitosha Mountain**, the snow-capped massif that rises above the southern edge of Sofia. Reaching more than 7,500 feet (2,286 m),

Vitosha is popular for summer hiking and hot springs, winter skiing and snowboarding. A year-round cable car and chairlift climb to two of the highest peaks. ■

Zagreb & Dubrovnik
Croatia

One of seven independent nations that emerged from the Yugoslav Wars of the 1990s, Croatia boasts two very different urban areas in ancient but artsy Zagreb and cool coastal Dubrovnik.

THE BIG PICTURE

Founded: Zagreb, 1242; Dubrovnik, seventh century A.D.

Population: Zagreb, 700,000, Dubrovnik, 65,000

Size: Zagreb, 72 square miles (186 sq km); Dubrovnik, 8 square miles (21 sq km)

Language: Croatian

Currency: Croatian kuna

Time Zone: GMT +1 and +2 (CET and CEST)

Global Cost of Living Rank: Zagreb, 140; Dubrovnik, unlisted

Also Known As: Zagreb—Little Vienna; Dubrovnik—Pearl of the Adriatic, Ragusa (ancient)

Croatia's vibrant capital city is just being discovered as a travel destination, while seaside Dubrovnik has been drawing oohs and ahhs since the Middle Ages.

ZAGREB

Of all the central European cities that have stepped into the limelight since the early 1990s, Zagreb is the most intriguing. Unlike Kraków, Prague, or Budapest, which became tourism darlings almost overnight, the Croatian capital took its time, cultivating a unique urban persona that combines vintage architecture and rich history with vibrant social life and a quirky modern art scene.

It helps that visitors arrive in Zagreb with few preconceived notions. During its tenure as part of Yugoslavia and the Austro-Hungarian Empire, the city was largely anonymous to the outside world, which makes exploring it today all the more exciting—everything feels fresh and new, even an old town that's been around for hundreds of years.

One of those rare places that avoided large-scale damage during the World Wars and Yugoslavia's breakup, the **Upper Town** (Gornji Grad) is where Zagreb was founded as a free royal city in medieval times. Much of the old town is now a pedestrian zone that fans out from **St. Mark's Square** and its entourage of historic structures: 13th-century **St. Mark's Church** with its colorful coat of arms roof and the mustard-colored **Hrvatski Sabor** (Croatian Parliament). On weekends between April and October, the square hosts a rousing changing of the guard ceremony by the **Royal Cravat Regiment** clad in 17th-century uniforms (including their namesake red or black cravats).

Just down the cobblestone road are two of Zagreb's most intriguing museums. From discarded rings and plush toys to wedding dresses and items stolen back from ex-lovers, the **Museum of Broken Relationships** takes a humorous and poignant look at bad romance. Meanwhile, the

A monument of politician Josip Jelačić, the viceroy of Croatia from 1848 to 1859, stands in the eponymous square.

Colorful buildings surround the city of Zagreb's Bana Jelačića Square.

Croatian Museum of Naïve Art exhibits amazing works by self-taught artists.

Pedestrians descend to the **Donji Grad** (Lower Town) via the antique **Uspinjača** funicular (opened in 1890) or meander down various streets and stairs. The area revolves around the broad **Bana Jelačića Square** (named for the national hero who sits astride the plaza's equestrian statue), the soaring **Cathedral of the Assumption**, and **Dolac Market** with its trademark red umbrellas, which primarily sells fruits and vegetable, but is also a great place for Croatian handicrafts and snacks.

The Lower Town also has its modern side, like the graffiti installations and summer concerts of the open-air **Art Park Zagreb**, the mysterious World War II-era **Grič Tunnel** (now used for underground exhibitions, concerts, and other events), and the

Nikola Tesla Technical Museum, with its science and transportation collection inspired by another Croatian national hero. After dark, locals and visitors alike make the trek up **Tkalčićeva**, a bar-lined pedestrian street that runs uphill from Dolac Market.

Be on the lookout for the city's abundant **murals**, from a giant blue whale and modern-day Gulliver to an incredibly lifelike urban waterfall and the innovative **"Medika Diving"** (named one of the world's top 100 murals). An entire wall outside the **Museum of Contemporary Art** in the Novi Zagreb district is covered in murals by OKO, one of the city's celebrated street artists.

Graffiti covers **"The Grounded Sun,"** a large bronze sphere on Bogovićeva Street that acts as the centerpiece of the **"Nine Views"** installation that includes all the planets in our solar

system, properly proportioned in relation to the sun and placed in locations around Zagreb that correspond to their distance from the sun.

Zagreb's hinterland flaunts its own adventures and charms. Half an hour west of the capital, **Samobor** is another free royal town established around the same time as Zagreb. However, it never grew into a big city and retains much of its medieval feel. **St. Anastasia Church** overlooks cobblestone **Kralja Tomislava Square** and its popular outdoor cafés, while looking on from the heights above town is ruined 13th-century **Samobor Castle**.

Medvednica (Bear Mountain) offers a huge outdoor escape just beyond the city's northern suburbs. Hiking and biking trails radiate from **Sljeme** summit (3,396 feet/ 1,035 m), which transforms into a popular skiing and snowboarding area in winter. The mountaintop also offers rustic hotels and restaurants. Farther down the slopes is **Medvedgrad Castle**, a partially restored

citadel built in the 13th century after the Mongols sacked Zagreb.

DUBROVNIK

When the makers of *Game of Thrones* were searching the world for the perfect medieval backdrop to play the fictional King's Landing in season two of the blockbuster fantasy, they looked no farther than Dubrovnik.

What sold them on the Croatian coastal town was the fact they barely had to change a thing: Encircled by thick walls and spangled with marble plazas, terra-cotta-roofed palaces, and flamboyant churches, Dubrovnik looks much the same today as 500 years ago when it rivaled Venice for domination of the Mediterranean Basin.

Needless to say, the TV show sparked a cottage industry of *GOT* Dubrovnik maps, booklets, and guided tours. Shooting took place at 16 locations in and around the old town, from **Lovrijenac Fort** (the Red Bastion) and **Minčeta Tower** (House of Undying) to the exquisite 14th-century **Dominican Monastery** and the heavily fortified 15th-century **Ploče Gate**.

Ploče is one of only four gates that breach the famed **City Walls**. The gates were totally repaired after sustaining damage during the 1990s Siege of Dubrovnik, and a walk around the ramparts is the first thing that every first-time visitor should undertake. The main entrance to the walls is a steep staircase that starts from **Paska Miličevića Plaza** just inside the **Pile Gate**.

Old town Dubrovnik's iconic seawalls and red-roofed buildings served as fictional King's Landing in HBO's *Game of Thrones*.

Two other landmarks are located on the square. Water pours forth from human and animal faces arrayed around the base of **Large Onofrio's Fountain**, which is also a pretty good place to sit and soak up the atmosphere. Built in the 14th century, the adjacent **Franciscan Church and Monastery** harbors a lovely cloister and a library-museum with sacred treasures and more than 100 rare incunabula (early printed books).

The plaza also marks the western end of the **Stradun**, a limestone-paved avenue that cuts straight across the old town. The street is flanked by cafés and specialty shops like **Uji**, which offers Croatian olive oil, truffle oil, wines, and other gourmet foods. The narrow lanes off Stradun have their own allures: incredible ice cream at **Dolce Vita**, images of conflict at the small **War Photo Limited** museum, and the baroque **Dubrovnik Synagogue**.

At the far end of Stradun, **Luža Square** is surrounded by 16th-century **Sponza Palace**, the flamboyant Venetian-style **Church of St. Blaise**, and **Dubrovnik Bell Tower**. The nearby **Rector's Palace**, a Gothic-Renaissance creation, houses the city's **Cultural History Museum** and a collection that runs a broad gamut from antique furnishings and vintage coins to paintings, sculptures, and weapons.

Beyond the walls, Dubrovnik offers half a dozen **Adriatic beaches** ideal for swimming, kayaks, and paddleboards. Ferries flit over to heavily wooded **Lokrum Island** (another *GOT* location), while the **Love Stories Museum** offers a romantic foil to Zagreb's broken-heart collection.

The **Dubrovnik Cable Car** makes its way up **Mount Srđ** with its aerial views of the old town and a **Homeland War Museum**

Stairs lead the way past ancient buildings and onto bustling streets in Dubrovnik.

detailing the Siege of Dubrovnik during the Yugoslav Wars. You can also reach the summit by car or by trekking the zigzag **Staza Prema Utvrdi Imperial** trail.

Gruz Harbor, the city's modern port, is the place to catch ferries to the nearby **Elaphites** (Deer Islands):

From south to north, **Koločep**, **Lopud**, and **Šipan** offer seaside villages, ancient ruins, small farms, charming little hotels, and beaches that are rarely crowded. An almost total lack of vehicles means the archipelago is also walker and biker friendly. ■

GATHERINGS

• **Snow Queen Trophy:** Sljeme mountain ski resort hosts this January grand prix slalom competition that features top professional women and men skiers from around the world.

• **Feast of St. Blaise:** Dubrovnik's third-century patron saint is feted on February 3 with a holy mass in his namesake church followed by a procession of holy relics, a medieval gun salute, and residents wearing folk costumes.

• **Zagrebdox:** Documentaries come into focus during a February/March festival that screens more than 100 films from around the planet; zagrebdox.net/en.

• **Cest is d'Best:** Musicians, acrobats, dancers, clowns, fire-eaters, jugglers, and other buskers descend on Zagreb in late May and early June for this global showcase of street performance; cestisdbest .com/en.

• **Dubrovnik Summer Festival:** Launched in 1950, this two-month-long event brings drama, dance, music, and opera to 70 venues around the Renaissance city in July and August; dubrovnik-festival.hr/en.

• **International Puppet Theatre Festival:** Puppet masters from around the world work their magic in Zagreb during this September event that was started in 1968; pif.hr/en.

Edinburgh
Scotland, United Kingdom

Staging a prestigious festival, rebranding its fabulous museums, or flaunting the best of Scottish culture, the "Athens of the North" is primed to take its place among the great capital cities of Europe.

THE BIG PICTURE

Founded: sixth century A.D.

Population: 525,000

Size: 46 square miles (120 sq km)

Language: English

Currency: Pound sterling

Time Zone: GMT +0 and +1 (GMT and BST)

Global Cost of Living Rank: Unlisted

Also Known As: Din Eidyn (ancient Celtic), Auld Reekie (Old Smokey), Edina, Athens of the North, Festival City

Whether or not the people of Scotland eventually decide to break away or stay in the United Kingdom is a moot point. Either way, Edinburgh feels like the capital of an independent nation.

Maybe that's because it was a national capital for more than a thousand years, the power center of Scottish kingdoms that defied the Romans, Vikings, and Anglo-Saxons. It wasn't until 1707 that Scotland was finally fused with England, and Edinburgh ceased being a national capital.

But the feeling never went away. And it's still there today in institutions that reflect a nation rather than a state or province. The incredible **National Museum of Scotland** safeguards iconic Scottish artifacts like the Hunterston brooch, the Lewis chessmen, the St. Ninian's Isle silver cache, the Whitecleuch chain, various Pictish stones, and the Jacobite flag from the Battle of Culloden.

In addition to the main branch, there's Scotland's **National War Museum** inside Edinburgh Castle and Scotland's **National Museum of Flight** at East Fortune airfield on the eastern outskirts of town—where visitors can explore a British Airways Concorde supersonic passenger jet. Edinburgh is also home to the **Scottish National Gallery**

with its European old masters and local virtuosos like Henry Raeburn and Alexander Nasmyth. The nearby **National Portrait Gallery** and **National Gallery of Modern Art** round out the great Scottish art collections in Edinburgh.

The rebranding of several of these institutions as "national" coincided with the upsurge in Scottish nationalism after the turn of the 21st century and the unveiling of the outrageously modern **Scottish Parliament Building** in Edinburgh's Holyrood area in 2004. Free guided and self-guided tours are offered Monday through Saturday whenever the Scottish Parliament (founded in 1999) is in session.

Opposite the futuristic parliament is the **Palace of Holyroodhouse**, a reminder of days when Scotland had its own king. Erected in the 12th century, the ruined abbey served as a meeting place and royal residence for the likes of Robert the Bruce, Bonnie Prince Charlie, and Mary Queen of Scots. The adjacent baroque residence—with its sumptuous **State Apartments** and **Throne Room**—arose during the Scottish Restoration of the late

Edinburgh Castle is an iconic fortress, built in the 11th century on Castle Rock.

The Dugald Stewart Monument is a memorial to the Scottish philosopher that stands on Calton Hill overlooking the city.

17th century and is still the official resident of the British monarch in Scotland.

The British monarchs also keep their yacht in Edinburgh, or at least the one they tool around in. **The Royal Yacht *Britannia*** is permanently berthed in **Leith** harbor on the **Firth of Forth** along with the former racing yacht *Bloodhound*. Make like a queen and take high tea aboard the *Britannia*, take a stroll along the picturesque **Water of Leith**, and raise a glass at ancient waterfront pubs like the **Malt & Hops** (opened in 1749) before returning to the city center.

Many a Scottish monarch may have lived at Holyroodhouse, but their power base was always **Edinburgh Castle**. The volcanic hilltop that anchors the citadel was most likely fortified by Celtic tribes

during Roman times before taking on its current Gothic form in the Middle Ages. In addition to hosting the **Royal Edinburgh Military**

Tattoo event, the castle features many permanent attractions, from the 12th-century **St. Margaret's Chapel** (Edinburgh's oldest

The Elephant House café is famous as the spot J. K. Rowling often sat writing early drafts of the Harry Potter series.

surviving building), the enormous **Mons Meg** siege gun, and the aforementioned War Museum to tearooms, a whisky and gourmet food shop, and a small cemetery for pets that have lived in the castle.

The cobblestone road that connects Edinburgh Castle and Holyroodhouse is the celebrated **Royal Mile**, which is both the most beloved and kitschy street in all of Scotland. Many of the stately

Georgian buildings are occupied by touristy restaurants, pubs, and souvenir shops dispensing kilts, clan badges, tartans, and other Scottish wares. However, the Royal Mile is also flanked by architectural landmarks and cultural attractions.

Right outside the castle's main gate is the **Witches' Well**, a small fountain and plaque marking the spots where hundreds of people accused of witchcraft were burned at

the stake between the 15th and 18th centuries. Across the street, the **Scotch Whisky Experience** explains everything you probably ever wanted to know about Scotland's national drink with several tours that include expert tasting sessions. Down the road, **Camera Obscura & World of Illusions** sounds like a cheesy tourist trap, but it's actually a historic Victorian-era concoction that now features five floors of optical illusions plus a roof terrace with panoramic city views.

Descending from Castlehill, the High Street portion of the Royal Mile is dominated by **St. Giles' Cathedral**, a hulking Gothic structure where John Knox kicked off the Protestant Reformation in Scotland. The cityscape looked quite a bit different back in those days, as evidenced by the **Mary King's Close**, a warren of medieval lanes where the black plague once flourished. Now buried beneath street level, this underground city is allegedly *very* haunted.

Several side streets off the Royal Mile beg investigation. Duck three blocks down George IV Bridge to the **Elephant House**, the café where then unknown author J. K. Rowling invented Harry Potter. Murder

GATHERINGS

• **Edinburgh Science Festival:** The world's oldest public science fest (founded 1988) presents more than 200 events around town over Easter week that revolve around discussion, debate, experimentation, and discovery; sciencefestival.co.uk.

• **Edinburgh International Children's Festival:** Youth-oriented theater, dance, music, arts, and crafts from around the globe are the focus

of this merry May jamboree; imaginate.org.uk/festival.

• **Edinburgh International Festival:** First staged in 1947, the event has grown into one of the world's premier showcases for performing arts; runs concurrently over the entire month of August with the Royal Edinburgh Military Tattoo, Edinburgh Festival Fringe, and Edinburgh Art Festival; eif.co.uk.

• **Scottish International Storytelling Festival:** Rooted in Scottish tradition,

this October event celebrates oral tradition via songs, tall tales, and good old-fashioned storytelling; www.sisf.org.uk.

• **Hogmanay:** Edinburgh rings in the new year with a three-day bash (December 30–January 1) that features traditional Scottish dancing and music, a torchlight parade, and a chance to "Auld Lang Syne" in the place where it was written; edinburghshogmanay.com.

mystery writers Ian Rankin and Alexander McCall Smith were also habitués. Another detour leads past the old **Scotsman newspaper building** (now a boutique hotel) and to **North Bridge** with its lofty city views.

The eastern stretch of the Royal Mile offers the **Scottish Storytelling Centre**—where traditional theater, music, and storytelling events are staged—and the offbeat but interesting **People's Story Museum** of Scottish social history inside the 16th-century **Canongate Tolbooth**. Economist Adam Smith and other famous Scots are buried in the **Canongate Kirkyard** cemetery.

Beyond the medieval warren of the old town, Edinburgh is surprisingly well endowed with green spaces like **Princes Street Gardens**, the **Royal Botanical Garden**, and the sprawling **Pentland Hills**

ALTER EGO

GLASGOW

With more than twice as many people as Edinburgh, Scotland's largest city has been the nation's economic powerhouse since the industrial revolution, when it emerged as a shipbuilding, heavy engineering, and textile center.

Though maritime industry remains a **Glasgow** specialty, the economy has diversified into fields like bioscience, video game development, and, increasingly, tourism.

In addition to its grandiose Victorian architecture—personified by **George Square** and **Kelvingrove Art Gallery**—the city flaunts astonishing modern structures like the **Glasgow Science Centre**, the **SEC Armadillo** (aka Clyde Auditorium), and the **Riverside Museum** of transportation and technology.

Glasgow also offers a robust pub culture and live music scene. And this being Scotland, the city boasts its fair share of whisky distilleries.

Regional Park beyond the city's southern fringe.

Once a private playground for the monarchy, **Holyrood Park** reaches a literal peak at **Arthur's Seat**, an 823-foot (251 m) volcanic plug. There's another great view from **Calton Hill**, strewn with

tributes to poet Robert Burns, philosopher Dugald Stewart, and other luminaries, as well as the Parthenon-like skeleton of the unfinished **National Monument of Scotland**, which much like Scotland's ambition for nationhood is still a work in progress. ∎

Historic Royal Mile is the main thoroughfare of the old town of the city.

Porto
Portugal

Sprawling along the Douro River in Portugal's picturesque wine country, the nation's northern metropolis basks in both sunshine and an appreciation for the finer things in life, especially eating, art, architecture, and outdoor urban adventures.

THE BIG PICTURE

Founded: 275 B.C.

Population: 1.3 million

Size: 307 square miles (795 sq km)

Language: Portuguese

Currency: Euro

Time Zone: GMT 0 and +1 (WET, WEST)

Global Cost of Living Rank: Unlisted

Also Known As: Oporto, Portus Cale (ancient)

The quintessential Mediterranean city, Porto is known for fine wine and romantic song, a sun-splashed climate that begs a day at the beach, and a red-tile skyline rising high above a boat-filled river flanked by castles and cathedrals.

Ironically, one of Iberia's oldest cities isn't anywhere near the Mediterranean; rather, it is on the Atlantic coast around 200 miles (320 km) north of Lisbon. Prince Henry the Navigator, explorer extraordinaire Ferdinand Magellan, blue-and-white azulejo tiles, port wine, and even the country's name are among Porto's contributions to Portuguese history and culture.

As it has been for more than 2,000 years, the **Ribeira** waterfront along the **Douro River** is still the city's heart and soul, a place to promenade at sunset, sample *bacalhau* (salted codfish), *tripas* (tripe), grilled sardines, and other local seafood favorites, and listen to street musicians singing melancholy fado songs.

First-time visitors can learn more about Porto on a **six bridges cruise** from the Ribeira quay or by visiting the **Museu da Cidade** in the 14th-century **Casa do Infante** near the waterfront. Beyond the museum, the **Praça do Infante Dom Henrique** and its equestrian statue of Prince Henry the Navigator is flanked by three of the city's most distinctive buildings.

The solemn Gothic façade of the **Igreja de São Francisco** church gives way to a flamboyant baroque interior and creepy catacombs. Once funded by the city's wealthy merchants, the neighboring **Palácio da Bolsa** houses the glass-domed **Pátio das Nações** and the exquisite golden **Salão Árabe**. The huge red metallic structure is the Victorian-era **Mercado Ferreira Borges**, which has morphed in recent years from a good market into the **Hard Club**, a cutting-edge cultural space with concerts and exhibitions.

Riding the heights behind the waterfront is the **Baixa** neighborhood, which blends downtown functions like banking and shopping with a potent nightlife scene that includes late-night cafés, cocktail bars, and music clubs. Pedestrians can scale the hills behind the waterfront via steep medieval streets, alleys, or stone staircases; in a supersize elevator called the **Lift Lada Ribeira**; or by riding the vintage **Funicular dos Guindais** (opened in 1891).

The interior of São Bento Railway Station features typical azulejo tiles.

Porto's colorful houses sit on the banks of the Douro River, where traditional *rabelo* boats are moored.

Although uptown is dominated by the medieval **Sé Cathedral** and the baroque **Torre dos Clérigos** campanile (240 steps to the top), Baixa is also renowned for art nouveau treasures like the **Livraria Lello** bookstore and the **Avenida dos Aliados**, as well as the beaux arts **São Bento Railway Station** with its stunning azulejos-covered vestibule.

Porto's architectural evolution continues at the **Serralves**, an expansive garden and contemporary art collection that revolves around a streamlined art deco building. The grounds also harbor an **art house cinema** and **treetop canopy walk**. West of Serralves, the city meets the sea, with half a dozen beaches between the Douro and the **Porto de Leixões**.

The south bank is reached via the aforementioned six bridges, including the modern **Ponte da Arrábida**, which features a vertiginous **Porto Bridge Climb** attraction. The historic **Ponte de Dom Luís I** links Ribeira and **Vila Nova de Gaia**, a south bank location renowned for caves or cellars where port wine is aged and stored. **Caves Ferreira** (founded in 1751) is one of the oldest, **Graham's Port Lodge** boasts the best view, and **Espaço Porto Cruz** offers hip, Gen-X style port tastings at its rooftop bar. ◼

LOCAL FLAVOR

• **éLeBê Entreparedes:** Fifty-year-old family recipes highlight a gastronomic trip that features *porco preto* (black pork) from Alentejo, *papas de sarrabulho* (potato-and-meat stew) from Minho, and Madeira-style tomato soup with poached eggs; *Rua de Entreparedes 37, Bolhão;* en.elebe.pt.

• **Café Santiago:** This tiny eatery arguably makes Porto's best *francesinha*, a sandwich that is stuffed with ham and linguica sausage, smothered in melted cheese and secret sauce, and garnished with a fried egg and french fries; *Rua Passos Manuel 226, Baixa;* cafesantiago.pt/en.

• **The Yeatman:** Chef Ricardo Costa earned the twin Michelin stars of this classy establishment by creating out-of-this-world seasonal tasting menus complemented by Portugal's best wine cellar; *Rua do Choupelo, Vila Nova de Gaia;* the-yeatman-hotel.com/en/food/restaurant.

Dublin

Ireland

From emotive music and illuminated manuscripts to single malt whiskey and Georgian mansions, the Irish capital flaunts multiple personalities: a city that makes it easy to fall head over heels for but eternally difficult to define.

Dublin has always been a hard city to put a finger on—not just for overseas visitors but also for the Irish themselves. "If I can get to the heart of Dublin," wrote hometown hero James Joyce, "I can get to the heart of all the cities of the world." Scores of other authors, poets, and songwriters have also struggled to interpret the city.

To some, Dublin is a city of pubs serving dark beer and a golden libation that derives its name from the old Gaelic term *uisce beatha* ("water of life"). To others, it's an erudite metropolis that nurtured scores of great writers and thinkers. To others still, it's the Irish equivalent of Nashville, the melodious confluence of "trad" music with modern rock, punk, and New Age. As Joyce and others recognized, there's no single Dublin.

Many of those with a thirst to slake head for **Temple Bar**, the rollicking epicenter for Dublin nightlife on the south bank of the **River Liffey**. The neighborhood is indeed old, the first outside the city walls during medieval times. But its guise as Ireland's pub hub is recent, traceable to the 1990s, when government

THE BIG PICTURE

Founded: 988

Population: 1.2 million

Size: 123 square miles (319 sq km)

Language: English, Irish

Currency: Euro

Time Zone: GMT +0 and +1 (GMT and IST)

Global Cost of Living Rank: 39

Also Known As: Baile Átha Cliath (Irish), the Big Smoke

redevelopment funds pumped new life into what had been one of Dublin's most decrepit districts.

Temple Bar's watering holes range from the historic **Palace Bar** (opened in 1823) and **Temple Bar Pub** (established in 1840) to newfangled hot spots like the **Liquor Rooms** speakeasy and **Porterhouse** craft brewery. There are also notorious spots: **Darkey Kelly's Bar** is located on the site of an 18th-century brothel whose namesake was burned at the stake for witchcraft and is now considered Ireland's first serial killer.

The nearby **Irish Whiskey Museum** provides an excellent overview of the amber nectar, and several of the factories are within short walking distance of Temple Bar, including the **Jameson Distillery** across the river in Smithfield and the **Teeling Whiskey Distillery** in the Liberties. The popular **Guinness Storehouse** is also nearby for those who prefer a totally different Irish libation.

A trail through musical Dublin also starts in Temple Bar, with the **Irish Rock 'n' Roll Museum**, a must-see for those who grew up on the tunes of U2, Thin Lizzy, Bob Geldof, Sinéad O'Connor, the Corrs, the Pogues, the Cranberries, and other great acts from the Emerald Isle.

Temple Bar in Dublin is one of the most iconic drinking wells in the city.

The Samuel Beckett Bridge crosses the River Liffey to the convention center and the Docklands.

Catch the stars of the future at the many Temple Bar pubs that offer live music, or join a **Traditional Irish Musical Pub Crawl** through the neighborhood (nightly April to October, three times a week the rest of the year). Famous names appear onstage at the brilliantly restored **Olympia Theatre**, a former Victorian-era music hall with plush red seats and elaborate golden decorations.

Crank back the musical clock even further by snapping a selfie in front of the **Molly Malone statue** of the fishmonger immortalized by the old Irish folk song. A block away (past the **Visit Dublin** tourist office) is **Grafton Street**, a pedestrian precinct renowned for both its shopping and street performers. Among those who have busked along Grafton are U2 lead singer Bono and Academy Award–winning singer and songwriter Glen Hansard. Tucked down a side street is a bronze statue of Thin Lizzy front man **Phil Lynott**

beside legendary music club **Bruxelles**, where rock-and-roll royalty like the Rolling Stones and Guns N' Roses have partied over the years.

Around the corner from Grafton Street, the excellent **Little Museum of Dublin** offers a permanent fan-curated exhibit called "U2: Made in Dublin" that delves into the 40-year history of a band with

deep roots in the city. As honorary "Freemen of Dublin," band members Bono and the Edge are entitled to the ancient right to pasture sheep on **St. Stephen's Green** across the street from the museum. Dublin is flush with U2 sights, from Mount Temple Comprehensive School where the band formed in 1976 to the Bonavox (Good Voice) hearing

---

ALTER EGO

BELFAST

Long renowned for the Troubles that plagued all of Northern Ireland, **Belfast** settled into an era of relative calm since the peace process of the late 1990s. Tourism has boomed, from around half a million annual overnight visitors during the Troubles to around three million in recent years.

They come to see vintage attractions like **St. George's Market** (opened in 1896) and the **Ulster**

Museum with its Spanish Armada artifacts, and new diversions like the **W5** interactive science discovery center, the futuristic *Titanic* **Belfast museum** on the site of the shipyard where the ill-fated ocean liner was built, and the new *Game of Thrones* **Studio Tour** in Banbridge.

The city's dark past also generates tourist attractions like the notorious **Crumlin Road Gaol** and the graffiti-covered "Peace Walls" that once divided the Catholic and Protestant communities.

aid store at 9 North Earl Street that inspired lead singer Paul Hewson's famous nom de plume.

Erudite Dublin starts on Grafton Street—at **Bewley's Grafton Street Café** (opened in 1927), where many a famous Irish writer lingered back in the day—and continues nearby at learned institutions like **Trinity College**. Authors Oscar Wilde, Bram Stoker, Samuel Beckett, and Jonathan Swift are but a few of the famed alumni. The university's prize possession is also literary: the ninth-century **Book of Kells** and other historical documents on display in the **Old Library.**

Despite its age—Trinity was founded by order of Queen Elizabeth I in 1592—many of the college buildings date from Georgian days when central Dublin was radically remade with wider streets and larger, British-inspired buildings. So pervasive was the change that Trinity and the surrounding area are often called Georgian Dublin. Among its landmarks are **Dublin Castle** (seat of British rule until independence in 1922), **Leinster House** (where the

Dublin has a strong literary history, which is on display in the Long Room at the Old Library of Trinity College.

Oireachtas, or Irish Parliament, convenes), and the **General Post Office** (GPO), epicenter of the 1916 Easter Rising against British rule and now home to a permanent exhibit on Irish nationalism called **"GPO Witness History."**

The museum cluster around parliament house is largely Victorian. The National Gallery of Ireland boasts the traditional Dutch and Italian masters, as well as a large collection of Indigenous art. However, the real riches are found inside the **National Museum of Ireland, Archaeology**, which safeguards the

Broighter Hoard, **Ardagh Chalice**, and other ancient Celtic treasures, as well as the 2,200-year-old **Gallagh Man** and other naturally mummified bog bodies.

Merrion Square, the most elegant of Dublin's many green spaces, is another product of the Georgian age. The author of *The Importance of Being Ernest* and *The Picture of Dorian Gray* grew up in the **Oscar Wilde House** on the square's northwest side. Poet and playwright **W. B. Yeats** lived on the opposite side of the square, in the Georgian manse at No. 82.

Yeats helped found the **Abbey Theatre** on the north side of the River Liffey. Now the national theater of Ireland, the playhouse offers a year-round slate of serious drama. The **Northside** area also harbors the **Dublin Writers Museum** with its prized first editions, personal mementos, and excellent bookshop, as well as the **James Joyce Centre**, which pays homage to both the novelist who so vividly described the city of his birth in tales like *Ulysses* and *Dubliners,* and the fictional characters who populated those stories. ∎

Baltic Cities
Vilnius, Lithuania; Riga, Latvia; Tallinn, Estonia

Stepping out from behind the Iron Curtain, the Baltic capitals have transformed into three of Europe's most intriguing destinations, cities with a shared heritage but divergent paths to the future.

Find colorful handmade Easter eggs at markets throughout Vilnius.

VILNIUS

Population: 515,000
Size: 85 square miles (220 sq km)

Vilnius first rose to fame as the capital of the powerful Kingdom of Lithuania that once stretched from the Baltic to the Black Sea as Europe's largest medieval state.

Eight hundred years later, it still bears all the trappings of a royal city, including the lavish **Palace of the Grand Dukes** who ruled Lithuania during its golden age and the hilltop **Gediminas Castle** with its exhibits on the nation's rich history. A statue of **Mindaugas**, the first king, dominates the plaza in front of the **National Museum**.

Founded in 1579, **Vilnius University** is one of Europe's oldest institutions of higher learning. The campus is spread across 13 interconnected courtyards, including an aptly named **Grand Courtyard** flanked by the wondrous Renaissance-era **VU Library** and **St. John's Church** with its classic Gothic facade.

Vilnius Cathedral features a freestanding **bell tower** that visitors can summit via a narrow wooden stairway. With its flamboyant redbrick facade, **St. Anne's Church** is even more astounding.

The city's Jewish population—one of Europe's largest until World War II—is evoked by the **Tolerance Center**, the restored **Choral Synagogue**, and the **Vilna Gaon Museum of Jewish History**. Located in the former KGB headquarters and prison, the **Museum of Occupations and Freedom Fights** is dedicated to all those who perished during the Nazi and Soviet occupations of Lithuania.

On the east side of the **Vilnia River** is the city's celebrated **"Republic of Užupis"**—a free-spirited neighborhood that declared its independence in 1998. Užupis's cobblestone streets are flanked by cafés and coffee shops, art and craft galleries, tattoo parlors, and little boutiques.

RIGA

Population: 605,000
Size: 100 square miles (259 sq km)

Latvia's capital is known for its photogenic blend of Gothic and art nouveau architecture. But what's not widely known is Riga's emergence as a leading foodie destination.

Fresh fish from the nearby Baltic and **Daugava River** is complemented by dairy, meats, and produce from Riga's rich hinterland at top-shelf restaurants like **Vincents** (with van Gogh–inspired decor) and the artful **Galerija Istaba**.

The city also boasts one of Europe's best (and biggest) food halls. **Riga Central Market** has been around since the Middle Ages, but it's now housed in five humongous structures originally used as zeppelin hangars by the German military during World War I. In addition to hundreds of fresh food stalls, there's a new **Gastro Market** with trendy little bars and eateries.

The House of the Brotherhood of the Blackheads is one of the most iconic buildings in all of Old Riga.

Riga's sweet tooth comes into tasty focus at the **Laima chocolate factory**, which offers a behind-the-scenes glimpse into the production process and a shop where visitors can load up on dragées, zephyrs, and caramels. On the left bank of the Daugava River, the **Kalnciema Kvartāls** market on Saturdays offers local food, drink, and entertainment in a courtyard surrounded by the historic wooden buildings of the **Kalnciema Quarter**.

Between all those eats, the **Vecriga** (Old City) flaunts Gothic masterpieces like **Riga Cathedral**, **St. Peter's Church**, and the **House of the Brotherhood of the Blackheads**, a medieval guild hall where the very first Christmas tree was allegedly erected, decorated, and then set ablaze in 1510.

Among the city's many art nouveau treasures are the apartment buildings along **Alberta Iela**, **Riga Synagogue** (Peitav Shul), and the **Cat House** with its famous black-cat motifs.

TALLINN

Population: 437,000
Size: 61 square miles (158 sq km)

Tallinn is totally wired. But in a good way. One of the first Eastern European cities to boom after the fall of the Iron Curtain, the metropolis transformed itself into a Baltic version of Silicon Valley. Skype is just one of the innovations that has emerged from here.

Estonia's capital also started to party hearty, a reputation that attracted young people from around Europe to its many lively clubs, bars, and underground dance scene.

The **Old City** is ground zero for Tallinn nightlife, hangouts like **Club Privé** and **Club Hollywood** along Müürivahe Street or **Sigmund Freud** and **Frank Underground** on nearby Sauna Street. During the summer, an outdoor party scene also revolves around two retro railcars at **Peatus** and the harborside **PADA cultural garden.**

By daylight, the Old City turns Gothic—the architecture that is. One of the best preserved medieval cities in northern Europe, the historic core was declared a UNESCO World Heritage site in 1997.

Raekoja Plats, the main square, is engulfed by classic Gothic buildings like the 13th-century **Town Hall** and **Revali Raeapteek pharmacy**, founded in 1422 and still an active drugstore (as well as a medieval pharmacy museum). The 407-foot (124 m) spire of 12th-century S**t. Olaf's Church** rises high above the cobblestone streets.

Danish knights intent on Christianizing Estonia built hilltop **Toompea Castle** with its solid medieval walls. Below the castle is a remaining portion of the **city walls**, and **Kiek in de Kök** tower that offers a military museum and entrance to the underground **Bastion Passages**.

Tallinn's 21st-century attractions include the **KUMU** art museum, boho bars, workshops, and the flea market at **Telliskivi Creative City**; dining and drinking in the refurbished **Rotermann Quarter**; and **Lennusadam** maritime museum in a repurposed seaplane hangar on the waterfront. ■

The Viru Gate marks the entrance to Old City Tallinn.

ACKNOWLEDGMENTS

All of the books in the 5,000 Ideas series are a team effort, and *100 Cities* is no exception. From the very start, I worked closely with senior editor Allyson Johnson to craft the outline and individual urban chapters. Meanwhile, Kay Hankins (designer and photo editor), Sanaa Akkach (art director), Susan Blair (director of photography), and Judith Klein (senior production editor) reviewed thousands of photos, developed the design, and created the book's amazing good look. Last but not least, fact-checker Lindsay Smith made sure that all the facts, figures, and information were correct and up to date. As she's done on all the 5,000 Ideas books, Julia Clerk undertook much of the research and performed a first-draft proofread of every chapter.

Thanks must also go to the many friends and colleagues who helped me discover these amazing urban spaces by traveling alongside or letting me crash on their couches. Thanks to photographer Ian Lloyd in Sydney, as well as our adventures in Auckland, Luang Prabang, and Antananarivo; Roberta "Bunny" Dans and Andrew Thomas in Manila and Singapore; Gigi and David Gourlay in Jakarta and Dubai; Debbie Howard and Bob Dieter in Tokyo; Nevada Weir in Hanoi and Ho Chi Minh City; Carolina Chia and David Levy in New York City; Shannon and Chelsea Yogerst in Boston; my aunt Eunice in Las Vegas; Duncan Beniston in Savannah and Charleston; Maribeth Mellin in Mexico City and Oaxaca; Marissa Silvera in Antigua and León; Patricia Ortiz in Lima, Cusco, and Buenos Aires; Karim El Minabawy in Cairo and Luxor; Aki Bockelmann in Johannesburg and Istanbul; Gabi Kurzenberger in Munich, Frankfurt, and Jerusalem; Mike and Caddie Chaddick in Budapest and Prague; Gaston Fasnacht in Zurich; Joanna and Markus Breuer in Innsbruck; Rebecca Morales in San Juan; Ruth and Junior Scott in Kingston.

ABOUT THE AUTHOR

During three decades as an editor, writer, and photographer, Joe Yogerst has lived and worked in Asia, Africa, Europe, and North America. His writing has appeared in *Condé Nast Traveler,* CNN Travel, *Islands* magazine, the *International New York Times* (Paris), *Washington Post, Los Angeles Times*, and *National Geographic Traveler*. He has written for 34 National Geographic books, including the best-selling *50 States, 5,000 Ideas* and the sequel *100 Parks, 5,000 Ideas.* His first U.S. novel, a murder mystery titled *Nemesis,* was published in 2018. Yogerst is the host of a National Geographic/ Great Courses video series on America's state parks.

ILLUSTRATIONS CREDITS

Cover: Domingo Leiva/Getty Images (BACKGROUND), sculpies/Shutterstock (UP LE), ira008/Shutterstock (UP RT), Patrick Foto/Shutterstock (LO LE), John J. Miller Photography/Getty Images (LO CT), Dieter Meyrl/Getty Images (LO RT); Spine, Antonino Bartuccio/Sime/eStock Photo; Back cover, Reinhard Schmid/Huber/eStock Photo; 2–3, Benny Marty/Shutterstock; 4, Tupungato/Shutterstock; 6, Wang Ying/Xinhua via Getty Images.

United States, Canada & The Caribbean
10–11, F. M. Kearney/Design Pics/robertharding;12, Tasos Katopodis/UPI/Alamy Stock Photo; 13, Matthew Carreiro/Shutterstock; 14, AlexCorv/Shutterstock; 16, Rolf_52/Alamy Stock Photo; 17, Robert Alexander/Getty Images; 18, Sean Pavone/Alamy Stock Photo; 19, Gino's Premium Images/Alamy Stock Photo; 20, Mario Tama/Getty Images; 21, trekandshoot/Shutterstock; 22, arkanto/Shutterstock; 23, Min C. Chiu/Shutterstock; 24, Susanne Kremer/Huber/eStock Photo; 25, Joseph Sohm/Shutterstock; 26, Richard T. Nowitz/robertharding; 27, Andrea Izzotti/Shutterstock; 28, SeaRick1/Shutterstock; 29, Shunyu Fan/Getty Images; 30, Carrie Yuan Images/Offset; 31, Sooksan Kasiansin/Alamy Stock Photo; 32, Joel Lerner/Xinhua via Getty Images; 33, Sean Pavone/Getty Images; 34, Giovanni Simeone/Sime/eStock Photo; 36, Jennifer Wright/Alamy Stock Photo; 37, f11photo/Shutterstock; 38, Laura Zeid/eStock Photo; 39, Edgar Bullon/Alamy Stock Photo; 40, picturelibrary/Alamy Stock Photo; 41, Chon Kit Leong/Alamy Stock Photo; 42, Jose Fuste Raga/robertharding; 43, Thierry Tronnel/Corbis via Getty Images; 44, Lorne Chapman/Alamy Stock Photo; 45, John Gress Media Inc/Shutterstock; 46, helen vt/Shutterstock; 47, Kit Leong/Shutterstock; 48, Marc Bruxelle/Shutterstock; 50, TRphotos/Shutterstock; 51, randy andy/Shutterstock; 52, Lenush/Shutterstock; 53, Toms Auzins/robertharding; 54, Robert Alexander/Getty Images; 55, Traveller70/Shutterstock; 56, Dina Rudick/The Boston Globe via Getty Images; 57, Dominionart/Shutterstock; 58, John Gress Media Inc/Shutterstock; 59, Barry Chin/The Boston Globe via Getty Images; 60, Toms Auzins/Shutterstock; 61, Kamira/Shutterstock; 62, Sean Pavone/Shutterstock; 63, Bill Grant/Alamy Stock Photo; 64, Nino Marcutti/Alamy Stock Photo; 65, Matt Parry/robertharding; 66, Ed Hasler/robertharding; 68, Jim Nix/robertharding; 69, ELEPHOTOS/Shutterstock; 70, richardamora/Shutterstock; 71, Xinhua/Alamy Stock Photo; 72, Jordan Banks/robertharding; 74, Marco Bicci/Shutterstock; 75, Ovidiu Hrubaru/Shutterstock; 76, hillsn_1992/Shutterstock; 77, Luca Venturelli/Shutterstock; 78, Mladen Antonov/AFP via Getty Images; 79 (UP), Malcolm Schuyl/robertharding; 79 (LO), John de la Bastide/Shutterstock.

Latin America
80–1, Francisco Guasco/EFE/lafototeca; 82, Irina Brester/Alamy Stock Photo; 83, Lucas Vallecillos/Alamy Stock Photo; 84, McClatchy-Tribune/Alamy Stock Photo; 85, Walter Shintani Images/Offset; 86, MaxPhotoArt/Alamy Stock Photo; 87, eskystudio/Shutterstock; 88, Bisual Photo/Alamy Stock Photo; 89, Brian Overcast/Alamy Stock Photo; 90, Santiago Salinas/Shutterstock; 91, Ecuadorpostales/Shutterstock; 92, Angela Meier/Shutterstock; 93, Sergio Pitamitz/robertharding; 94, wtondossantos/Shutterstock; 95, bonandbon/Shutterstock; 96, Metin Celep/Shutterstock; 97, Antonio Salaverry/Shutterstock; 98, mdm7807/Shutterstock; 99, Karol Kozlowski/robertharding; 100, Cavan-Images/Shutterstock; 101, Iurii Dzivinskyi/Shutterstock; 102, reglain/Stockimo/Alamy Stock Photo; 103, Ronaldo Almeida/Shutterstock; 104, Tacito.fotografia/Shutterstock; 105, Luis War/Shutterstock; 106, Luiz Barrionuevo/Shutterstock; 107, art of line/Shutterstock; 108, Matthew Williams-Ellis/robertharding; 109, Yadid Levy/robertharding; 110, Ivo Antonie de Rooij/Shutterstock; 111, Location South/Alamy Stock Photo; 112, Vadim Zignaigo/Shutterstock; 113, Alexandre Fagundes De Fagundes/Dreamstime; 114, Karol Kozlowski/robertharding; 116, Lucas Vallecillos/robertharding; 117, Roberto Galan/Alamy Stock Photo; 118, Roberto Galan/Alamy Stock Photo; 119, Francesco Palermo/agefotostock/Alamy Stock Photo; 120, videobuzzing/Shutterstock; 121, marcosdominguez/Shutterstock; 122, Mark Green/Shutterstock; 124, Al Argueta/Alamy Stock Photo; 125 (UP), PixieMe/Shutterstock; 125 (LO), R. de Bruijn_Photography/Shutterstock.

Australia & Oceania

126–7, Ingus Kruklitis/Shutterstock; 128, Page Light Studios/Shutterstock; 129, Christian Mueller/Shutterstock; 130, Phillip B. Espinasse/Shutterstock; 131, Phillip B. Espinasse/Shutterstock; 132, Keitma/Alamy Stock Photo; 133, Richard Taylor/Sime/eStock Photo; 134, Mo Wu/Shutterstock; 135, Benny Marty/Shutterstock; 136, Jam Travels/Shutterstock; 137, James D. Morgan/Getty Images; 138, Saeed Khan/AFP via Getty Images; 140, Karind/Alamy Stock Photo; 141, Michael Williams/Dreamstime; 142, marcobrivio.photo/Shutterstock; 143, Julius Chang/Alamy Stock Photo; 144, TravelCollection/Image Professionals GmbH/Alamy Stock Photo; 145, David Wall/Alamy Stock Photo; 146, Olivier Goujon/robertharding; 147 (LE), Michael Runkel/imageBROKER/Alamy Stock Photo; 147 (RT), maloff/Shutterstock.

Africa

148–9, Gary Yeowell/Getty Images; 150, DEA/V. Giannella/Getty Images; 151, mbrand85/Shutterstock; 152, Martine Wallenborn/Shutterstock; 153, Grant Duncan-Smith/Shutterstock; 154, Andrea Armellin/Sime/eStock Photo; 156, ICP/incamerastock/Alamy Stock Photo; 157, Andrea Armellin/Sime/eStock Photo; 158, Rich T Photo/Shutterstock; 159, Chen Cheng/Xinhua/Alamy Live News; 160, Chen Cheng/Xinhua/Alamy Live News; 161, Jeffrey Greenberg/Universal Images Group via Getty Images; 162, Lyu Shuai/Xinhua/Alamy Live News; 163, Bildagentur-online/Universal Images Group via Getty Images; 164, Nick Brundle Photography/Shutterstock; 165, Joseph Sohm/Shutterstock; 166, Mohamed el-Shahed/AFP via Getty Images; 167, Khaled Desouki/AFP via Getty Images; 168, Mohamed el-Shahed/AFP via Getty Images; 169, FCerez/Shutterstock; 170, Giuseppe Cacace/AFP via Getty Images; 171, Emily Marie Wilson/Alamy Stock Photo; 172, Michael Runkel/robertharding; 173, Oguz Dikbakan/Shutterstock; 174, Balate Dorin/Shutterstock; 175, Matej Kastelic/Shutterstock; 176, Zouhair Lhaloui/Alamy Stock Photo; 178, Ppictures/Shutterstock; 179 (UP), OHudecek/Shutterstock; 179 (LO), Lucky-photographer/Shutterstock.

Middle East & Central Asia

180–1, hadynyah/Getty Images; 182, DedMityay/Shutterstock; 183, Captured Blinks/Shutterstock; 184, Rahmaniyas/Shutterstock; 185, Manto Konikkara/Shutterstock; 186, Finn stock/Shutterstock; 187, Sea Salt/Shutterstock; 188, Ludovic Marin/AFP via Getty Images; 189, Balate Dorin/Shutterstock; 190, DyziO/Shutterstock; 191, Diy13/Getty Images; 192, Massimo Ripani/Sime/eStock Photo; 193, Ruslan Gilmanshin/Alamy Stock Photo; 194, dragoncello/Alamy Stock Photo; 196, Jacobs photography/Shutterstock; 197, James Kerwin/robertharding; 198, Oleksandr Rupeta/Alamy Stock Photo; 199, Dynamoland/Alamy Stock Photo; 200, Ilia Torlin/Dreamstime; 201, Larisa Dmitrieva/Alamy Stock Photo; 202, Jack Guez/AFP via Getty Images; 204, Dmitriy Feldman svarshik/Shutterstock; 205, mbrand85/Shutterstock; 206, trabantos/Shutterstock; 207, Roman Sigaev/Shutterstock; 208, Mohamed Abdelrazek/Alamy Stock Photo; 209 (UP), Fedor Selivanov/Shutterstock; 209 (LO), Frank Fell/robertharding.

Eastern & Southern Asia

210–1, Damien Douxchamps/robertharding; 212, Sergey Bogomyako/Shutterstock; 213, marcociannarel/Shutterstock; 214, Sergey Pozhoga/Shutterstock; 215, Travel Wild/Getty Images; 216, travellinglight/Alamy Stock Photo; 217, Michihiko Kanegae/robertharding; 218, Uino/Shutterstock; 220, f11photo/Shutterstock; 221, David Guttenfelder/National Geographic Image Collection; 222, Galih Yoga Wicaksono/Shutterstock; 223, Wolfgang Kaehler/LightRocket via Getty Images; 224, i viewfinder/Shutterstock; 225, Jan Wlodarczyk/Alamy Stock Photo; 226, Frank Fell/robertharding; 227, dowell/Getty Images; 228, Markus Mainka/Shutterstock; 229, Frank Fell/robertharding; 230, nantonov/Getty Images; 231, GG6369/Shutterstock; 232, Michael McInally/Eclipse Sportswire/CSM/Alamy Live News; 233, Zen S Prarom/Shutterstock; 234, Julien Jean Zayatz/Shutterstock; 235, Rob Atherton/Shutterstock; 236, Yooran Park/Dreamstime; 237, Artur Widak/NurPhoto via Getty Images; 238, Valery Sharifulin/TASS/Alamy Stock Photo; 239, Nhut Minh Ho/Shutterstock; 240, Peerapong Bovornlohachai/Shutterstock; 241, Sean Pavone/Shutterstock; 242, Toms Auzins/robertharding; 243, Fabrizio Troiani /Alamy Stock Photo; 244, TY Lim/Shutterstock; 245, neo2620/Shutterstock; 246, leonovo/Alamy Stock Photo; 247, Tjetjep Rustandi/Alamy Stock Photo; 248, Tim Whitby/Alamy Stock Photo; 249, Matyas Rehak/Shutter-

Europe

INDEX

Boldface indicates illustrations.

A

Abu Dhabi, United Arab Emirates 184

Addis Ababa, Ethiopia **162,** 162–163, **163**

Agra, India **269,** 270–271

Alexandria, Egypt **170,** 171

Algiers, Algeria 172–173, **173**

Amman, Jordan **204,** 204–205, **205**

Amsterdam, Netherlands 362–365
  Amsterdam Music Project 365
  canal system **363,** 363–364
  De Poezenboot (The Catboat) 365
  gatherings 363
  hidden treasures 365
  Micropia 365
  museums **362,** 363, 364, 365, **365**
  Nieuwe Spiegelstraat 365
  Red Light District 363
  Rembrandt sites 364
  Royal Palace 362
  Vondelpark **364**

Andalusia, Spain
  Córdoba **352,** 352–353, **353,** 355
  Granada **354, 355,** 355–356
  Sevilla 355, **356,** 356–357, **357**

Angkor, Cambodia 243

Antananarivo, Madagascar 178–179, **179**

Antigua, Guatemala 124, **124**

Archaeology see Urban relics

Argentina see Buenos Aires, Argentina

Art festivals see Gatherings

Artbeat
  Athens, Greece 296
  Auckland, New Zealand 143
  Berlin, Germany 327
  Buenos Aires, Argentina 99
  Chicago, Illinois, U.S.A. 33
  Copenhagen, Denmark 341
  Dublin, Ireland 383
  London, England, United Kingdom 309
  Mexico City, Mexico 83
  Mumbai, India 267
  New Orleans, Louisiana, U.S.A. 37
  Seattle, Washington, U.S.A. 31
  Singapore 251
  Stockholm, Sweden 291
  Toronto, Ontario, Canada 45
  Vienna, Austria 322

Aswan, Egypt 164–165, **165**

Athens, Greece 294–297
  Acropolis 294–295, **295, 297**
  Agora 295
  artbeat 296
  Ermou Avenue **296,** 297
  Filopappou Hill 294
  Hellenic Parliament **294,** 297
  Kolonaki district 294
  Lycabettus Hill 294
  Monastiraki area 295–296
  museums 294, 295, 296, 297, **297**
  National Garden 296–297
  Parthenon ruins, Acropolis 294, **295**
  Plaka area 296
  Sparta 295

Auckland, New Zealand 142–145
  Aotea Square 144
  artbeat 143
  bungee jumping 144, **145**
  central business district (CBD) **143,** 144
  Karekare Beach **144**
  lay your head 145
  museums 142, **142,** 144–145
  North Island day trips 144
  parks and reserves 143, 144
  Sky Tower **143,** 144
  waterfront 142–143

Australia
  Canberra 137
  Melbourne **132,** 132–135, **133, 134, 135**
  Sydney **126–127, 136,** 136–141, **137, 138, 140, 141**

Austria
  Innsbruck 323, **323**
  Vienna **320,** 320–323, **321, 322**

B

Background check (history and culture)
  Algiers, Algeria 173
  Beijing, China 226
  Boston, Massachusetts, U.S.A. 57
  Cape Town, South Africa 153
  Dubai, United Arab Emirates 183
  Hong Kong, China 233
  Istanbul, Turkey 193
  Johannesburg, South Africa 161
  Las Vegas, Nevada, U.S.A. 52
  London, England, United Kingdom 311
  Melbourne, Australia 133
  Nashville & Memphis, Tennessee, U.S.A. 17
  New York, New York, U.S.A. 65
  Paris, France 285
  Rio de Janeiro, Brazil 105
  San Francisco, California, U.S.A. 71
  Tokyo, Japan 219
  Vienna, Austria 321
  Zagreb, Croatia 371

Bagan, Myanmar 249

Ballestas Islands National Reserves, Peru 123

Baltic cities
  Riga, Latvia 384–385, **385**
  Tallinn, Estonia 385, **385**
  Vilnius, Lithuania 384, **384**

Baltimore, Maryland, U.S.A. 15

Bangkok, Thailand 240–245
  Angkor 243
  BTS Skytrain 245
  Chang Chui Plane Night Market 245
  Chao Phraya River 240, 243
  Chinatown 243
  floating markets **243,** 244, 245
  gatherings 241
  Grand Palace 240
  lay your head 244
  Lumphini Park 240, **240**
  markets 241, 243, **243, 244,** 244–245
  Muay Thai (Thai boxing) 243, 244
  museums 240, 241, 243, 244, **245**
  Papaya Studio 245
  Royal City (Phra Nakhon) 240–241, **241,** 243
  Samut Prakan Ancient City 245
  Sanam Luang royal parade ground 241
  Siam area 244
  Silom Road skyscrapers 243–244
  temples 240–241, **241, 242,** 243
  Wat Pho 240–241
  Wat Traimit 243

Barcelona, Spain 304–307
  Barri Gòtic 304–305, 306, **307**
  El Born neighborhood 306–307
  El Raval quarter 306
  El Recinte Modernista de Sant Pau 305
  fútbol (soccer) sites **304,** 305–306
  gatherings 307
  Gaudí sites 304, **305, 306**
  hidden treasures 305
  La Rambla 307
  Miró sites 304–305
  MUHBA Turó de la Rovira 305
  Museu de la Xocolata 305
  museums 304, 305, 306, 307
  Parc del Laberint d'Horta 305
  Plaça de Catalunya 306

Beijing, China 224–227
  background check 226
  central business district (CBD) **224,** 226
  Forbidden City 224
  gatherings 225

Great Wall of China 225, 227, **227**
hutong (alleys) 226
museums 224, 225, 226, 227
Olympic Village 227
performing arts 225–226
798 Art District 227
Summer Palace 226, **226**
Temple of Heaven 226
Tiananmen Square 225, **225**
Belfast, Northern Ireland, United
    Kingdom 381
Belgium **280–281, 366,** 366–367,
    367
Berlin, Germany 324–329
    artbeat 327
    Berlin Wall 67, 324, 327
    Brandenburg Gate 324, **325**
    Checkpoint Charlie 327
    East Germany sites 327
    gatherings 328
    Holocaust Memorial 327, **328**
    Jewish Berlin 328–329
    museums 325, **326,** 327, 329,
        329
    Palaces and Parks of Potsdam 329
    Potsdamer Platz 327–328
    Sanssouci 329
    Scheunenviertel 328–329
    Spandauer Vorstadt 328–329
    Tiergarten 328
    Unter den Linden 324–325
    World War II sites 327, 329
Bhutan
    Paro 265, **265**
    Thimphu **264,** 264–265
Bloemfontein, South Africa 156
Books *see* Artbeat
Boston, Massachusetts, U.S.A. 56–59
    Back Bay 57, 58
    background check 57
    Beacon Hill 57–58
    Black Heritage Trail 57–58
    Cambridge 58
    Charlestown 58–59
    downtown **58**
    Fenway Park 58
    First Church of Christ, Scientist 58
    Freedom Trail 56–57
    lay your head 59
    Lexington and Concord 59
    museums **57,** 58, 59, **59**
    North End 56–57
    North Shore 59
    Public Garden **56,** 57
    South Boston 59
Brazil
    Rio de Janeiro **100,** 100–105, **101,**
        102, 103, 104, 105
    Salvador **94,** 94–95, **95**
    São Paulo 101

British Columbia, Canada
    Vancouver **38,** 38–39, **39**
    Victoria 39
Bruges, Belgium **280–281, 366,** 366–
    367, **367**
Budapest, Hungary **318,** 318–319, **319**
Buenos Aires, Argentina 96–99
    artbeat 99
    Avenida Alvear 99
    Calle Florida pedestrian zone 97
    local flavor 97
    museums 96, 97, 98
    Obelisco **97,** 97–98
    Plaza de Mayo **96,** 96–97
    Plaza General San Martín 97
    Puerto Madero 98
    Recoleta Cemetery 99, **99**
    Recoleta neighborhood 99
    Reserva Ecológica Costanera Sur 98–99
    soccer 99
    tango **96,** 99
    Teatro Colón 98, **98**
Bulgaria **368,** 368–369, **369**

C

Caesarea National Park, Israel 201
Cairo, Egypt 166–171
    Cairo Tower **166,** 170
    Citadel 170–171
    Coptic Quarter 171
    downtown Cairo 169–170
    Gezira 170
    lay your head 169
    local flavor 167
    markets **148–149,** 171
    mosques **169,** 170, 171
    museums 166, 167, 169–170, 171, **171**
    New Cairo City 171
    Pyramids of Giza 166–167, **167**
    Sphinx 167, **167**
    Step Pyramid of Saqqara **168,** 169
    Tahrir Square 169–170
California, U.S.A.
    Los Angeles **20,** 20–25, **21, 22, 23,**
        24, 25
    San Diego 22
    San Francisco **70,** 70–75, **71, 72, 74,**
        75
Cambodia 243
Canada
    Montréal, Québec **40,** 40–43, **41, 42,**
        43
    Québec, Québec **46,** 46–49, **47, 48**
    Toronto, Ontario **44,** 44–45, **45**
    Vancouver, British Columbia **38,**
        38–39, **39**
    Victoria, British Columbia 39
Canberra, Australia 137
Cape Town, South Africa 152–157
    background check 153

Blaauwberg Nature Reserve 155
Bo-Kaap neighborhood **152,** 155
Boulders Beach penguins **156,** 157
Camissa Township Tours 155
Cape Peninsula 155–157, **156**
Cape Point lighthouse **154,** 156
Castle of Good Hope 153
City Bowl 153
Coffee Beans Routes (tour company)
    155
guided tours 155
hidden treasures 155
Kirstenbosch National Botanical
    Garden 157, **157**
Mandela sites 152–153, 155
museums 152, 153, 155
Norval Foundation 155
Robben Island 152–153
shopping 155
Signal Hill 155
Table Mountain 152, **153**
Tygerberg Nature Reserve 155
Victoria & Albert Waterfront 152
vineyards 157
west coast 155
Woodstock 157
Caribbean cities 78–79
    Georgetown, Guyana 79, **79**
    Kingston, Jamaica **78,** 78–79
    Port of Spain, Trinidad and Tobago
        79, **79**
Casablanca, Morocco 177
Chan Chan, Peru 121
Chengdu, China 278, **278**
Chicago, Illinois, U.S.A. 32–35
    artbeat 33
    Chicago River 32, **34**
    Des Plaines River Trail 35
    Forest Preserves of Cook County 35
    Garfield Park Conservatory 35
    hidden treasures 35
    Indiana Dunes 35
    Lake Shore Drive 32, 35
    Michigan Avenue 32–33, 35
    Miracle Mile 32–33
    museums **32,** 33, 35
    parks 33, **33,** 35
    skyscrapers 32–33, **34**
Chile *see* Santiago, Chile
China
    Beijing **224,** 224–227, **225, 226, 227**
    Chengdu 278, **278**
    Guilin 278–279, **279**
    Hangzhou 231
    Hong Kong **232,** 232–235, **233,**
        234, 235
    Shanghai **228,** 228–231, **229, 230, 231**
    Xi'an 279, **279**
Copenhagen, Denmark **340,** 340–341,
    341

Córdoba, Spain **352,** 352–353, **353, 355**
Croatia
Dubrovnik **372,** 372–373, **373**
Zagreb **370,** 370–372, **371**
Cuba
Havana **60,** 60–61, **61**
Santiago 61
Cusco, Peru 112–115
Cabildo del Cusco 112
gatherings 115
Iglesia de Santo Domingo 115
Inca foundations 115
La Merced 112–113
Machu Picchu 112, 113, **114**
markets **112**
museums 112
Plaza de Armas 112–113, **113,** 115
Sacred Valley day trips 115
San Blas neighborhood 115
Cyprus **206,** 206–207, **207**
Czechia **292,** 292–293, **293**

D

Delhi, India 268–270
Agrasen Ki Baoli **271**
Chandni Chowk Spice Market **268,** 269
Gandhi sites 269
Jama Masjid 268–269
local flavor 270
Lotus Temple **270**
Mehrauli Archaeological Park 270
museums 269, 270
Qutb Minar Complex 270
Rajpath 270
Rashtrapati Bhavan 270
Red Fort 268–269
Swaminarayan Akshardham 270
Denmark **340,** 340–341, **341**
District of Columbia see Washington, D.C., U.S.A.
Doha, Qatar **2–3,** 208, **208**
Dubai, United Arab Emirates 182–185
Al Fahidi Historical District 184–185
aquariums 182, **182,** 183
Atlantis Dubai resort **182,** 183
background check 183
beach zone 185
Burj Khalifa 182
Dubai Creek 184
Dubai Frame 183, **183**
Green Planet bio-dome 183–184
markets 184
museums 183, 184, 185
Palm Jumeirah archipelago 182–183
Rub' al Khal (Empty Quarter) 185
Sheikh Zayed Grand Mosque **185**
Shindagha Historic District 185

Ski Dubai 183
theme parks 185
water taxis and cruises 184, **184**
Dublin, Ireland 380–383
artbeat 133
Georgian Dublin 383
lay your head 383
literary Dublin **382,** 383
museums 380, 381, 383
musical Dublin 380–381, 383
River Liffey **381**
Temple Bar **380,** 380–381
Trinity College **382,** 383
Dubrovnik, Croatia **372,** 372–373, **373**

E

Ecuador see Quito, Ecuador
Edinburgh, Scotland, United Kingdom 374–377
Calton Hill **375,** 377
Edinburgh Castle **374,** 375–376
Elephant House café **376,** 376–377
gatherings 376
local flavor 375
museums 374, 376, 377
Palace of Holyroodhouse 374–375
parks and gardens 377
Royal Mile 376–377, **377**
royal yachts 375
Scottish Parliament Building 374
Egypt
Alexandria **170,** 171
Aswan 164–165, **165**
Cairo **148–149, 166,** 166–171, **167, 168, 169, 171**
Luxor **164,** 164–165, **165**
Pyramids of Giza 166–167, **167**
England see London, England, United Kingdom
Ephesus, Turkey 195
Estonia 385, **385**
Ethiopia
Addis Ababa **162,** 162–163, **163**
Lalibela 163
Events see Gatherings
Évora, Portugal 349

F

Fatehpur Sikri, India 269
Festivals see Gatherings
Fiji 146–147, **147**
Finland **358,** 358–359, **359**
Florence, Italy **4,** 316, 316–317, **317**
Food see Gatherings; Local flavor
France
Paris **284,** 284–289, **285, 286, 288, 289**
Versailles 289

Frankfurt am Main, Germany **336,** 336–338, **337**

G

Gatherings (festivals and events)
Amsterdam, Netherlands 353
Bangkok, Thailand 241
Barcelona, Spain 307
Beijing, China 225
Berlin, Germany 328
Bruges, Belgium 367
Budapest, Hungary 319
Córdoba, Spain 355
Cusco, Peru 115
Edinburgh, Scotland, United Kingdom 376
Granada, Spain 355
Honolulu, Hawaii, U.S.A. 129
Jaipur, India 273
Las Vegas, Nevada, U.S.A. 51
Los Angeles, California, U.S.A. 21
Melbourne, Australia 134
Montevideo, Uruguay 111
Montréal, Québec, Canada 43
Munich, Germany 315
Sevilla, Spain 355
Stuttgart, Germany 339
Tokyo, Japan 221
Zagreb & Dubrovnik, Croatia 373
Geneva, Switzerland **282,** 282–283, **283**
Georgetown, Guyana 79, **79**
Georgia (republic) **196,** 196–197, **197**
Germany
Berlin 324–329, **325, 326, 328, 329**
Frankfurt am Main **336,** 336–338, **337**
Hamburg **324,** 325
Munich **314,** 314–315, **315**
Stuttgart 338–339, **339**
Glasgow, Scotland, United Kingdom 377
Granada, Spain **354, 355,** 355–356
Great Zimbabwe, Zimbabwe 159
Greece
Athens **294,** 294–297, **295, 296, 297**
Sparta 295
Thessaloniki 297
Guadalajara, Mexico 116–119
Barranca de Oblatos-Huentitán 117
baseball 118
Calle Independencia **118**
Centro Histórico 119
folk dance **80–81,** 116
hidden treasures 117
local flavor **118,** 119, **119**
lucha libre (wrestling) 118
mariachi music 116, 117
markets and shops 118–119, **119**
museums 117–118, 119
Panteón de Belén 117

Plaza de Armas **117**
rodeo 116–117
soccer 118
tequila **116,** 117–118
Tlaquepaque 118
Zona Rosa 119
Guatemala 124, **124**
Guilin, China 278–279, **279**
Guyana 79, **79**
Gyeongju, South Korea 259

H

Haifa, Israel 200–203
Báb shrine **202**
Bat Galim 201
Caesarea National Park 201
Druze settlements 203
German Colony neighborhood 201
Glashanim Beach **200**
Hadar 200–201
lay your head 203
Mount Carmel **201,** 203
museums 200–201, 203
Wadi Nisnas 200
Hamburg, Germany **324,** 325
Hangzhou, China 231
Hanoi, Vietnam **238,** 238–239
Havana, Cuba **60,** 60–61, **61**
Hawaii, U.S.A.
Hilo 131
Honolulu **128,** 128–131, **129, 130, 131**
Helsinki, Finland **358,** 358–359, **359**
Hidden treasures
Amsterdam, Netherlands 365
Barcelona, Spain 305
Cape Town, South Africa 155
Chicago, Illinois, U.S.A. 35
Guadalajara, Mexico 117
Helsinki, Finland 359
Hong Kong, China 234
Kathmandu, Nepal 263
Kuala Lumpur, Malaysia 277
Kyoto, Japan 215
Los Angeles, California, U.S.A. 25
Marrakech, Morocco 175
New York, New York, U.S.A. 67
Nicosia, Cyprus 207
Paris, France 287
Philadelphia, Pennsylvania, U.S.A. 27
Prague, Czechia 293
Reykjavik, Iceland 361
Samarkand, Uzbekistan 199
Tbilisi, Georgia 197
Tokyo, Japan 217
Ulaanbaatar, Mongolia 257
Venice, Italy 331
Hilo, Hawaii, U.S.A. 131

History *see* Background check; Urban relics
Ho Chi Minh City, Vietnam 238–239, **239**
Hollywood *see* Los Angeles, California, U.S.A.
Hong Kong, China 232–235
background check 233
Central District 232
Happy Valley Racecourse **232,** 234
hidden treasures 234
Hong Kong Island **233,** 234
Hong Kong Wetland Park 234
International Commerce Centre 234
Kowloon 234–235
Lantau island 235
MacLehose Trail 234
markets 235, **235**
museums 232, 234, 235
parks and trails 234
Ping Shan Heritage Trail 234
Sheung Wan neighborhood 233
Ten Thousand Buddhas Monastery 234, **234**
transportation 232–233, 234, 235
UNESCO Global Geopark 234
Wan Chai 233–234
waterfront walkway 233
West Kowloon Cultural District 234
Honolulu, Hawaii, U.S.A. 128–131
Ala Moana Bowls surf spot **130**
Arts District 131
Chinatown 131
Diamond Head **129,** 130
gatherings 129
'Iolani Palace 128
museums 128–129, 130, **131**
Pearl Harbor Visitor Center 130
Queen Kapi'olani Regional Park 130
road trips 131
U.S.S. *Arizona* Memorial **128,** 130
Waikiki Beach 129–130
Hotels *see* Lay your head
Huế, Vietnam 239
Hungary **318,** 318–319, **319**

I

Iceland **360,** 360–361, **361**
Illinois *see* Chicago, Illinois, U.S.A.
India
Agra **269,** 270–271
Delhi **268,** 268–270, **270, 271**
Fatehpur Sikri 269
Jaipur **272,** 272–273, **273**
Mumbai **266,** 266–267, **267**
Indian Ocean cities
Antananarivo, Madagascar 178–179, **179**
Port Louis, Mauritius 178, **178**
Victoria, Seychelles 179, **179**

Indiana Dunes, Indiana, U.S.A. 35
Indonesia
Jakarta 223
Yogyakarta **222,** 222–223, **223**
Innsbruck, Austria 323, **323**
Intangible Cultural Heritage *see* UNESCO Intangible Cultural Heritage
Ireland *see* Dublin, Ireland
Israel
Caesarea National Park 201
Haifa **200,** 200–203, **201, 202**
Jerusalem **186,** 186–191, **187, 188, 190, 191**
Tel Aviv 189, **189**
Istanbul, Turkey 192–195
background check 193
Beyoğlu 195
Blue Mosque **180–181,** 195
Bosporus ferries 192–193, **193**
Hagia Sophia 193, **194,** 195
Levent and Maslak districts 195
museums 195
Sufi Mevlevi Order **192**
Suleymaniye Mosque **193**
Sultanahmet 193, 195
Topkapi Palace 195
Troy 195
Italy
Florence **4, 316,** 316–317, **317**
Pompeii 343
Rome **342,** 342–347, **343, 344, 346, 347**
Venice **330,** 330–333, **331, 332, 333**

J

Jaipur, India **272,** 272–273, **273**
Jakarta, Indonesia 223
Jamaica **78,** 78–79
Japan
Kyoto **210–211, 212,** 212–215, **213, 214, 215**
Nara **212,** 213
Tokyo **216,** 216–221, **217, 218, 220, 221**
Jerusalem, Israel 186–191
Arab Souk 186–187
Armenian Quarter 186, 189–190
Christian Quarter 186
Church of the Holy Sepulchre 187, **190**
Damascus Gate 186
Dome of the Rock 186, **187,** 189
downtown Jerusalem 191
Givat Ram neighborhood 191
Jaffa Road 191
Jewish Quarter 186, 189
Kidron Valley 190
lay your head 191
local flavor 187

Montefiore windmill 191, **191**
museums 189–191
Muslim Quarter 186, 187
Old City 186–187, 189
ramparts 186–187
Rehavia-Talbiya area 191
shops and markets **186,** 186–187
Temple Mount 189
Via Dolorosa (Stations of the Cross)
  187, 189
Western Wall **187, 188,** 189
Johannesburg, South Africa 158–161
apartheid history **160,** 160–161
background 161
Constitutional Court 158–159
gold-rush history 158, 159–160
Great Zimbabwe 159
Maboneng neighborhood 161
Mandela sites 158, **159,** 160
markets **158,** 161
museums 160, **160,** 161
Old Fort 158
Parktown neighborhood 159
Soweto 160–161
Sterkfontein Caves 160, **161**
Jordan
Amman **204,** 204–205, **205**
Petra 205

K

Kansas City, Missouri, U.S.A. 62–63,
  63
Kathmandu, Nepal **262,** 262–263, **263**
Kenya **150,** 150–151, **151**
Kingston, Jamaica **78,** 78–79
Kraków, Poland **334,** 334–335, **335**
Kuala Lumpur, Malaysia 274–277
architecture 274, **274,** 276, **277**
Batu Caves **275,** 277
Bukit Bintang neighborhood
  276–277
Butterfly Park 276, **276**
Dataran Merdeka 274
hidden treasures 277
Kanching Rainforest Waterfall 277
KL Forest Eco Park 277
Krau Wildlife Reserve 277
lay your head 276
Malacca 275
markets 274, 275
museums 276
parks and gardens 275–276, 277
Petronas Twin Towers 276
Taman Tugu 277
tin history 274, 277
Kyoto, Japan 212–215
Gion district 212
hidden treasures 215
Imperial Palace 214
local flavor 214

markets 215
museums 215
Nara **212,** 213
Nijō Castle 214
parks and gardens 213–214, **214**
temples **210–211,** 212–214, **213,**
  215, **215**
workshops and classes 215
Yokai (Monster) Street 215

L

Lalibela, Ethiopia 163
Laos **246,** 246–247, **247**
Las Vegas, Nevada, U.S.A. 50–53
background check 51
casino-hotels 50–52, **51**
day trips 52–53
Elvis Presley sites 50–51
Fremont Street 52
gatherings 51
museums 50, 51, **52**
Red Rock Canyon National
  Conservation Area 52–53, **53**
Vegas Strip **50, 51,** 51–52
Latin American cities
Antigua, Guatemala 124, **124**
León, Nicaragua 124–125
Paramaribo, Suriname 125
Latvia 384–385, **385**
Lay your head (unique lodgings)
Auckland, New Zealand 145
Bangkok, Thailand 244
Boston, Massachusetts, U.S.A. 59
Cairo, Egypt 169
Dublin, Ireland 383
Florence, Italy 317
Frankfurt am Main, Germany 337
Haifa, Israel 203
Jerusalem, Israel 191
Kuala Lumpur, Malaysia 276
Lima, Peru 123
Luxor & Aswan, Egypt 165
Mexico City, Mexico 84
Montréal, Québec, Canada 41
Nashville & Memphis, Tennessee,
  U.S.A. 19
Québec, Québec, Canada 47
Quito, Ecuador 91
Rome, Italy 346
Salvador, Brazil 95
San Francisco, California, U.S.A. 74
Santiago, Chile 108
Shanghai, China 229
Singapore 255
Sydney, Australia 140
León, Nicaragua 124–125, **125**
Lima, Peru 120–123
Chan Chan 121
churches 120–121
day trips 123

lay your head 123
local flavor 120, 121, 123
Malecón 121, 123
Miraflores neighborhood 121
museums 120, **120,** 121, 123
Plaza de Armas 120, **121, 122**
Lisbon, Portugal 348–351
Alfama district 348
Bairro Alto 350
Baixa district 349
Elevador de Santa Justa 349–350
Évora 349
fado music **348,** 348–349
local flavor 348, 351
museums 349, 350–351, **351**
São Jorge Castle **349**
Sé de Lisboa Cathedral 348, **350**
Tram 28 351
Lithuania 384, **384**
Local flavor (restaurants)
Buenos Aires, Argentina 97
Cairo, Egypt 167
Delhi, India 270
Edinburgh, Scotland, United
  Kingdom 375
Guadalajara, Mexico 119
Jerusalem, Israel 187
Kyoto, Japan 214
Lima, Peru 123
Lisbon, Portugal 351
London, England, United Kingdom
  313
Luang Prabang, Laos 247
Madrid, Spain 301
Melbourne, Australia 135
Montréal, Québec, Canada 42
Porto, Portugal 379
Québec, Québec, Canada 49
Quito, Ecuador 92
Rio de Janeiro, Brazil 103
Rome, Italy 345
San Francisco, California, U.S.A. 73
San Juan, Puerto Rico, U.S.A. 77
Santiago, Chile 107
Seoul, South Korea 260
Singapore 253
Sofia, Bulgaria 369
Sydney, Australia 139
Tokyo, Japan 217
Venice, Italy 333
Lodgings *see* Lay your head
London, England, United Kingdom
  308–313
artbeat 309
background check 310
Buckingham Palace **308,** 311
Bushy Park 308
Chelsea 311
Docklands 312–313
East End 312

Eel Pie Island 309
Greenwich 313
local flavor 313
museums 312, **312**, 313, **313**
Richmond-upon-Thames 309
River Thames 308–309, 311, 312–313
Royal Botanic Gardens, Kew 311
Tower of London **309,** 312
Trafalgar Square 311
Twickenham 308–309
Westminster Abbey **310,** 311
Westminster Pier 311
Longyearbyen, Norway 299
Los Angeles, California, U.S.A. 20–25
Bel Air 20
Beverly Hills 20, 23
California Institute of Abnormal Arts 25
Colorado Boulevard 24–25
downtown 22
gatherings 21
Getty Center 20
Griffith Park **21,** 25
hidden treasures 25
Hollywood Boulevard 21
Hollywood Forever Cemetery 25
Hollywood Sign **24,** 25
Hollywood Walk of Fame 21, **22**
Long Beach 24
Los Angeles County Museum of Art **23**
Malibu 20, 24
Miracle Mile 23
museums 22, 23, 24, 25
Pacific Coast Highway (PCH) 24
Pasadena 25
Phantasma Gloria at RandylandLA 25
Santa Monica 23–24
Sunset Boulevard 20–22
UCLA 20
underground tunnels 25
Venice Beach 23–24, **25**
Walt Disney Concert Hall **20,** 22
Wilshire Boulevard 22–23
Louisiana, U.S.A. **36,** 36–37, **37**
Luang Prabang, Laos **246,** 246–247, **247**
Luxor, Egypt **164,** 164–165, **165**

M
Machu Picchu, Peru 113, **114**
Madagascar 178–179, **179**
Madrid, Spain 300–303
art 302, 303
Barrio de Las Letras (Literary Quarter) 303
Cibeles Fountain **301**
Gran Via 301–302
local flavor 300, **300,** 301, **302**

markets 303
museums 302
Palacio Real 300
parks and gardens 300, 303
Plaza de España 300–301
Plaza Mayor 300
Retiro Park **303**
soccer 303
theaters 302
Toledo 302
Malacca, Malaysia 275
Malaysia
Kuala Lumpur **274,** 274–277, **275, 276, 277**
Malacca 275
Manila, Philippines **236,** 236–237, **237**
Map 8–9
Marrakech, Morocco 174–177
Bab Debbagh 175
Cactus Thiemann 175
gardens **174,** 175, 177
Gueliz neighborhood 177
Hassan II Mosque **176**
hidden treasures 175
Hivernage neighborhood 177
Jewish community 175
Koutoubia Mosque 174
markets 174, 175, **175,** 177
Medersa Ben Youssef 177
Medina Quarter 174, 175, **175**
Musée de la Palmeraie 175
museums 175, 177
Palmeraie 177
royal landmarks 174–175
Maryland, U.S.A. 15
Massachusetts *see* Boston, Massachusetts, U.S.A.
Mauritius 178, **178**
Melbourne, Australia 132–135
background check 133
Carlton Gardens 135
Docklands 134
ethnic neighborhoods 132
Federation Square 133, **133**
gatherings 135
Healesville Sanctuary 135
local flavor 134
Melbourne Zoo 135
museums 132, 133, 134, 135
Phillip Island wildlife 135, **135**
Queen Victoria Market 132, **132**
Royal Botanic Gardens Victoria 135
Southbank 134
St. Kilda Beach **134,** 135
Victorian structures 135
Memphis, Tennessee, U.S.A. 16, **18,** 18–19, **19**
Mexico
Guadalajara **80–81,** 116–119, **117, 118, 119**

Mexico City **82,** 82–87, **83, 84, 85, 86, 87**
Monte Albán 89
Oaxaca **88,** 88–89, **89**
Teotihuacan 87
Tequila **116,** 117–118
Mexico City, Mexico 82–87
Alameda Central 85
artbeat 83
Avenida Madero 84
Casa de la Maniche 82
Centro Histórico 84
Chapultepec Park 85, 87
Coyoacán neighborhood 82, **87**
food scene **84**
lay your head 84
museums **82,** 82–85, **83,** 87
Palacio Nacional de Mexico 84
Paseo de la Reforma 85
Polanco 87
San Ángel 83
Tenochtitlan 83–84
Torre Latinoamericana 84
trajineras (boats) **86,** 87
Zócalo plaza 84, **85**
Zona Rosa 85
Middle East cities
Abu Dhabi, United Arab Emirates 184
Doha, Qatar **2–3,** 208, **208**
Dubai, United Arab Emirates **182,** 182–185, **183, 184, 185**
Muscat, Oman 209, **209**
Riyadh, Saudi Arabia 208–209, **209**
Missouri, U.S.A.
Kansas City 62–63, **63**
St. Louis **62,** 62–63
Mongolia **256,** 256–257, **257**
Monte Albán, Mexico 89
Montevideo, Uruguay **110,** 110–111, **111**
Montréal, Québec, Canada 40–43
Boulevard Saint-Laurent 40–42
gatherings 43
Jean-Talon Market 42
La Grande Roue Ferris wheel 40, **41**
lay your head 41
local flavor 42, **43**
Mile End 41–42
Mont-Royal 42–43
museums 40, 43
Notre-Dame Basilica 40, **42**
Parc Jean-Drapeau 43
Vieux-Montréal (Old Montréal) **40,** 40–41
Morocco
Casablanca 177
Marrakech **174,** 174–177, **175, 176**
Movies *see* Artbeat
Mumbai, India **266,** 266–267, **267**

Munich, Germany **314,** 314–315, **315**
Muscat, Oman 209, **209**
Museums *see* Hidden treasures
Music *see* Artbeat; Gatherings
Myanmar
    Bagan 249
    Yangon **248,** 248–249, **249**

N

Nairobi, Kenya **150,** 150–151, **151**
Nara, Japan **212,** 213
Nashville, Tennessee, U.S.A. **16,** 16–18, **17**
Nepal **262,** 262–263, **263**
Netherlands *see* Amsterdam, Netherlands
Nevada, U.S.A.
    Las Vegas **50,** 50–53, **51, 52, 53**
    Reno 53
New Caledonia 147, **147**
New Mexico, U.S.A.
    Santa Fe **54,** 54–55, **55**
    Taos Pueblo 55
New Orleans, Louisiana, U.S.A. **36,** 36–37, **37**
New York, New York, U.S.A.
    background check 65
    Bronx 65
    Brooklyn 65, 67, 69
    Central Park 65
    Ground Zero Museum Workshop 67
    Harlem 68
    hidden treasures 67
    Hudson River Park **65**
    Mmuseumm 67
    museums **64,** 67, 68, 69, **69**
    9/11 Memorial and Museum **64,** 69
    One World Trade Center **65,** 69
    A Piece of the Berlin Wall 67
    Queens 65, 68, 69
    skyline **10–11**
    skyscrapers 67–68, 69
    Staten Island 65
    Statue of Liberty **6,** 67, 285
    Third Rail Projects 67
    Times Square **66**
    Triborough Bridge **10–11**
New Zealand
    Auckland **142,** 142–145, **143, 144, 145**
    Queenstown 145
Nicaragua 124–125
Nicosia, Cyprus **206,** 206–207, **207**
Northern Ireland, United Kingdom 381
Norway
    Longyearbyen 299
    Oslo **298,** 298–299, **299**
Nouméa, New Caledonia 147, **147**

O

Oaxaca, Mexico **88,** 88–89, **89**
Oklahoma City, Oklahoma, U.S.A. 63
Old Delhi, India **268,** 268–269
Oman 209, **209**
Ontario, Canada **44,** 44–45, **45**
Oregon, U.S.A. 29
Oslo, Norway **298,** 298–299, **299**

P

Papeete, Tahiti 146, **146**
Parades *see* Gatherings
Paramaribo, Suriname 125, **125**
Paris, France 284–289
    Arc de Triomphe 287–288
    background check 285
    Champs-Élysées 286–287
    Eiffel Tower **284,** 287
    hidden treasures 287
    Île de la Cité 284–285
    Jardin des Plantes 287
    La Défense, Île-de-France **288**
    La REcyclerie 287
    Le Mur des Je t'aime ("I Love You Wall") 287
    Louvre **285,** 288, **289**
    Montmartre 288
    Montparnasse 288–289
    Musée des Arts Forains 287
    museums 285, **285,** 287, 288, **289**
    Notre-Dame Cathedral 284
    Opéra Garnier **286**
    Palais de Justice 285
    Seine **284,** 289
    Tour Saint-Jacques 285
    Versailles, Palace of 289
Paro, Bhutan 265, **265**
Pennsylvania, U.S.A. **26,** 26–27, **27**
Peru
    Chan Chan 121
    Cusco **112,** 112–115, **113, 114**
    Lima **120,** 120–123, **121, 122**
    Machu Picchu 113
    national reserves 123
Petra, Jordan 205
Philadelphia, Pennsylvania, U.S.A. **26,** 26–27, **27**
Philippines **236,** 236–237, **237**
Poland
    Kraków **334,** 334–335, **335**
    Warsaw 335
Pompeii, Italy 343
Port Louis, Mauritius 178, **178**
Port of Spain, Trinidad and Tobago 79, **79**
Portland, Oregon, U.S.A. 29
Porto, Portugal **378,** 378–379, **379**
Portugal
    Évora 349

Lisbon **348,** 348–351, **349, 350, 351**
Porto **378,** 378–379, **379**
Prague, Czechia **292,** 292–293, **293**
Puerto Rico, U.S.A. **76,** 76–77, 77

Q

Qatar **2–3,** 208, **208**
Québec, Canada
    Montréal **40,** 40–43, **41, 42, 43**
    Québec **46,** 46–49, **47, 48**
Québec, Québec, Canada 46–49
    Basse-Ville (Lower Town) 46, 47, 49
    Fairmont Le Château Frontenac 47, **47**
    festivals 46
    Grand Allee 49
    Haute-Ville (Upper Town) 46–47
    Hôtel de Glace (Ice Hotel) 46, **46**
    lay your head 47
    local flavor 49
    murals 49
    museums 49
    Old Port 49
    performing arts 49
    Plains of Abraham 49
    Saint-Jean-Baptiste neighborhood 49
    Saint-Roch area 49
    seasonal activities 46
    shopping 49
    Umbrella Alley **48**
Queenstown, New Zealand 145, **145**
Quito, Ecuador 90–93
    Barrio Guápulo 93
    Barrio La Floresta 92
    Barrio Mariscal 93
    Basílica del Voto Nacional 91, **92**
    Casa de la Cultura Ecuatoriana (CCE) 91–92
    Cayambe Coca Ecological Reserve 90
    Centro Histórico 90, **90**
    chocolate makers 93
    Ciudad Mitad del Mundo 93
    Iglesia de la Compañía de Jesús 91
    Iglesia de San Francisco **93**
    Iglesia y Convento de San Francisco 91
    lay your head 91
    local flavor 92
    monument of the winged Virgin **91**
    museums 91–92
    parks 93
    Pichincha volcano 90, 93
    Plaza Grande 90

R

Reno, Nevada, U.S.A. 53
Restaurants *see* Local flavor
Reykjavik, Iceland **360,** 360–361, **361**
Riga, Latvia 384–385, **385**

Rio de Janeiro, Brazil 100–105
  background check 105
  Boulevard Olimpico **102,** 105
  Christ the Redeemer statue **101,** 103–104
  Copacabana Beach 100, 105, **105**
  Guanabara Bay 104, 105
  Ipanema beach 101, 103
  local flavor 103
  museums **103,** 104, 105
  Selarón Staircase **100,** 105
  stadiums 105
  Sugarloaf peaks 103, **104**
  Tijuca Forest National Park 104
Riyadh, Saudi Arabia 208–209, **209**
Rome, Italy 342–347
  lay your head 346
  local flavor 345, **346**
  museums 342, 345, 346, 347
  Pompeii 343
  Roman Forum ruins 345, **347**
  Spanish Steps **342,** 345
  Vatican City **343, 344,** 346–347

S

Salvador, Brazil **94,** 94–95, **95**
Samarkand, Uzbekistan **198,** 198–199, **199**
San Diego, California, U.S.A. 22
San Francisco, California, U.S.A. 70–75
  Alcatraz Island 71, **72,** 74
  background check 71
  cable cars **72,** 74
  California Academy of Sciences 75
  Castro District 74–75
  China Basin 70
  Chinatown **70,** 73
  Dogpatch 71
  downtown 74
  Fisherman's Wharf 74
  Golden Gate Bridge **71,** 73
  Golden Gate Park 71, **71,** 75
  lay your head 74
  local flavor 73
  Market Street 74–75
  Mission Bay 70–71
  Mission District 75
  museums 70, 71, 73, 74, 75
  Napa Valley **75**
  Naval Air Station Alameda 73
  North Beach 73
  Pier 39 sea lions **74**
  Presidio 71, 73
  Salesforce Tower 74
  San Francisco-Oakland Bay Bridge 73
  Treasure Island 73
San Juan, Puerto Rico, U.S.A. **76,** 76–77, **77**

Santa Fe, New Mexico, U.S.A. **54,** 54–55, **55**
Santiago, Chile 106–109
  Avenida Apoquindo 106–107
  Barrio Brasil 109
  Barrio el Golf neighborhood 106–107
  Barrio Yungay 109
  Bellavista 109
  Centro Historico 108–109
  Cerro San Cristóbal **108,** 109
  Cerro Santa Lucia 108
  Gran Torre Santiago (Costanera Center) 106
  Las Condes district 106
  Lastarria neighborhood 107–108
  lay your head 108
  local flavor 107
  museums 107, 108, 109
  National Zoo **106,** 109
  Neptune Fountain 108, **109**
  Parque Forestal 108
  Plaza de las Armas **107,** 108
  Rio Mapocho 107
  Vitacura neighborhood 106–107
Santiago, Cuba 61
São Paulo, Brazil 101
Saudi Arabia 208–209, **209**
Scotland, United Kingdom
  Edinburgh **374,** 374–377, **375, 376, 377**
  Glasgow 377
Seattle, Washington, U.S.A. 28–31
  Amazon Spheres **28,** 31
  artbeat 31
  Bainbridge Island 31
  Ballard Locks **30**
  Central District 29–30
  Chinatown Historic District 29
  day trips 30–31
  gold-rush history 29
  Lake Washington Ship Canal 30–31
  museums 28, 30, 31
  Pike Place Market 28, **31**
  Pioneer Square 28–29
  Seattle Center 28
  Space Needle 28, **29**
  waterfront 28, 30–31
Seoul, South Korea 258–261
  Bukchon 261
  Changdeokgung Palace 260
  Cheonggyecheon stream 261
  Deoksugung Palace 260
  Dongdaemun Design Plaza (DDP) 261
  Gangnam District **258,** 258–259
  Gyeongbokgung Palace **259,** 260
  Gyeongju 259
  Insadong 261
  Itaewon 259
  local flavor 260, 261

museums 258, 259, 261
  Myeong-dong district 260, **261**
  N Seoul Tower 260, **260**
  Namsan Mountain 260
  Namsangol Hanok Village 261
  Palace District 260–261
  sports scene 259
  Starfield COEX Mall 258, **258**
  templestay programs 259, 261
Sevilla, Spain 355, **356,** 356–357, **357**
Seychelles 179, **179**
Shanghai, China 228–231
  Bund waterfront area 228–230, **229**
  French Concession 230–231
  lay your head 229, 230
  Longjing village of Hangzhou **231**
  Lujiazui business district (Pudong) 228, **229**
  M50 Arts District 231
  museums 228, 230
  Nanjing Road East 230
  Old City district **230,** 231
  People's Park 230
  performing arts 230, 231
  Shanghai Disneyland Park 231
  Shanghai Transrapid maglev train 228, **228**
  Yuyuan Bazaar 231
  Yuyuan Garden **230,** 231
Singapore 250–255
  artbeat 251
  Chinatown 253
  churches 254
  Fort Canning Hill 254–255
  Helix Bridge 251, **251,** 253
  Jurong Bird Park **250**
  Kampong Glam 253
  lay your head 255
  Little India 253–254
  local flavor 253, **254**
  museums **251,** 254, **255**
  nature reserves 250
  Orchard Road 255
  outlying islands 250
  Singapore Zoo 251
  Supertree Grove 251, **252**
Sofia, Bulgaria **368,** 368–369, **369**
South Africa
  Bloemfontein 156
  Cape Town **152,** 152–157, **153, 154, 156, 157**
  Johannesburg **158,** 158–161, **159, 160, 161**
South Korea
  Gyeongju 259
  Seoul **258,** 258–261, **259, 260, 261**
South Pacific cities
  Nouméa, New Caledonia 147, **147**
  Papeete, Tahiti 146, **146**
  Suva, Fiji 146–147, **147**

Spain
  Barcelona **304,** 304–307, **305, 306, 307**
  Córdoba **352,** 352–353, **353,** 355
  Granada **354, 355,** 355–356
  Madrid **300,** 300–303, **301, 302, 303**
  Sevilla **356,** 356–357, **357**
  Toledo 302
Sparta, Greece 295
Sports events *see* Gatherings
St. Louis, Missouri, U.S.A. **62,** 62–63
Stockholm, Sweden **290,** 290–291, **291**
Stone Town, Zanzibar, Tanzania 151
Stuttgart, Germany **338,** 338–339, **339**
Suriname 125
Suva, Fiji 146–147, **147**
Sweden **290,** 290–291, **291**
Switzerland
  Geneva **282,** 282–283, **283**
  Zurich 283
Sydney, Australia 136–141
  Anzac Memorial, Hyde Park South 139, **140**
  beaches 136, **136**
  central business district (CBD) 140–141
  Circular Quay 139, 140
  Darling Harbour 141
  ferries and water taxis 139
  Hyde Park 139–140, **140**
  lay your head 140
  local flavor 139
  museums 139, 140, 141, **141**
  national parks 136, 139
  neighborhoods 141
  Opera House **126–127,** 137, **138,** 139
  The Rocks area 139
  Sydney Harbour Bridge 137, **137,** 139

T

Tahiti 146, **146**
Taj Mahal, Agra, India **269,** 270–271
Tallinn, Estonia 385, **385**
Tanzania 151
Taos Pueblo, New Mexico, U.S.A. 55
Tbilisi, Georgia **196,** 196–197, **197**
Tel Aviv, Israel 189, **189**
Television shows *see* Artbeat
Tennessee, U.S.A.
  Memphis 16–19, **18, 19**
  Nashville **16,** 16–19, **17**
Teotihuacan, Mexico 87
Tequila, Mexico **116,** 117–118
Thailand *see* Bangkok, Thailand
Theater *see* Artbeat
Thessaloniki, Greece 297
Thimphu, Bhutan **264,** 264–265

Tokyo, Japan 216–221
  Akihabara ("Electric City") 219–220
  Asakusa 219
  background check 219
  Chatei Hatou 217
  cherry blossoms **217,** 221
  gatherings 221
  Ginza shopping district 217, 219
  Gōtokuji Temple 217
  hanamachi (geisha district) 219
  Harajuku 220–221
  hidden treasures 217
  museums 216, 219, 220, 221
  restaurants 217, 219, **221**
  Sensōji Temple 216, 219, **220**
  Shibuya **216,** 221
  Shinjuku 220
  Shinto Hie Shrine **218**
  Shiro-Hige's Cream Puff Factory 217
  subway and rail system **216,** 216–217, 219, 220–221
  Sugamo neighborhood 217
  Sumida River 219
  sumo matches 219
  Tsukiji 219
  2D Café 217
  Ueno Park 220
Toledo, Spain 302
Toronto, Ontario, Canada **44,** 44–45, **45**
Trinidad and Tobago 79, **79**
Troy ruins, Turkey 195
Turkey
  Ephesus 195
  Istanbul **180–181, 192,** 192–195, **193, 194**
  Troy 195
TV shows *see* Artbeat

U

Ulaanbaatar, Mongolia **256,** 256–257, **257**
UNESCO Global Geopark, Hong Kong, China 234
UNESCO Intangible Cultural Heritage
  Bruges, Belgium **366,** 367
  Guadalajara, Mexico 116–117
  Marrakech, Morocco 174, **175**
  Salvador, Brazil 94
  Vienna, Austria 320, 322
UNESCO World Heritage sites
  Addis Ababa, Ethiopia 163
  Algiers, Algeria 172, **172**
  Amman, Jordan 205
  Amsterdam, Netherlands 363–364
  Angkor, Cambodia 243
  Antananarivo, Madagascar 179
  Antigua, Guatemala 124, **124**
  Barcelona, Spain 304, 305
  Beijing, China 224, 226, 227

  Berlin, Germany 329
  Chan Chan, Peru 121
  Córdoba, Spain 352–353, **353**
  Guadalajara, Mexico 119
  Gyeongju, South Korea 259
  Hamburg, Germany **324,** 325
  Hangzhou, China 231
  Havana, Cuba 60, **61**
  Helsinki, Finland 359
  Huế, Vietnam 239
  Johannesburg, South Africa 160
  Kathmandu, Nepal 262, 263, **263**
  Kraków, Poland **334,** 334–335, **335**
  Kyoto, Japan **210–211,** 212
  León, Nicaragua 124–125
  Lima, Peru 120
  London, England, U.K. 311
  Luang Prabang, Laos **246,** 246–247, **247**
  Machu Picchu, Peru 113, **114**
  Mumbai, India 266, **267**
  Paramaribo, Suriname 125
  Pompeii, Italy 343
  Port Louis, Mauritius 178
  Québec, Québec, Canada 46–47
  Quito, Ecuador **90,** 90–93, **91, 92, 93**
  Riyadh, Saudi Arabia 208, 209
  Samarkand, Uzbekistan 198
  Tallinn, Estonia 385, **385**
  Taos Pueblo, New Mexico, U.S.A. 55
  Tel Aviv, Israel 189
  Teotihuacan, Mexico 87
  Tequila, Mexico 117–118
  Toledo, Spain 302
  Versailles, Palace of, France 289
United Arab Emirates
  Abu Dhabi 184
  Dubai **182,** 182–185, **183, 184, 185**
United Kingdom
  Belfast, Northern Ireland 381
  Edinburgh, Scotland **374,** 374–377, **375, 376, 377**
  Glasgow, Scotland 377
  London, England **308,** 308–313, **309, 310, 312, 313**
United States
  Baltimore, Maryland 15
  Boston, Massachusetts **56,** 56–59, **57, 58, 59**
  Chicago, Illinois **32,** 32–35, **33, 34**
  District of Columbia **12,** 12–15, **13, 14**
  Hilo, Hawaii 131
  Honolulu, Hawaii **128,** 128–131, **129, 130, 131**
  Kansas City, Missouri 62–63, **63**
  Las Vegas, Nevada **50,** 50–53, **51, 52, 53**

Los Angeles, California **20,** 20–25, **21, 22, 23, 24, 25**
Memphis, Tennessee 16–19, **18, 19**
Nashville, Tennessee **16,** 16–19, **17**
New Orleans, Louisiana **36,** 36–37, **37**
New York, New York **6, 10–11, 64,** 64–69, **65, 66, 68, 69**
Oklahoma City, Oklahoma 63
Philadelphia, Pennsylvania **26,** 26–27, **27**
Portland, Oregon 29
Queens, New York 65, 68, 69
Reno, Nevada 53
San Diego, California 22
San Francisco, California **70,** 70–75, **71, 72, 74, 75**
San Juan, Puerto Rico **76,** 76–77, **77**
Santa Fe, New Mexico **54,** 54–55, **55**
Seattle, Washington **28,** 28–31, **29, 30, 31**
St. Louis, Missouri **62,** 62–63
Taos Pueblo, New Mexico 55
Washington, D.C. **12,** 12–15, **13, 14**
Urban relics (archaeological sites)
Angkor, Cambodia 243
Bagan, Myanmar 249
Caesarea National Park, Israel 201
Chan Chan, Peru 121
Évora, Portugal 349
Fatehpur Sikri, India 269
Great Zimbabwe, Zimbabwe 159
Gyeongju, South Korea 259
Huế, Vietnam 239
Lalibela, Ethiopia 163
Machu Picchu, Peru 113, **114**
Malacca, Malaysia 275
Monte Albán, Mexico 89
Nara, Japan 213
Petra, Jordan 205
Pompeii, Italy 343
Sparta, Greece 295
Stone Town, Zanzibar, Tanzania 151
Taos Pueblo, New Mexico, U.S.A. 55
Teotihuacan, Mexico 87
Toledo, Spain 302
Troy, Turkey 195
Versailles, Palace of, France 289
Uruguay **110,** 110–111, **111**
Uzbekistan **198,** 198–199, **199**

V

Vancouver, British Columbia, Canada **38,** 38–39, **39**
Vatican City, Rome, Italy **343, 344,** 346–347
Venice, Italy 330–333
    Basilica San Marco **331,** 332
    Burano Island 331
    canals **332**
    Carnival **330**
    Doge's Palace 332
    Drogheria Màscari 331
    Enoteca Millevini 331
    glassblowing 332
    hidden treasures 331
    John Cabot House 331
    lacemaking 331
    Libreria Acqua Alta 331
    local flavor 333
    Murano Island 331–332
    museums 331, 332, 333, **333**
    Piazza San Marco 330, **330, 331,** 332
    Piscina di Sacca Fisola 331
    Renaissance structures 332–333
    Santa Maria e San Donato 331–332
    Torcello 330–331
Versailles, Palace of, France 289
Victoria, British Columbia, Canada 39
Victoria, Seychelles 179, **179**
Vienna, Austria 320–323
    artbeat 322
    background check 321
    Beethoven sites 321–322
    Belvedere Gardens 323
    churches 320
    coffeehouse culture 320, **320**
    gardens 323
    Hofburg 322–323
    Maria-Theresien-Platz 323
    Mozart sites 321
    museums 320, 322, 323
    music history 320–321
    performing arts 320
    Prater amusement park 323
    Schloss Belvedere **321**
    Schloss Schönbrunn 323
    Spanish Riding School 322
    Strauss sites 321–322, **322**
    tram system 323
    Vienna Boys' Choir 322–323
Vietnam
    Hanoi **238,** 238–239
    Ho Chi Minh City 238–239, **239**
    Huế 239
Vilnius, Lithuania 384, **384**

W

Warsaw, Poland 335
Washington, D.C., U.S.A. 12–15
    Arlington National Cemetery 15
    C&O Canal 15
    Connecticut Avenue 15
    Dupont Circle 15
    Embassy Row 15
    Georgetown 15
    monuments and memorials 12–13, **13**
    museums 13, **14,** 15
    National Mall **12,** 13
    parks and gardens 15
    Pennsylvania Avenue 12
    Smithsonian museums 13, **14**
Washington, U.S.A. *see* Seattle, Washington, U.S.A.
World Heritage sites *see* UNESCO World Heritage sites

X

Xi'an, China 279, **279**

Y

Yangon, Myanmar **248,** 248–249, **249**
Yogyakarta, Indonesia **222,** 222–223, **223**

Z

Zagreb, Croatia **370,** 370–372, **371**
Zanzibar, Tanzania 151
Zimbabwe 159
Zurich, Switzerland 283

100
CITIES
5000
IDEAS

Since 1888, the National Geographic Society has funded more than 14,000 research, conservation, education, and storytelling projects around the world. National Geographic Partners distributes a portion of the funds it receives from your purchase to National Geographic Society to support programs including the conservation of animals and their habitats.

Get closer to National Geographic Explorers and photographers, and connect with our global community. Join us today at nationalgeographic.org/joinus

For rights or permissions inquiries, please contact National Geographic Books Subsidiary Rights: bookrights@natgeo.com

The information in this book has been carefully checked and to the best of our knowledge is accurate. However, details are subject to change, and the publisher cannot be responsible for such changes, or for errors or omissions. Assessments of sites, hotels, and restaurants are based on the author's subjective opinions, which do not necessarily reflect the publisher's opinion.

Library of Congress Cataloging-in-Publication Data

Names: Yogerst, Joseph R., author.
Title: 100 cities, 5,000 ideas : where to go, when to go, what to see, what to do / Joe Yogerst.
Other titles: One hundred cities, five thousand ideas
Description: Washington, D.C : National Geographic, [2022] | Includes index. | Summary: "Discover the world's 100 best cities to explore - including amazing skylines, mouthwatering bites, and pure fun - in this smart and inspiring travel resource"-- Provided by publisher.
Identifiers: LCCN 2021053125 | ISBN 9781426221675 (Paperback)
Subjects: LCSH: Urban tourism. | Cities and towns.
Classification: LCC G156.5.U73 Y64 2022 | DDC 910.9173/2--dc23/eng20220429
LC record available at https://lccn.loc.gov/2021053125

ISBN: 978-1-4262-2167-5

Printed in South Korea

22/SPSK/1